1989

University of St. Franci
GEN 576.15 F604
Ecolog...

C0-AWF-005
...801 00076374 4

ECOLOGY OF MICROBIAL COMMUNITIES

SYMPOSIA OF THE
SOCIETY FOR GENERAL MICROBIOLOGY*

* Published by the Cambridge University Press, except for the first Symposium, which was published by Blackwell's Scientific Publications Limited.

ECOLOGY OF
MICROBIAL COMMUNITIES

EDITED BY
M. FLETCHER, T. R. G. GRAY AND J. G. JONES

FORTY-FIRST SYMPOSIUM OF
THE SOCIETY FOR GENERAL MICROBIOLOGY
HELD AT
THE UNIVERSITY OF ST ANDREWS
APRIL 1987

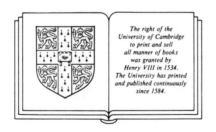

*The right of the
University of Cambridge
to print and sell
all manner of books
was granted by
Henry VIII in 1534.
The University has printed
and published continuously
since 1584.*

Published for the Society for General Microbiology

CAMBRIDGE UNIVERSITY PRESS
CAMBRIDGE
LONDON   NEW YORK   NEW ROCHELLE
MELBOURNE   SYDNEY

LIBRARY
College of St. Francis
JOLIET, ILL.

Published by the Press Syndicate of the University of Cambridge
The Pitt Building, Trumpington Street, Cambridge CB2 1RP
32 East 57th Street, New York, NY 10022, USA
10 Stamford Road, Oakleigh, Melbourne 3166, Australia

© The Society for General Microbiology Limited 1987

First published 1987

Printed in Great Britain at The Bath Press, Avon

British Library Cataloguing in Publication Data
Society for General Microbiology, (*Symposium:
41st: 1987: University of St. Andrews*)
Ecology of microbial communities: forty-first
symposium of the Society for General
Microbiology held at the University of St.
Andrews, April 1987. – (Symposia of the
Society for General Microbiology)
1. Microbial ecology   2. Biotic communities
I. Title   II. Fletcher, M.   III. Gray, T. R. G.
IV. Jones, J. G.   V. Series
576'.15   QR100

Library of Congress Cataloguing-in-Publication Data

Society for General Microbiology.
Symposium (41st: 1987: University of St Andrews)
Ecology of microbial communities.

(Symposia of the Society for General Microbiology; 41)
"Forty-first Symposium of the Society for General
Microbiology, held at the University of St. Andrews, April 1987."
Includes index.
1. Microbial ecology – Congresses.   I. Fletcher,
Madilyn.   II. Gray, T. R. G. (Timothy R. G.), 1937–
III. Jones, J. G. (J. Gwynfryn)   IV. Society for General
Microbiology.   V. Series: Society for General
Microbiology. Symposium. Symposia; 41.
QR1.S6233 no. 41 [QR100]   576 s [576'.15] 86-28344

ISBN 0 521 33106 4

BS

5 76.15

F604

CONTRIBUTORS

ANDERSON, J. M. Wolfson Ecology Laboratory, Department of Biological Sciences, University of Exeter, Exeter EX4 4PS, UK

ANDERSON, K. L. Department of Microbiology, Montana State University, Bozeman, MT 59717, USA

AZAM, F. Institute of Marine Resources, Scripps Institute of Oceanography, University of California, La Jolla, CA 92093, USA

BATESON, M. M. Department of Microbiology, Montana State University, Bozeman, MT 59717, USA

BODDY, L. Department of Microbiology, University College, Newport Road, Cardiff CF2 1TA, UK

BROCK, T. D. Department of Bacteriology, University of Wisconsin – Madison, Madison, WI 53706, USA

CHO, B. C. Institute of Marine Resources, Scripps Institute of Oceanography, University of California, La Jolla, CA 92093, USA

DOWSON, C. G. School of Biological Sciences, University of Bath, Claverton Down, Bath BA2 7AY, UK

FLETCHER, M. Center of Marine Biotechnology, The University of Maryland, Central Administration, Adelphi, Maryland 20783, USA

GRAY, T. R. G. Department of Biology, University of Essex, Colchester, Essex CO4 3SQ, UK

HAMILTON, W. A. Department of Genetics and Microbiology, Marischal College, University of Aberdeen, Aberdeen AB9 1AS, UK

JACKSON, G. A. Institute of Marine Resources, Scripps Institution of Oceanography, La Jolla, CA 92093, USA

JONES, J. G. Freshwater Biological Association, The Ferry House, Ambleside, Cumbria, LA22 0LP, UK

KARL, D. M. Department of Oceanography, University of Hawaii, Honolulu, HI 96822, USA

LYNCH, J. M. Glasshouse Crops Research Institute, Littlehampton, West Sussex, BN17 6LP, UK

134994

NEDWELL, D. B. Department of Biology, University of Essex, Colchester, Essex, CO4 3SQ, UK

PARKES, R. J. Scottish Marine Biology Association, Dunstaffnage Marine Research Laboratory, P.O. Box 3, Oban, Argyll PA34 3AD, UK

POINDEXTER, J. S. The Public Health Research Institute of The City of New York, Inc., New York, NY 10016, USA

RAYNER, A. D. M. School of Biological Sciences, University of Bath, Claverton Down, Bath BA2 7AY, UK

TAYNE, T. A. Department of Microbiology, Montana State University, Bozeman, MT 59717, USA

TETT, P. School of Ocean Sciences, University College of North Wales, Marine Science Laboratories, Menai Bridge, Gwynedd LL59 5EH, UK

WARD, D. M. Department of Microbiology, Montana State University, Bozeman, MT 59717, USA

CONTENTS

EDITORS' PREFACE

The ultimate aim of the study of microbial ecology is to understand the interactions within microbial communities and between them and their environments. Two microbiologists who pioneered the early study of microbial ecology, Sergei Winogradsky and M. W. Beijerinck, developed the enormously useful technique of enrichment culture and observed microbial transformations of elements such as nitrogen, sulphur and iron. This provided the basis for understanding processes which occur in the natural environment. Following these early discoveries, and with the development of the pure culture technique by Robert Koch, microbiology became firmly entrenched in the laboratory, and very little of our knowledge about microbial processes was obtained directly from the field. Instead, microbiological studies in the field were often descriptive, and studies on the physiology of microbes from natural environments were performed only after sampling material and attempting (not necessarily successfully) to culture resident microorganisms in the laboratory. In the 1950s and 1960s microbial ecologists became more and more aware of the limitations imposed by laboratory culture and searched for ways of examining the physiological processes of microbes *in situ*. Considerable advances were made through the application of microanalytical and radiotracer techniques, but by far most studies were still done in the laboratory, and usually with pure cultures. Such studies clearly were of enormous importance, as they provided detailed information on the potential activities of microorganisms isolated from natural environments. However, they threw little light on the actual activities of microbes *in situ*. Nor did they indicate how the various microbes within the mixed assemblages found in natural communities interacted with one another and thus determined the net transformations occurring in their soil, water or animal or plant host environments.

We now live in a very exciting time when the study of the ecology of microbial communities is coming of age. There are more and more microbial ecologists who are coming to grips with the complexity of natural microbial communities and who are attempting to unravel and describe the interactions and transformations occurring in such communities *in situ*. This has been made possible through the development of advanced methods in the laboratory, followed by their modification and adaptation to field situations. A powerful battery of techniques is being applied, including more advanced developments in microanalytical, radiotracer, and microscopical techniques, as well as spin-offs from the ongoing revolution in microbial genetics and molecular biology. Combined with these practical innovations is an increase in intuitive insight into ecological processes among microbiologists. This awareness has arisen only through many years of observing microbes in their natural habitats and recognizing how critically important are those habitats in determining the nature of the inhabiting community. Consequently, we are beginning to perceive the true complexity of microbial communities *in situ* and have some hope of understanding some of the many facets of their ecology.

In recognition of the considerable progress made in the study and understanding of microbial ecology over the past 20 or so years, the Ecology Group for the Society of General Microbiology proposed a main society symposium, which described some of these advances and their significance. The papers in this volume are based on those presented at that symposium, the 108th Ordinary Meeting, held in April 1987 at St Andrews, Scotland. We wish to express our thanks to the Society for mounting the meeting and especially to the authors, for coming in from the field long enough to prepare their manuscripts.

THE STUDY OF MICROORGANISMS *IN SITU:* PROGRESS AND PROBLEMS

THOMAS D. BROCK

Department of Bacteriology, University of Wisconsin–Madison, Madison, WI 53706, USA

It is perhaps appropriate that I should write this chapter. Twenty years ago I published a book – *Principles of Microbial Ecology* (Brock, 1966) – which dealt for the first time in an integrated way with the field of microbial ecology. Although that book was flawed, it did provide a focus on microbial ecology and perhaps helped to strengthen the weak ties holding fields such as soil microbiology, aquatic microbiology, and pathogenic microbiology together. In my preface to this book I said:

The task of microbial ecology is to provide an understanding of the place of microorganisms in nature and in human society. In the past, microbial ecology has been fragmented into a number of subfields, such as soil microbiology, food microbiology, marine microbiology, medical microbiology, etc. I believe it is now possible, and desirable, to integrate these fields, which is what I have tried to do in this book.

Where have we come in those twenty years? Depending on one's viewpoint: far, but not nearly far enough; or not very far at all.

During the past twenty years, the real field of ecology, *macroecology*, that is, has grown up and matured. The field of microbiology has changed beyond all recognition. But microbial ecology? True, new journals have appeared: *Microbial Ecology, Applied and Environmental Microbiology* (General Microbial Ecology section), *FEMS Microbiology Ecology, Advances in Microbial Ecology*. Symposia have been held, including Modern Methods in Microbial Ecology (Uppsala, 1972) and the various International Symposia on Microbial Ecology (1st, New Zealand; 2nd, Warwick; 3rd, East Lansing; 4th, Yugoslavia).

Success? Perhaps. But I worry that microbial ecology has become too inwardly directed, too parochial in its outlook. It has been my hope that microbial ecology will not become too narrow in its viewpoint. I have always believed that microbial ecology should be part of microbiology, and that microbial ecology papers should be refereed by microbiologists rather than by microbial ecologists. However, papers in the field of microbial ecology were pushed out of the *Journal of Bacteriology* and discouraged in *Archives of Microbiology*. Only the *Journal of General Microbiology*, of general microbiology journals, still has an ecology section, albeit rather small.

What is microbial ecology? An acceptable definition is that *microbial ecology is the study of the relationships of microorganisms with their natural environments*. The operative word here is 'natural'. What we are concerned about, or should be concerned about, is the study of microorganisms *in situ*. The words *in situ* were in the title of this paper as given to me by the organizers of this symposium, and that is the title I will write to.

Unfortunately, much of what passes for microbial ecology does not fit this definition. Further, many studies that pretend to be ecological are still using antiquated, discredited, or meaningless methods. Many studies are unfocussed, or do not deal with important questions. In many studies, a clear question is not formulated.

Even worse, the idea is about that microbial ecology is easy – that if you can't do chemistry, molecular biology, or genetics, you can at least do microbial ecology. WRONG! Microbial ecology is not easier than those other things, but harder. Because microbial ecology deals with complex systems, it requires skill, insight, imagination, attendance to quantitative detail, even mathematical insight, that far surpasses what is needed for genetics and molecular biology.

THE CURRENT STATE OF MICROBIAL ECOLOGY

Before I began to write this chapter, I did a statistical survey of current research in microbial ecology by classifying research papers published in two key journals, *Applied and Environmental Microbiology* (General Microbial Ecology section, GME) and *Microbial Ecology* (ME). I looked at each paper published in the GME in 1985 and all papers published in ME in 1983, 1984, and 1985. I was particularly interested in the *methods* used, since methods most clearly reveal the real nature and significance of a study. I summarize my results in Table 1. Let me explain how I have categorized the research papers in this table, since this explanation will reveal not only my biases, but will permit me to make some general comments about various approaches to microbial ecology.

Table 1. *Categories of microbial ecology papers published in two leading journals*

Category	GME (%)	ME (%)
Meritorious studies		
Microenvironment	1	0
Integrated field/laboratory	2	1
Model systems	8	11
Biomass/activity	25	23
Direct microscopy	3	8
Questionable studies		
Pure cultures	41	23
Mixed cultures	12	7
Viable counts	7	20
Species diversity	0	7

GME: *Applied and Environmental Microbiology*; papers of 1985 listed in the General Microbial Ecology section ($n = 137$).
ME: *Microbial Ecology*; all papers of 1983–5 ($n = 86$).
Categories are partly arbitrary; see text for discussion.

Pure culture studies

This was a shocker to me! Many of the papers were not ecological at all, but dealt with studies in pure culture! In GME, 41% of the papers dealt with studies of pure cultures under highly controlled conditions in the rich culture media used by physiologists. These were, not model system studies in which the culture attempted to mimic nature (such studies are discussed later). No, these were studies which if they had been done with *Escherichia coli* would have been considered physiological (and might have been rejected by appropriate journals as being trivial). Why, then, did their authors, and the editors of the journals, deem them ecological?

What makes a pure culture study ecological, as far as the editors of these journals were concerned? An interesting question. These pure culture studies dealt with photosynthetic bacteria, methanogens, iron-oxidizing bacteria, nitrifiers, and thermophiles. So that seems to be it – if a study deals with a pure culture of an unusual organism, not *Escherichia coli* or *Bacillus subtilis*, then it is ecological. WRONG! These studies are physiological studies, possibly even poor ones, masquerading as ecology. It may even be the kind of study that would be *rejected* by the *Journal of Bacteriology* or the *Archives of Microbiology*. Even in ME, that paragon of a microbial ecology journal, 23% of the papers dealt with pure culture studies. Please note: I am not saying there is anything wrong with pure culture studies *per se*. I have published lots of pure culture studies myself. I am just saying that if they do not involve attempts to duplicate critical features of the natural environment, they are hardly microbial ecology.

Mixed culture studies

These studies, in my way of thinking, are even more suspect than pure culture studies. I include here those studies in which some sample of the natural habitat is placed under artificial culture conditions in the laboratory, allowed to grow, and then transferred periodically. Finally, at some ill-defined stage, a study is done with this mixed culture of unknown provenance. Yes, such a study might be called an *enrichment culture* study, but an enrichment culture is only a mixed culture on its way to becoming a pure culture. The mixed culture studies I am considering here *never* lead to pure cultures! The initial inoculum from nature, composed of who knows how many organisms, is gradually altered by successive transfer until little is left of nature, but the part that is left *is* a mystery to us.

Fortunately, such mixed culture studies constituted only 12% of papers in GME and 7% of papers in ME.

What kind of 'microbial ecologist' would do a study of this sort? Well, rumen microbiologists and sewage microbiologists, especially!

Studies involving viable counts

The viable counting procedure, as far as microbial ecology is concerned, was discredited 20 years ago, if not before. Why? There are many reasons. Dormancy, for example; a dormant cell, doing nothing in the environment, will show up in a viable count. The viable count is notorious for its selectivity toward unicellular organisms, yet in many environments, filamentous organisms are the rule. The viable count shows a colony of 100 cells as a single colony-forming unit (calling it a CFU does not legitimize it), yet microscopy shows us that in nature many organisms are found not as single cells, but in colonial masses. Finally, even the most efficient plating medium is not 100% efficient. Viable counts usually run 1% or less of direct microscopic counts. Even *Escherichia coli* under the best laboratory conditions rarely plates at 100% efficiency.

Twenty years ago I had several pages in my book about the meaninglessness of the viable count, of which the following sums it up:

The most serious difficulty ... is that [viable counts] are (1) selective for unicellular organisms or organisms which produce many propagules and (2) they are selective for organisms which spread rapidly on the surface of the medium, either because of motility or because of growth by extension of filaments. Alternatively, [viable counts] are selective against organisms which cannot grow well against the resistance which is offered by the solid surface. Finally, an organism which was dormant in the ecosystem may still produce a colony. Thus, the organisms counted on the agar plate are not necessarily the ones which are functioning in the ecosystem.

Was nobody reading? Surely everybody knows all this. No one does viable counts in microbial ecology anymore. Well, 7% of the papers in GME and 20% (yes, 20%) of the papers in ME used viable counting procedures as their main technique. In the mid 1980s!

Species diversity and similar mumbo jumbo
There is another area that seems to attract some microbial ecologists, and this involves the determination of species diversity in natural systems (sometimes called community structure or guild structure). In this approach, one isolates pure cultures from nature and then categorizes them taxonomically, either to species or to other meaningless grouping such as OTU (operational taxonomic unit). Then, by use of a formula derived from information theory, one determines a diversity index.

Diversity of what? Not of the natural population, since we have already shown above that culture procedures do not provide representative samples of the natural populations. (As we all know, 'species' is barely a useful concept in bacteriology depending on how well studied a genus is, the number of described species will vary radically). Even if we could culture everything, and even if species did have some real meaning, for what purpose would one determine species diversity? Here is one idea: 'species diversity of a community is directly correlated with the stability of that community'. WRONG! Not in microbial ecology! This is a good example of an idea taken over from macroecology which is wrong and/or meaningless in a microbial ecology context. In microbial systems, population stability is determined by environmental stability. Oak forests may resist environmental onslaughts, but microbial populations do not. Microbial populations respond to environmental variability by *changing*. In fact, an inability to understand this essential difference between microbial populations and populations of higher organisms is at the base of one of the current fashionable arguments between microbial ecologists and macroecologists about the dangers of release of genetically engineered organisms in the environment.

For some reason, GME published no papers on species diversity, but 7% of the papers published in ME dealt with this infamous subject.

But there is Hope!

I have only dealt with 61% of the papers in GME and 57% of the papers in ME. There are some papers left. There is hope for the future.

Model systems

A number of papers deal with one of the most promising areas of microbial ecology, the model. True, model systems frequently involve use of pure cultures, but in a much different way to that discussed above. The goal of the model system is to mimic in the laboratory the natural environment (or some part of it) and to analyze the behavior of an organism (for instance, Wiggins & Alexander, 1985). The good model system uses well-defined cultures that are known to be significant in the habitat of interest, low-density populations (ecologically relevant densities, usually $1\,000\,000$ cells ml^{-1} or below), low nutrient concentrations, slow growth rates (ecologically relevant growth rates – *Escherichia coli* has been estimated to double about twice a day in the intestine), or in some other way mimic nature.

Alternatively, the model system uses a powerful and realistic mathematical approach, attempting to understand the behavior of organisms in their natural environments by describing in mathematical terms their physical and chemical environments (Schmidt, Simkins & Alexander, 1985).

Another type of model attempts to understand some process occurring in nature, using a pure culture as a tool, such as the work by Jenneman, McInerney & Knapp (1985) on the penetration of bacteria through sandstone.

About 10% of the papers in each journal dealt with model systems.

Direct microscopic studies

I am not discussing in this section the direct microscopic count, which falls under the biomass topics discussed later. I am talking about microscopic study of organisms directly in their natural environments.

The cell is small, and therefore its environment is also small. Some aspects of the environment, such as temperature and pressure, are the same at both the macro- and microenvironmental level, so that a measurement by traditional methods can be expected to have microbiological meaning. Yet for many other factors, a microenvironment may differ greatly from its surrounding macroenvironment. A single soil crumb, for instance, may contain numerous microenvironments differing in moisture content, nutrient concentration, pH, or other factors. *Chemical analysis of the microenvironment requires the use of the microscope.* (Brock, 1966)

If one accepts these words, then one realizes that *any* study in microbial ecology must begin with the microscope. Only by the use of some sort of microscope can we really see organisms in their actual environments. The light microscope is, or course, the simplest type of microscope for ecological studies, and is still, over 100 years after Louis Pasteur & Robert Koch introduced it to our field, the most important tool of the microbial ecologist. But the electron microscope, especially the scanning electron microscope, plays an increasingly important role. However, we are not talking about just taking pretty pictures, as some workers do, but using the microscope intelligently to characterize the environment. A few papers are making use of some microscopic procedures in an ecologically useful way. The thrust of such papers is frequently morphological, and currently the widest such studies are with animal-associated microbes (Czolij, Slaytor & O'Brien, 1985). Especially with the use of the scanning electron microscope, one can see organisms in place and can discern important ecological features.

We all recognize the limitations of microscopic study. Identification of microorganisms is usually difficult and their activities cannot be determined without additional studies. Even the use of the fluorescent antibody technique, one of the more attractive approaches to direct microscopy, has serious limitations. But microscopy does constitute the base, the essential foundation, on which any study in microbial ecology must build.

How important is microscopy in microbial ecology? As an analogy, suppose an animal ecologist did a wide sweep of habitats from the California desert to the High Sierra (or from the Mediterranean Sea to the top of Mont Blanc), mixed all of the animals collected together, and then carried out some kind of analysis. What ecological significance would such a study have? Yet, many microbial ecologists take samples in precisely this way (sediment cores, for instance, or soil or fecal samples). Without knowing where these organisms lived, can we make any sense out of such analyses?

Microscopy, admittedly, is difficult. This is, of course, why it is not done more. Is this any excuse? If nothing more, microscopic studies will help keep the microbial ecologist thinking about the essence of the problem – that microbes are small, and their environments are also small. As examples of the possibilities of direct microscopic analysis, consider the paper by Heldal, Norland & Tumyr (1985) using X-ray microanalysis in a transmission electron microscope to measure the dry matter and elemental content of individual bacteria, or the use of image-analyzed epifluorescence microscopy for the detection, enumeration, and sizing of planktonic bacteria (Sieracki, Johnson & Sieburth, 1985).

Microscopic activity studies

There is one way in which a limitation of microscopy can be overcome – to use a microchemical of microautographic procedure to study the activities of microorganisms *in situ*. In my survey, only *one* paper of over 200 papers published in GME or ME used this technique (and this paper was only a Note: Bern, 1985). I developed the technique of microscopic autoradiography over 20 years ago, have published extensively on it (Brock & Brock, 1966, 1967a, 1967b; Kelly & Brock, 1969; Munro & Brock, 1968), including a detailed methods paper (Brock & Brock, 1968), and the technique has been used through the years by a number of workers (for instance, Knoechel and Kalff, 1976a, 1976b). The utility of the technique has been widely recognized by 'aquatic' types, and it has even been reintroduced independently, several times. Why hasn't the technique become more popular with microbial ecologists? Limitations it certainly has, but the main reason for its lack of popularity, I believe, is that there is the perception that it is too much work! Inertia, ignorance, laziness; on such qualities is much of microbial ecology built.

Biomass / activity studies

But things are not all bad. A number of papers have approached studies in microbial ecology in a modern manner, attempting to understand natural systems by: (1) quantifying microbial biomass directly; and (2) measuring microbial activity with radioisotopic or chemical methods. Although these studies frequently ignore the microenvironment, they do provide a solid basis for advances in microbial ecology. A full quarter of the papers in GME and 23% of those in ME made use of such techniques.

True, it is primarily the 'aquatic' types who quantify and measure activity (Novitsky & Karl, 1985; Simon, 1985). Indeed, the techniques employed were first developed by marine scientists over thirty years ago and were gradually incorporated into microbial ecology. The initial studies by both aquatic scientists and microbial ecologists were on microscopic algae (phytoplankton); these organisms were adaptable to such studies because of their relatively large size and taxonomic recognizability under the microscope, and because they carried out as one of their major metabolic activities the process of photosynthesis, which can be readily measured in natural environments with carbon-14. The same techniques have been applied, albeit with some difficulty, to the bacteria. The introduction of epifluorescence microscopy for bacterial counting and the use of isotope techniques on a 'whole system sample', have permitted some of the most interesting studies in microbial ecology.

True, there are still extensive controversies about the validity of some of these methods. Most of the controversy centers around 'short-cut' methods which attempt to circumvent some of the required complexities of analysis. Also, many of these studies would repel the taxonomically trained bacteriologist, since the bacteria are often lumped together as one large 'black box'. Meaningless concepts, such as 'bacterial production' and 'heterotrophic potential' are employed, thus attempting to legitimize the study. Nevertheless, these studies are on the right track, and we must be prepared for some weak studies in any field.

Integrated field / laboratory studies
My favorite studies, because they come closer to the central problem of microbial ecology, are those which integrate field and laboratory study. Again, we are talking primarily about aquatic microbial ecology, primarily in reasonably well-defined habitats – a lake, an estuary, a salt marsh, a hot spring. A full three papers in GME (for example, Paerl et al., 1985) and one whole paper in ME dealt with such studies. True, if I had looked in a 'real' aquatic journal, such as Limnology and Oceanography, I would have seen a lot more such papers. But we are considering here the state of microbial ecology, not that of limnology / oceanography.

PROGRESS IN MICROBIAL ECOLOGY

Clearly, I do not give much of microbial ecology research strong marks. Yet, there is hope, and there is a future. Microbial ecology will not go away, any more than will the problems which need microbial ecology for a solution.

Indeed, serious students of microbial ecology can be encouraged by the dramatic developments in basic biological research which have led to our current excitement about genetics and molecular biology. A new era is at hand, an era in which new insights will come from new methods.

What are the important areas of research where microbial ecology will make a contribution? I list them here and discuss them briefly in what follows: (1) microbial evolution and phylogeny; (2) population genetics of microorganisms; (3) pollution control; (4) public health microbiology; and (5) the roles of microorganisms in global ecosystems.

Microbial evolution and phylogeny

One of the most exciting areas of current research deals with the comparative study of the relationships among organisms, using molecular techniques. New approaches of molecular biology have permitted a facile comparison of gene sequences, making it possible to discern at the most basic level genetic differences between organisms. This has led to the development of rational phylogenies of organisms (Schleifer & Stackebrandt, 1985).

Yet, these studies have been done in the almost complete absence of ecological interpretation. If we know anything about evolution, we know that organisms do not evolve in a vacuum. They evolve because of natural selection, arising either from changes in environment or through competitive pressures from other organisms. How much more satisfying would be our understanding of microbial evolution if we could relate it to microbial ecology. With the exception of the work of Norman Pace (Stahl *et al.*, 1985), there seems to be no work attempting to relate the new molecular phylogeny studies to the environment. However, even the work of Stahl *et al.*, interesting as it is, is essentially another approach at 'species diversity'. Although Stahl *et al.* have stated, 'In principle, phylogenetic placement can be interpreted in terms of physiology', this must be true only in the grossest terms. Regulation and dormancy are two phenomena that affirm that what an organism *can* do is not the same as what it *does* do. (I do not discuss the obvious fact that the Stahl / Pace approach only permits characterization of the dominant members of the community. Perhaps the direct cloning of ribosomal RNA genes will increase the sensitivity of the technique.)

Population genetics of microbes

Recent concern about the fate in the environment of genetically modified organisms has focussed attention on how limited is our knowledge of microbial population genetics. An incisive review of the population genetics of *Escherichia coli* (Hartl & Dykhuizen, 1984) reveals the current state of knowledge of this most studied bacterium. Unfortunately, *E. coli* is not a favorable organism for *direct* studies in microbial ecology, since it lives in nature either with a congeries of other organisms, or is present in microenvironments inaccessible to accurate sampling and experimentation. Therefore, most of our knowledge of the population genetics of *E. coli* has been obtained indirectly from studies on the genetic structure of pure-culture isolates. One area of current interest, the exchange of genetic elements between strains, has been little studied, although the frequency in nature of genetic exchange appears low in *E. coli*.

The population structure of prokaryotes can be describe in 'terms of a "commonwealth" of clones, consisting of largely independent cell lineages in terms of chromosomal genes, but sharing among themselves an enormous diversity of extrachromosomal genetic elements' (Hartl and Dykhuizen, 1984).

One area of population genetics where *E. coli* is the pre-eminent organism is in experimental studies of mutation / selection in laboratory models, using chemostats. Studies dealing with the selective effects of temperate bacteriophage and of transposable elements have shown that these elements are favorable for growth and reproduction of the bacterial host. *Why* prophages and transposons should be favorable for bacterial growth is not known. Studies on the selective effect of enzyme isomers (isozymes) at a number of gene loci have shown that in most cases electrophoretic alleles are selectively neutral or nearly neutral, although they possess a potential for selection that permits them to respond to natural selection pressures. A number of studies on the evolution of new metabolic pathways in *E. coli* and other bacteria attest to the importance of mutation in the evolution (and hence, by implication, in the ecology) of microbes.

When we consider the complexities revealed by a study of a single *E. coli* strain in a controlled laboratory condition, we can marvel at the bravery of those who set up mixed culture model systems!

Microbes in pollution control

The removal of toxic wastes, either in the environment or in controlled treatment systems, is another major area in which microorganisms are involved and microbial ecology has significance. The evolution of microorganisms capable of degrading xenobiotics is well established, and we can anticipate that a microbial system capable of degrading any randomly chosen compound might eventually evolve.

How long will it take for an organism capable of degrading a particular xenobiotic to evolve? Is a single organism enough? Are there any environmental limitations (temperature, aerobiosis, pH, nutrients, water activity, etc.)? Simply to ask these questions exposes the depth of our ignorance. Although the genetic engineer may well be able to construct an organism capable of handling a particular compound, this engineer will need a microbial ecologist if the organism is to be released into the environment and actually *work*.

Although the use of genetic engineering for the development of new organisms for pollution control is currently in vogue, we really don't know enough about evolutionary processes to be certain that genetics will always make it possible. Is it true that constructing a new organism in the laboratory is quicker and more effective than isolating an organism from nature? What if the xenobiotic can only be degraded by a system of coupled organisms? Perhaps the genetic engineer might still then play a major role, providing that first hard-to-obtain organism, but the genetic engineer will than have to work hand-in-glove with the microbial ecologist, in order to get the whole system to work.

But however the organism is obtained, whether from nature or from the laboratory, it is clear that it is not sufficient that the organism or consortium of organisms act on the xenobiotic in the laboratory. Action must occur in the world at large. If we have learned nothing else from research in microbial ecology, we have learned that microbes do not do the same things in the laboratory that they do in nature.

So the pollution-control researcher is immediately faced with ascertaining the environmental requirements of the organisms, and the *likelihood* that these requirements will be *met* in the habitat of concern.

An interesting series of studies that relate to this question is that from Martin Alexander's laboratory (Schmidt, Simkins & Alexander, 1985; Simkins & Alexander, 1985) which deals with the kinetics of biodegradation. These authors come to grips with the whole question of the concentration-dependence of biodegradation rate, the relevance of the Monod equation for predicting rate, the use of various kinetic equations for *predicting* the ultimate fate of a biodegradable xenobiotic. One would hope that studies of this type, when used to model the biodegradation process quantitatively at environmentally relevant concentrations, will point the way toward the development of a predictive microbial ecology.

Ultimately, what will be needed is the development of quantitative models that will not only describe the rates of growth and activity of biodegrading microorganisms, but which will also describe these processes under a wide range of environmental conditions. Then, one will have to assess the environmental conditions in the habitat of interest, at the *microenvironmental* level. If such a two-phased approach were used, one should be able to predict the likelihood that an organismal construct might actually do what is intended.

There is no reason to reinvent the wheel each time a new compound is to be tried in a new environment. We *must* develop general principles that will be widely applicable!

Public health microbiology
At this point I would like to turn briefly to an old field that needs some real help from microbial ecology. In public health microbiology we are concerned with how various microbial pathogens are released, spread, survive, and reinfect new hosts. We have lots of qualitative information on these processes, but precious little quantitative information. Release and spread have been studied widely, but rarely quantitatively. Survival has been poorly studied, except for the coliform, that surrogate of the enteric pathogen. Reinfection, the most difficult part of the sequence to approach, has been studied almost not at all.

How many viable pathogens are needed to initiate an infection? One? A few? Many? And what is the probability of infection? These are microbial ecology questions that should not be left to the medical microbiologist or epidemiologist.

We need quantitative information – and mathematical models – to describe these processes.

The roles of microorganisms in global ecosystems

The field of biogeochemistry deals with the chemical transformations in the earth that are carried out by living organisms. Microorganisms are the prime biological agents in many transformations, and an understanding of their roles in biogeochemical processes is of great importance.

Unfortunately, from a microbial ecology viewpoint, much of microbial biogeochemistry is treated as a 'black box', with little reference to specific organisms. This is certainly acceptable for many studies, as it is for the biomass / activity studies discussed earlier. However, it leaves us embarrassingly in the dark (I was tempted to say 'in the black') about the roles of specific microbes. For example, sulfate reduction occurs in marine sediments and is an important process in anaerobic decomposition, but what sulfate-reducing organisms are involved?

As a further example of what I mean about the importance of knowing the organisms involved in biogeochemical processes, and why a 'black box' approach is not enough, consider the role of microorganisms in anaerobic carbon breakdown: organic carbon $\rightarrow CH_4$. This biogeochemical process, so central to global carbon cycling and so widely studied in sanitary engineering, was for many years (and still often is) treated as a single process without consideration of the involvement of specific microbes. We know, of course, that the key microbes in this process, the methanogens, only use a limited array of substrates, and that the methanogens thus depend on other carbon-metabolizing anaerobic organisms for the production of the substrates they need. Therefore, although it is crucial to understand the ecology of the methanogens, it is equally important to understand the organisms the methanogens depend on, the fermentative and biodegradative organisms. Indeed, do we even know which group of organisms is rate-limiting? To ask such a question is to expose our ignorance.

Obviously, modelling biogeochemical processes will only be possible if we understand the kinetic parameters of the individual groups, and the coefficients involved in the transfer of molecules from one group to another. We may not have to analyze *each* individual organism, but we will have to analyze each group. So far, little work has been done on microorganisms other than the methanogens.

WHAT DO WE NEED IN THE FUTURE?

I have emphasized in this chapter the importance of the quantitative approach, and the vital need for more research at the microenvironmental level. Unfortunately, these areas are the most difficult to study.

It is much easier to dump a radioisotope into a sample and blindly follow its fate, than it is to study the system microscopically.

It is much easier to do a viable count than a direct microscopic count.

It is much easier to work with a pure culture than a natural system.

It is much easier to work in the laboratory than in the field.

It is much easier to do taxonomy on the easily studied organisms than on the difficult ones, although the latter may be more important.

It is much easier to avoid mathematical modelling and to restrict oneself to qualitative studies.

In 1966, I wrote:

Microbial ecology has the potential of becoming the most sophisticated branch of ecology for two reasons: (1) The tremendous store of fundamental knowledge on the genetics, physiology, and biochemistry of microorganisms provided by microbiologists is now ripe for ecological exploitation. (2) Microbial ecology can become a meaningful experimental science, since in many cases a simple test tube or flask can be converted into a precise, reproducible, and meaningful ecosystem. Just as it is possible to do meaningful microecological experiments in the laboratory, so is it also possible to make meaningful ecological observations in the field. Thus field and laboratory can complement each other. The microbial ecologist can have his roots in the soil, but his feet in the laboratory.

This statement is still true. But what has happened during those past twenty years? My book went through five printings, so it obviously sold fairly well. Did anyone *read* the book? Was no one listening? How about all those papers summarized in Table 1?

What is needed first for any successful *in situ* study in microbial ecology is the *will* to carry it out. Many excuses can be advanced for not studying microorganisms directly in their natural environments. But we should recognize these for what they are, *excuses*. The result of this timidity of microbial ecology is that our knowledge of *in situ* activities of microorganisms is extremely limited.

One final note: Although field microbial ecology is difficult, it *can* be fun! What is needed is an intelligent choice of the habitat to be studied, a habitat that is not only important and amenable to study, but exciting to visit. There is no reason why the microbial ecologist cannot carry on important studies in interesting and exciting locations. After all, isn't that one reason we went into microbial ecology in the first place, to get out into the fresh air and travel to interesting places?

REFERENCES

BERN, L. (1985). Autoradiographic studies of [*methyl*-³H]-thymidine incorporation in a cyanobacterium (*Microcystis wesenbergii*)–bacterium association and in selected algae and bacteria. *Applied and Environmental Microbiology*, **49**, 232–3.

BROCK, M. L. & BROCK, T. D. (1968). The application of microautoradiographic techniques to ecological studies. *Mitteilungen der Internationale Verein Limnoloqie*, No. 15.

BROCK, T. D. (1966). *Principles of Microbial Ecology*. Englewood Cliffs, NJ, Prentice-Hall.

BROCK, T. D. (1967a). Bacterial growth rate in the sea: direct analysis by thymidine autoradiography. *Science*, **155**, 81–3.

BROCK, T. D. (1967b). Mode of filamentous growth of *Leucothrix mucor* in pure culture and in nature, as studied by tritiated thymidine autoradiography. *Journal of Bacteriology*, **93**, 985–90.

BROCK, T. D. & BROCK, M. L. (1966). Autoradiography as a tool in microbial ecology. *Nature*, **209**, 734–6.

CZOLIJ, R., SLAYTOR, M. & O'BRIEN, R. W. (1985). Bacterial flora of the mixed segment and the hindgut of the higher termite *Nasutitermes exitiosus* Hill (Termitidae, Nasutitermitinae). *Applied and Environmental Microbiology*, **49**, 1226–36.

HARTL, D. L. & DYKHUIZEN, D. E. (1984). The population genetics of *Escherichia coli*. *Annual Review of Genetics*, **18**, 31–68.

HELDAL, M., NORLAND, S. & TUMYR, O. (1985). X-ray microanalytic method for measurement of dry matter and elemental content of individual bacteria. *Applied and Environmental Microbiology*, **49**, 1251–7.

JENNEMAN, G. E., MCINERNEY, M. J. & KNAPP, R. M. (1985). Microbial penetration through nutrient-saturated Berea sandstone. *Applied and Environmental Microbiology*, **49**, 383–91.

KELLY, M. T. & BROCK, T. D. (1969). Physiological ecology of *Leucothrix mucor*. *Journal of General Microbiology*, **59**, 153–62.

KNOECHEL, R. & KALFF, J. (1976a). The applicability of grain density autoradiography to the quantitative determination of algal species production: a critique. *Limnology and Oceanography*, **21**, 583–90.

KNOECHEL, R. & KALFF, J. (1976b). Track autoradiography: a method for the determination of phytoplankton species productivity. *Limnology and Oceanography*, **21**, 590–6.

MUNRO, A. L. S. & BROCK, T. D. (1968). Distinction between bacterial and algal utilization of soluble substances in the sea. *Journal of General Microbiology*, **51**, 35–42.

NOVITSKY, J. A. & KARL, D. M. (1985). Influence of deep ocean sewage outfalls on the microbial activity of the surrounding sediment. *Applied and Environmental Microbiology*, **49**, 1464–73.

PAERL, H. W., BLAND, P. T., BOWLES, N. D. & HAIBACH, M. E. (1985). Adaptation to high-intensity, low-wavelength light among surface blooms of the cyanobacterium *Microcystis aeruginosa*. *Applied and Environmental Microbiology*, **49**, 1046–52.

SCHLEIFER, K. H. & STACKEBRANDT, E. (1985). *Evolution of prokaryotes*. London, Academic Press.

SCHMIDT, S. K., SIMKINS, S. & ALEXANDER, M. (1985). Models for the kinetics of biodegradation of organic compounds not supporting growth. *Applied and Environmental Microbiology*, **49**, 323–31.

SIERACKI, M. F., JOHNSON, P. W. & SIEBURTH, J. McN. (1985). Detection, enumeration, and sizing of planktonic bacteria by image-analyzed epifluorescence microscopy. *Applied and Environmental Microbiology*, **49**, 799–810.

SIMKINS, S. & ALEXANDER, M. (1985). Nonlinear estimation of the parameters of Monod kinetics that best describe mineralization of several substrate concentrations by dissimilar bacterial densities. *Applied and Environmental Microbiology*, **50**, 816–24.

SIMON, M. (1985). Specific uptake rates of amino acids by attached and free-living bacteria in a mesotrophic lake. *Applied and Environmental Microbiology*, **49**, 1254–9.

STAHL, D. A., LANE, D. J., OLSEN, G. J. & PACE, N. R. (1985). Characterization of a Yellowstone hot spring microbial community by 5S rRNA sequences. *Applied and Environmental Microbiology*, **49**, 1379–84.

WIGGINS, B. A. & ALEXANDER, M. (1985). Minimum bacterial density for bacteriophage replication: implications for significance of bacteriophages in natural ecosystems. *Applied and Environmental Microbiology*, **49**, 19–23.

THE COMMUNITIES – SOLID PHASE SYSTEMS

SOILS AND SEDIMENTS
AS MATRICES FOR MICROBIAL GROWTH

D. B. NEDWELL AND T. R. G. GRAY

*Department of Biology, University of Essex, Colchester, Essex
CO4 3SQ, UK*

This review will highlight some of the factors affecting microbial growth that are characteristic of soils and sediments. The bulk of the volume of a soil consists of finely divided, inorganic materials which do not act as substrates for metabolism but nevertheless have a profound effect on the liquids, inorganic and organic solutes, gases and microorganisms that occupy the pores between the particles. In sandy materials, only about 30% of the volume is pore space, although the pores are relatively large in diameter. In clays, 50% of the volume is pore space, although the pores are much smaller in diameter. These pores are filled with liquids and gases to varying degrees. Heterotrophic microbial growth will only be sustained in these matrices if there is a sufficient input of organic material. Peculiarities of the supply, distribution and utilization of this organic material will also be considered.

One of the major consequences of the abundance of solids is that movements of organisms, water, gases and substrates are much more restricted than in aqueous environments. This leads to the development of environmental heterogeneity and it is with this aspect that we start.

THE DEVELOPMENT OF HETEROGENEITY IN SOILS AND SEDIMENTS

Gross heterogeneity

Soils arise from the laying down of weathered particles formed from parent rock material (Foth, 1984). If the rate of erosion is less than the rate of particle formation, a soil will develop on top of the parent material. The processes which lead to the formation of a soil will vary in their duration and intensity, resulting in the formation of different soil types. The primary factors that affect these processes will be the nature of the parent rock material, the climate and the types of macro- and microorganisms that become established. During the long process of soil formation, major climatic changes have occurred so that today's soils may reflect not only current factors and processes but those of preceding eras. Most soils have developed during the last hundred million years but in the temperate regions which have been subject to repeated glaciations over the last 1.5 million years the soils are relatively recent. However, even in these soils much of the boulder clay deposited by the glaciers contains previously weathered clays as well as newly weathered materials.

During the development of soils, the solid matrix is relatively stable although subject to periodic disturbance through the action of burrowing animals, root growth and cultivation. A consequence of stability is that vertical environmental gradients develop between the underlying parent material and the above-ground atmosphere. These gradients are caused by differences in the relative rates of production and consumption of organic matter at the surface and at depth in the soil, differences in the relative rates of production and removal by plant roots of inorganic materials, differences in the relative rates of production, consumption and diffusion of carbon dioxide, oxygen and other gases in the atmosphere and the soil, and the progressive chemical and physical interaction between the organic and inorganic solids and downward-percolating rain water (or water rising from the water table by capillary action). Over a period of time, interactions between these different gradients result in the formation of more or less distinct soil horizons, each providing fundamentally different environments for microbial growth. Fig. 1a and b show the variation in a number of important factors that would affect microbial growth in a podsol and an altosol, soil types from which are derived many of the present-day soils in Great Britain. The number of horizons in a soil profile changes with time. Surface horizons appear first during a few decades, the middle horizons next during the first 5000 years, and the deeper horizons over even longer periods (Fitzpatrick, 1971).

Erosion of terrestrial soils may cause movement of the particles through air or water to be deposited elsewhere as a sediment, e.g. as dunes or riverine, lacustrine, estuarine and marine deposits. Characteristically, these sediments are relatively homogeneous in respect of particle size at any one place, as the rate of sedimentation is dependent on particle size and the depositional environment. Small particles which remain in suspension even at low current velocities are transported for greater distances, hence the occurrence of finer muds on the upper reaches of tidal marshes. However, processes leading to erosion and transport will vary in intensity over periods of time so that layers of different-sized particles may be deposited on top of one another. In physically high-energy environments, surface materials may be subject to turbulence so that solids are constantly resuspended and deposited, together with any microorganisms that are in them. Away from the continental shelf, oceanic sediments are dominated by particulate material, principally carbonates and silicates, which are of biogenic rather than terrestrial origin and derived from planktonic microorganisms.

Typically, aquatic sediments have a surface aerobic layer of variable depth, below which the sedimentary environment is anaerobic (see also p. 35). The depth of the aerobic layer depends upon a balance between the rates of oxygen removal by the sedimentary microflora and diffusion of oxygen into the sediment from the overlying water. Factors which stimulate respiration, such as increased temperature and increased supplies of organic carbon, decrease the depth of the aerobic layer. Heterogeneity within sediments is associated principally with the establishment of vertical concentration gradients of solutes. These may be produced through removal of electron acceptors, e.g. oxygen, nitrate and sulphate, from the sediment during microbial respiration, resulting in solutes diffusing into the sediment from the overlying water. The products of detrital mineralisation, such as phosphate and ammonium ions, are usually higher in concentration in the sediment than the overlying water and so they diffuse upwards. Bottom sediments are thus both important sinks and sources of inorganic nutrients for the plankton.

The amount of organic material reaching aquatic bottom sediments is a function of the depth of the overlying water column. Microbial decay of organic material as it settles through the water column progressively reduces the total amount of surviving organic material with increased depth of water (Fig. 2), and also changes its quality as the more labile, degradable organic components are removed. In the deep ocean only a small fraction of the most refractory organic components of the detritus, initially derived from primary production in the euphotic zone, may survive to reach the sediment/water interface, and this is reflected in the trend of reduction of the organic carbon content of sediments as water depth increases (Fig. 3). The effect upon microbial activity is illustrated by the similar trend of decrease of oxygen uptake rates by sediments with increased water depth (Fig. 4), reflecting diminished respiratory

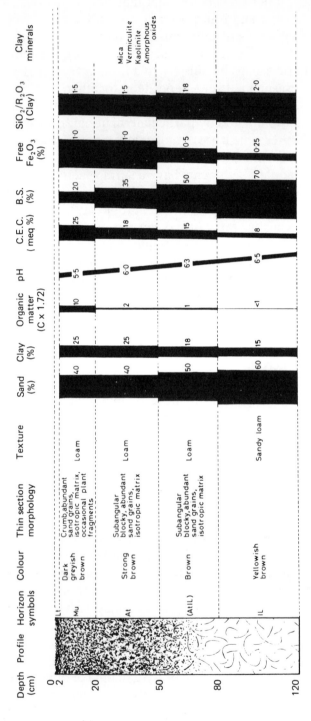

a

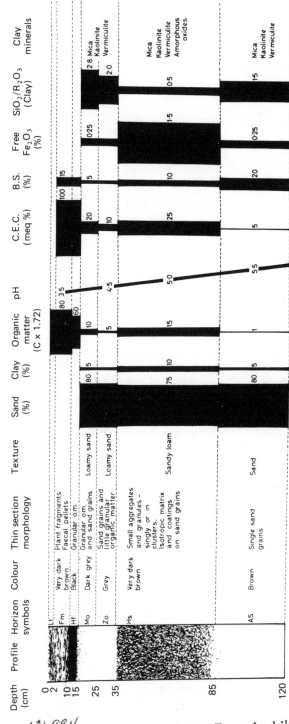

Fig. 1. A comparison of environmental data for the horizons of two soil profiles: (a) an altosol (brown earth), with Lt (litter), Mu, (mullon or A_1), At (alton or A_2) and IL (underlying material) horizons; and (b) a podsol, with Lt (litter), Fm (fermenton or F), Hf (humifon or H), Mo (modon or A_1), Zo (zolon or A_2), Hs (husequon or B_{hfe}) and AS (underlying material) horizons. G.E.C. = Cation Exchange Capacity expressed as meq per 100 g of soil of the <2 mm size fraction; B.S. in Base Saturation expressed as a % of G.E.C.; R = Fe + Al; o.m. = organic matter. (After Fitzpatrick (1971), with permission from Longmans).

b

131,994   College of St. Francis Library
Joliet, Illinois

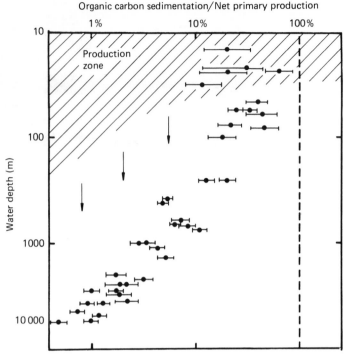

Fig. 2. Amounts of residual organic matter remaining with depth in the water column after settlement from the euphotic zone. Reprinted with permission from *Nature*, **288**, 262–3, © copyright 1980 Macmillan Journals Ltd.

activity of the microbial community as availability of organic electron donors decreases with increased water depth.

In historical times, soil and sediment development have been greatly influenced by man's activities. Man has altered natural vegetation patterns, producing large areas of heathland, savanna and prairie and thus modifying the soil, as well as improving soils through reclamation from the sea and devastating others through over-exploitation.

This degree of complexity at the gross level ensures that the data from laboratory model systems are of limited value. The comparison of the additions, losses, translocations and transformations of materials in a field soil with those in a column containing an artificial matrix through which a medium is flowing shows why this is so (Fig. 5a and b, adapted from Foth (1984)). Laboratory models are extremely useful tools for investigating individual, underlying processes in soil, but less useful for understanding how these different processes interact. They are also useful for studying short-term processes which

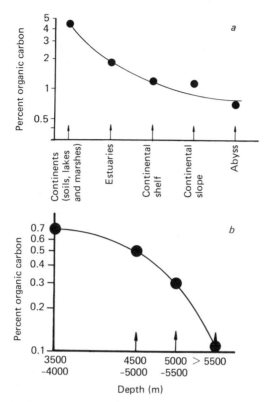

Fig. 3. Average organic carbon contents of sediments: (a) in major types of sedimentary environments; (b) in marine abyssal sediments. The depth axis is not to scale. (Redrawn from Vigneaux *et al.*, 1980.)

may or may not be related to long-term changes taking place in the natural environment.

Heterogeneity of microhabitats

In space

Although soil and sediment horizons each have characteristic properties, considerable variation in the environmental factors that affect microbial growth occurs from point to point. For heterotrophic bacteria, it is the distribution of their organic substrates which primarily determines the occurrence of pockets of activity, sometimes referred to as microhabitats. The majority of the organic substrates entering soil are either insoluble or are soluble but packaged in cells and tissues with insoluble boundaries. Before decomposition takes place, therefore, there will be limited dispersion of the substrate within

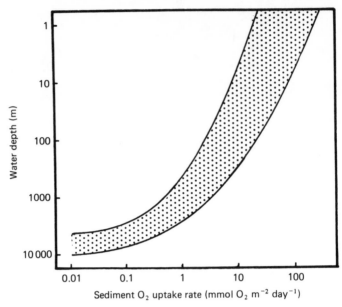

Fig. 4. Uptake of oxygen by microorganisms in marine bottom sediments in relation to depth of overlying water. (Redrawn from Jørgensen (1983), with permission from SCOPE.)

the inorganic soil matrix, and litter material often spends a considerable period of time at the surface in the O horizons where much comminution and decomposition takes place. Nevertheless, some soluble materials will diffuse into the soil water and be leached out of the surface horizons. Also, animals may transport and distribute larger pieces of organic material unevenly within the soil matrix.

Organic matter is also produced within the soil by plant roots. In 1 ha of a 50-year-old pine forest, the soil to a depth of 200 cm may contain about 4 tonnes of living roots less than 0.3 mm in diameter, about 50% of which are replaced each year. This biomass of roots may represent 50% of the mass of an intact tree (over 90% in the case of some grasses). At the soil–root boundary, microorganisms can colonize the surfaces of living cells and root hairs as well as crushed and partly disorganised cells. They can also penetrate and utilise the contents of living cells, exploit the mucigel and sloughed off cells, and use the soluble exudates that diffuse out from the younger parts of roots. Living roots raise the pH of the soil by as much as 1 pH unit for a few millimetres around the root and set up water and ion gradients as well.

The extent of these gradients, as well as gradients of soluble breakdown products of organic matter, will be affected not only by the diffusion coefficients of the different substances but also by the degree of discontinuity of the water films in the soil pores and the amount of downward mass flow of water. Solute diffusion decreases approximately linearly with volumetric water content so that it will be restricted at low matric potentials, potentials that will also restrict bacterial movement (Griffin, 1972) (see below). All these factors will be influenced by variations in the textural composition of the soil and rates of loss of water through evaporation or drainage. When the pore spaces between the solid particles are water-filled, then the pattern of decomposition of the substrate will change, with anaerobic processes predominating where rates of oxygen consumption exceed rates of supply of oxygen by diffusion or percolation. Thus, surrounding a solid substrate will be a volume of soil, changing in diameter with time, into which miocroorganisms can grow outwards to exploit dissolved nutrients and spread to other substrate pockets. Organisms with a limited ability to spread and disperse will be at a disadvantage, for the amount of substrate present in any one microhabitat will be small and readily exhaustible. Their only alternative strategy will be to reduce their metabolic rate by forming resting stages and to wait for fresh substrate to arrive or to be carried to substrate by burrowing animals.

An uneven distribution of substrate would be expected to influence the morphological types of microorganism that could grow well. In those types of aquatic environment in which the majority of substrate is soluble, small unicellular organisms with large surface:volume ratios are favoured. However, in soils and sediments only some of the substrate is soluble and hence occurs in the water films in the pore spaces. The remaining solid substrates may be adsorbed onto inorganic particles, or trapped inside clay lattices, or they may exist as large discrete particles. The decomposition of such solid substrates presents several interesting features:

1. they are usually small in quantity at any one point, so that rapid microbial growth will cease quickly when the growth-limiting nutrient is exhausted;
2. for rapid exploitation, penetration of the substrate is desirable; for slow exploitation, gradual erosion of the surface is sufficient;
3. individual substrate particles may be separated by considerable distances (in microbial terms), although individual substrate particles may be ingested or moved in other ways through animal activity;

(a)

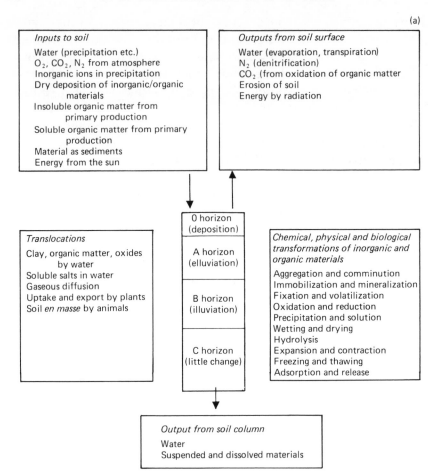

Fig. 5. A comparison of inputs, outputs, translocations and transformations taking place in (a) a soil (modified from Foth, 1984) and (b) a glass column supplied with a nutrient medium and containing a pure culture of a heterotrophic bacterium. Note that the time scale for the soil system is a long one, during which many environmental changes and fluctuations will occur. The time scale for a model system is generally short and environmental conditions are kept constant.

4. in soils with a clay content greater than 12–14%, aggregation of particles to form crumbs will occur; aggregate formation may entrap both substrate and microorganism, effectively isolating them from other parts of the soil;

5. humified materials are often deposited in particular horizons in a soil profile; organisms exploiting these substrates may need to adhere to the soil particles to avoid being separated and washed out of the soil.

(b)

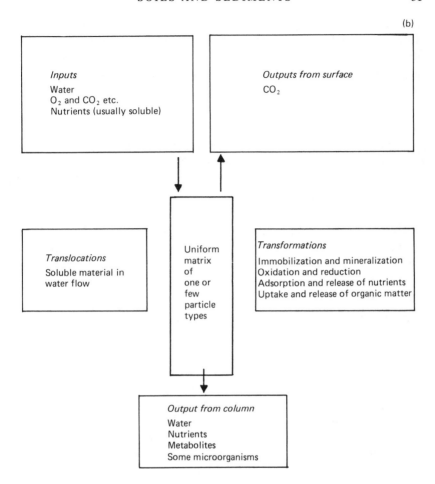

Thus, organisms inhabiting soil are often found to have developed one or more of the following characteristics: filamentous/mycelial/cord habit or small colonies of unicells; resistant wall structures; dormant propagules or shut-down cells; adhesive mechanisms; dimorphic or pleomorphic potential. Gray & Williams (1971a) proposed a series of growth patterns of microorganisms in soil which incorporated this diversity. These seven patterns were: (i) non-migratory unicells, (ii) migratory unicells, (iii) plasmodia, (iv) substrate-restricted hyphae, (v) locally spreading hyphae, (vi) mycelial strands/rhizomorphs, and (vii) diffuse spreading hyphae. These concepts have been refined and expanded by Cooke & Rayner (1984; see also this volume), partly in the realization of the pleomorphic potential of many organisms. Dowson, Rayner & Boddy (1986) have shown that wood blocks placed about 5 cm from an inoculated wood block cause marked changes in the form and growth characteristics of the mycelial network of two basidiomycetes, *Hypholoma fasciculare* and *Phanaerochaete velutina*. Initially, sparse exploratory mycelial systems were produced from the inoculated block. The direction of growth was changed toward the uninoculated block and further radial extension ceased following contact with it. Regression of the mycelial front not in contact with the block followed and at the same time fan-shaped systems of effuse mycelium spread over the uninoculated block.

It is clear from these experiments that fungi have the potential to change their growth pattern in response to the presence of substrates. The most dramatic change that can take place is in those dimorphic forms capable of existing as yeasts or mycelium. It is generally considered that yeasts are at a disadvantage in soils and sediments, as they are incapable of penetrating solids and have limited ability to disperse themselves by growth or spore production. They are rarely isolated from soils, except where they have found their way into the soil from surfaces of living leaves and fruit where they exploit exuded carbohydrates.

However, other unicellular colonial organisms such as bacteria do not seem to be at such a serious disadvantage for they can produce a sizeable biomass, e.g. Gray, Hissett & Duxbury (1974). But where are these other unicellular forms found? The majority of them probably occur on plant surfaces. In this case the surfaces are those of roots (Foster & Rovira, 1976) or hyphae (Siala & Gray, 1974; Fradkin & Patrick, 1982), where they utilise the nitrogen-rich, soluble exudates or metabolic byproducts of fungal decomposition of insoluble polymers. Those bacteria that are found on non-living, inorganic, solid particles occur in very small colonies, the majority of them on particles coated with iron and humified materials (Siala, Hill & Gray, 1974). They may well obtain many of their nutrients from water percolating through the soil. However, Griffin & Luard (1979) have pointed out that if a cell is on a solid surface, it and its progeny will be trapped by surface tension at one site unless the water film is at least as thick as the diameter of the cell. Furthermore, for movement to occur, the neighbouring pore system must be water-filled and the pore necks big enough to allow passage of the cell. Thus, if the matric potential of a soil is $-147\,kPa$ or less, pore necks of $1\,\mu m$ radius will be air-filled and restrict bacterial movement.

Movement of larger yeasts or fungal zoospores (radius 10 μm) would be prevented when the matric potential of the soil reached −14.7 kPa. In soils where pore neck sizes are highly variable, the probability of a continuous water-filled pathway existing is much reduced, so that even a matric potential of −100 kPa, equivalent approximately to a water content of 20% in a loam soil and 50% in a clay soil, will have a marked effect on bacteria and an even bigger effect on yeasts and zoospores. The activity of unicellular organisms is thus limited by exhaustion of nutrients following a decline in the matric potential. It is not surprising, therefore, that Chuang & Ko (1981) found that there is an inverse linear relation between propagule volume and abundance of microorganisms in soils (on a logarithmic basis), though this will be due partly to the impossibility of sustaining large numbers of large propagules on the energy available. Lundgren (1984) has also found that most soil bacteria are smaller than 1 μm and that 55% are smaller than 0.5 μm. Soil protozoans would also be limited by the number of available pore spaces. Darbyshire, Robertson & Mackie (1985) have measured the distribution of different-sized pores above 3 μm in diameter in agricultural soils which would allow the passage of protozoans, but data on the water status of these pores are not available. Filamentous organisms behave differently for they are active at lower values of matric potential. Their growth is prevented not indirectly by nutrient exhaustion but more directly by matric potential and water availability. It is noteworthy that matric potentials can be much lower than solute potentials in nature and that in some dry soils survival of xerotolerant forms only can be expected (Lanyi, 1979).

Despite the presence of pores through which oxygen can diffuse rapidly, anaerobic environments also occur in soil, especially when there is a high clay content and a tendency to produce aggregates. The finer pores inside the aggregates remain water-filled, even at quite low matric potentials, and Greenwood (1961) has estimated that soil aggregates with a diameter greater than approximately 3 mm (depending upon temperature, organic matter content and matrix texture) will have anaerobic centres. Larger crumbs will thus have an anaerobic core surrounded by an aerobic shell. However, microbial activity in the core may be negligible, as substrates which are rapidly exhausted there will not be replaced (Allison, 1968).

In some parts of the soil profile, notably the litter layers, organic substrates and aerobic environments are not discontinuous. In these habitats, spreading hyphae of fairy-ring type fungi are extremely common where growth, albeit slow, is reminiscent of that occurring on agar plates (Cooke & Rayner, 1984). However, in these environments, fungi may still not be distributed uniformly, even if they are indiscriminate in their choice of substrate. An excellent example of such a fungus is *Mycena galopus*, which has the potential to attack all the major constituents of plant litter and is known to colonize a wide variety of broad-leaved and coniferous litter types (Frankland, 1984). It grows in close proximity with other basidiomycetes such as *Marasmius, Collybia* and *Cystoderma*. In laboratory experiments, *Mycena galopus* was outcompeted by *Marasmius androsaceus* when growing on sterile spruce litter but, in the field, these two fungi were separated vertically, with the mean depth of origin of the fruit bodies of the former being 6 mm and of the latter, 1 mm (Fig. 6). It has been suggested by Newell (1984) that this distribution is

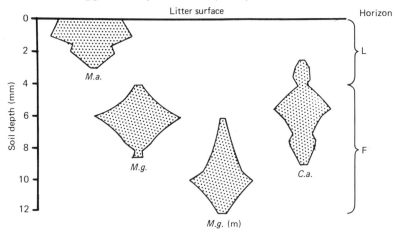

Fig. 6. The fruiting depths of *Marasmius androsaceus* (*M.a.*), *Mycena galopus* (*M.g.*) and *Cystoderma amianthinum* (*C.a.*) in a *Picea sitchensis* plantation. *M.g.* (m) = mixed species clumps of *M. galopus* with *M. androsaceus*. The width of the kite bars at any depth is proportional to the percentage number of basidiocarps originating at that depth. (After Frankland (1984), with permission of the British Mycological Society.)

caused partly by the collembolan mite *Onychiurus latus* which grazes differentially on these fungi, preferring *M. androsaceus* mycelium. These collembolans are most abundant at a depth of about 6 mm. Dix (1984) has also shown that *Mycena galopus* is less tolerant than *Marasmius androsaceus* of low water potentials, and thus the observed field distribution may be due to a combination of physical and biological factors. Frankland (1984) has also shown that *Mycena galopus* is not distributed at random horizontally over the forest floor: surprisingly, its position is related to the position of trees even though it is not mycorrhizal. Further studies on the distribution of this fungus may be assisted by the development of specific fluorescent antibodies that will enable its mycelium to be identified *in situ* in the absence of fruit bodies; such antisera are now being developed (Frankland *et al.*, 1981; Chard, Gray & Frankland, 1983, 1985*a,b*).

In waterlogged sediments the picture is very different, for here the environment is often anaerobic, fungi are absent and bacteria play a much greater role in decomposition. The diffusion coefficient of oxygen in water is about 10^4 times lower than in air and the concentration of oxygen in water is also low, there being only 0.28 mmoles l^{-1} at $20°C$ in freshwater and 0.23 mmoles l^{-1} at $20°C$ in seawater. Thus, it is easy to understand why oxygen-consuming, waterlogged systems become anaerobic below the surface. The depth of the surface aerobic layer is a function of several different factors. Increases in available organic matter and temperature enhance respiratory removal of oxygen and thus diminish the aerobic layer; increases in sediment particle size and porosity enhance transport rates and extend the depth of the aerobic layer. However, sulphate reduction can occur in anaerobic microenvironments, even within the generally aerobic surface layers of marine sediments, possibly associated with localised organically rich copepod faecal pellets where oxygen is consumed rapidly (Jørgensen, 1977). In a highly organic sediment with a restricted depth of aerobic layer, a significant proportion of the organic material deposited on the surface may pass through the aerobic zone and be mineralised anaerobically. In marine sediments, the major electron acceptor after oxygen removal is sulphate, which is present at approximately 20 mM concentration in seawater, compared to oxygen at approximately 0.23 mM concentration. Jørgensen (1982) has emphasised the importance of sulphate reduction in the mineralisation of organic material in those shallow-water marine sediments which have a large input of organic matter and a restricted aerobic layer. Approximately 50% of the total mineralisation of organic carbon is brought about by sulphate-reducing bacteria in these environments. By contrast, deep oceanic sediments have a low organic input, are at low temperature and have extensive aerobic layers in which the majority of organic carbon is mineralised aerobically. Sulphate reduction is therefore less important in these deep-water marine environments because of the extensive aerobic layer and it is of little quantitative significance in freshwater sediments as freshwater contains little sulphate. Nitrate reduction and carbon dioxide reduction (methane formation) are the most important respiratory processes in anaerobic freshwater sediments (Jones & Simon, 1980).

In time

Microenvironments are not only spatially heterogeneous but also heterogeneous with time. Bosatta & Berendse (1984) have reviewed the ways in which the nutrient composition of substrates can change. Thus, during litter decomposition a phase of nitrogen accumulation is succeeded by a period of nitrogen release, a change which probably also occurs with other mineral nutrients (Swift, Heal & Anderson, 1979). The change from one phase to the other depends upon a change in the carbon:element ratio of the litter. Bosatta & Berendse (1984) point out that carbon mineralisation is negatively correlated with nitrogen mineralisation in both laboratory and field experiments. Thus the common idea that the priming effect of nitrogen additions on soil nitrogen mineralisation is due to enhanced microbial activity may not be true. A qualitatively different response to the same perturbation can be expected depending upon whether the system is carbon- or nitrogen-deficient. In unperturbed systems, Bosatta & Staaf (1982) showed that the retention and release of nitrogen was regulated by decomposition rate and initial nitrogen concentration of the litter. Increased decomposition rates reduced the rates of nitrogen release per unit of carbon mineralised. Litters with a low initial nitrogen concentration immobilised more nitrogen but over a shorter period of time. Thus, in all litter types, carbon:nitrogen ratios converge as decomposition proceeds.

More recently, Bosatta & Agren (1985) have proposed a theory for the microbial decomposition of heterogeneous substrates in soil. They point out that decomposing organic matter is heterogeneous both because of its original composition and because of new compounds produced during decomposition and humification. Both the carbon left in the substrate and that returned by the microbes can be thought of as being of lower quality than that which was present originally, i.e. more refractory. This, as decomposition proceeds the substrate becomes progressively poorer in quality. Quantity and quality of soil organic matter are the result, therefore, of interactions between many processes with time constants ranging from seconds to thousands of years.

The patterns and rates of decomposition could be described, according to Bosatta & Agren (1985), in terms of a few critical functions of the microbial population (mean concentration of carbon in the microbial biomass, mean concentration of nitrogen in the microbial biomass, efficiency of carbon utilisation, the rate of substrate utilisation), the quality of the carbon resource, and the initial quality and nitrogen:carbon ratio of the litter. The most important functions were the microbial efficiency and the rate of substrate utilisation. Bosatta & Agren pointed out that if the degradation in substrate quality were more rapid than that in substrate quantity, then eventually a finite amount of undegradable material would remain undecomposed. For *all* the material to be degraded, the fall in quantity would need to be more rapid than that in quality which could be achieved by decreasing microbial efficiency as quality decreased. Application of theoretical models to experimental data obtained by McClaugherty *et al* (1985) for the decomposition of aspen, hemlock, white maple, sugar maple, red maple and white pine leaves over 2 years suggested that the efficiency of utilisation of litter varied from 0.19 (white pine) to 0.367 (aspen). The assumption that the efficiencies remained constant during decomposition was shown to fit the data, except for the final observations in the time series. These coincided with a second period of nitrogen mineralisation and a shift in extracellular enzyme activity. McClaugherty *et al.* ascribed this to a change from one microbial community to another with a different and lower efficiency. If this interpretation turns out to be correct, then these latter organisms may be regarded as truly autochthonous (Winogradsky, 1924).

Soils and some sediments may be subject also to seasonal changes. Soil is a poor conductor of heat so that large temperature changes in the atmosphere may not change soil conditions much below the surface layers except in extreme environments. Nevertheless, some other major seasonal changes in the environment will occur in sub-surface soil, mostly related to changes in the moisture content and consequent degree of aeration. The buffering effect of surface soils on sub-surface horizons is illustrated by data obtained from soils during forest fires, where surface temperatures of 850° C can be reached while soil at about 4 cm depth remains at 16° C (Hofmann, 1917). As soils are often dark in colour, direct insolation can also cause surface soils to increase in temperature well above air temperature. Antarctic fellfield soils can experience diurnal temperature changes of 40° C, although these effects are damped out with increasing depth (Fig. 7) and only longer-term seasonal temperature cycles are observed in deeper layers.

In intertidal sediments which are exposed to air, seasonal, diurnal

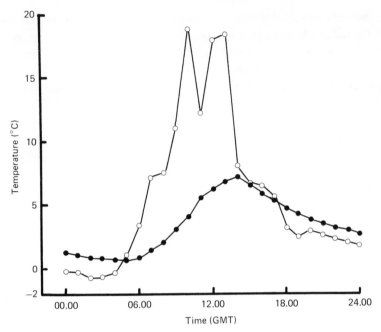

Fig. 7. Diurnal temperature changes at the surface (○) and at 10 cm (●) depth in an Antarctic fellfield soil during summer. (Unpublished data from D. W. H. Walton, British Antarctic Survey, Cambridge.)

and tidal cyclic changes occur and temperatures about 10° C above air temperature have been recorded in such surface muds in summer in Great Britain. However, water has a large thermal capacity, so that temperature changes in permanently submerged aquatic sediments are damped relative to those in air and these sediments are only affected significantly by seasonal changes: even these are small in deep oceanic environments. Other indirect seasonal changes can occur in deep sediments, however, because of the settling of organic detritus formed periodically in the euphotic zone. The microbial communities in different sediments may be influenced, therefore, by very different environmental regimes. When short-term variations, e.g. diurnal and tidal cycles, occur, speed of physiological response by a species to environmental change may be more important than its steady-state physiological competitiveness. Where only long-term variations occur, the reverse may be true.

Heterogeneity at the molecular level

The surface chemistry of soils has been reviewed extensively by Sposito (1984). Some of the solids present in soil exhibit considerable surface activity, notably clays and humic materials. Since clays are, by definition, finely divided particles, the surface area they possess is enormous. In a cubic metre of clay, each particle having a diameter of $2\,\mu m$, the total surface area would be about $3 \times 10^6 m^2$. On this surface, in contact with the soil water, are many chemically reactive units separated by fixed distances, interacting with one another as well as with molecules in the water film. The principal reactive units on clay surfaces are siloxane ditrigonal cavities, inorganic hydroxyl groups and a variety of organic groups in humic substances which are themselves complexed with clay.

The siloxane ditrigonal cavity (diameter 0.26 nm) is formed by six corner-sharing silica tetrahedra and is thus bordered by six oxygen atoms, each with a set of lone pairs of electron-orbitals. If the clay has not undergone isomorphous replacement in the underlying layer of the molecule, these cavities act as electron donors and form unstable complexes only with neutral dipolar molecules such as water. However, if isomorphous replacement of trivalent aluminium with divalent iron or magnesium has occurred, e.g. as in montmorillonite, an excess negative charge is generated which enables cations to be complexed. Isomorphous replacement of tetravalent silicon with trivalent aluminium, e.g. as in vermiculite, also generates an excess charge but as it is distributed over fewer surface oxygen atoms than in the previous example, even stronger complexes with cations and dipolar molecules occur. Inorganic hydroxyl groups are the most abundant and reactive groups but their properties are varied, depending upon the atoms with which they are coordinated. They are capable of complexing hydroxide anions, hydrogen ions, oxyanions such as HPO_4^{2-}, and metal cations.

Organic compounds in the soil solution can be bound to clay particles and themselves then bind protons. The mechanisms underlying the adsorption of soluble organic compounds to clay include cation exchange, protonation, anion exchange, water bridging, cation bridging, ligand exchange, hydrogen bonding and van der Waal's interactions (Mortland, 1970; Greenland, 1971). Generally, the quantity of dissolved organic material adsorbed decreases as the pH increases above 4.0 (Theng, 1979). Humic materials have a number of surface functional groups, including carboxyl, carbonyl, amino, imidazole, phenylhydroxyl, sulphydryl and sulphonic groups. In well oxidised organic matter, the carbonyl, carboxyl and phenylhydroxyl groups are the most significant (Stevenson, 1982), binding protons with increasing degrees of relative stability. The variability of the nature and position of the functional groups on the surface ensures that different interactions occur between the groups which, in turn, affect the stability of proton complexes.

The idea that enzymes may be immobilised by humic materials has been explored by Nannipieri, Ceccanti, Cervelli & Sequi (1978) and developed by Burns (1983). Recently, Serban & Nissenbaum (1986) suggested, on the basis of laboratory experiments, that extracellular enzymes such as peroxidase and catalase in the soil solution could become incorporated in the rigid, three-dimensional, macromolecular matrix of humic acid and that this rendered them resistant to decomposition by pronase and thermal denaturation but did not impair their activity. They suggested that humus-immobilised enzymes seemed to retain their activity more consistently than those bound to clay and might be responsible for the persistence of peroxidase-like activity in sediments 7 million years old.

The reactive groups at the soil surface all have profound effects on the water and its solutes surrounding the solid particles. Water is adsorbed and its structure altered in thin films. There is a more rigid configuration of water molecules interacting with fixed reactive groups on the clay surfaces and with cations in the water. These effects will be felt through several layers of water molecules so that adsorbed water extends for up to 3 nm around kaolinite, 5 nm around vermiculites and perhaps up to 10 nm around montmorillonite particles (Sposito, 1984). Adsorbed water has different solvent properties to bulk liquid away from the surface. There will be enhancement of complex formation between dissolved materials and between exchangeable cations and the siloxane ditrigonal cavity, thus retarding the development of the double diffuse layer. Acidity will also increase.

Microorganisms also possess fixed reactive groups on their surfaces, including COO^- and PO_4^{3-} groups (Rogers, 1979). The degree of attraction between particle and bacterium therefore depends upon a balance between electrostatic repulsion, electrostatic attraction and van der Waal's attractive forces. Marshall (1976) has shown that there are two distances of separation between surfaces at which net attraction will occur and that for adsorption to take place, the repulsion barrier between these two distances must be overcome. The degree of repulsion decreases with decreasing radius of curvature of the particles and increasing electrolyte concentration. Thus, as soils dry out and electrolyte concentration increases, the chance of adsorption will increase, so that bacteria, because of their relatively large size, could become coated with clay particles or become attached to larger humus-coated sand grains. Stabilisation of such an attachment may require the formation of stronger bonds, through the production of fimbriae, lipopolysaccharides and peptidoglycan in the cell wall or extracellular polymers and capsules which are often polysaccharides (Fletcher & Marshall, 1982).

This high degree of variability and reactivity in the matrix of soils and sediments thus ensures that all the components involved in metabolism (water, anions, cations, hydrogen ions, organic matter, enzymes, metabolites and cells) will be distributed non-uniformly in relation to particle surfaces. Furthermore, the environmental conditions surrounding cells will be quite different from those occurring in 'liquid water' (Stillinger, 1980). Unfortunately, non-destructive methods of measuring these environmental factors are not available and so the precise conditions under which much decomposition takes place remain a matter for conjecture.

The potential complexity of substrate–clay–cell interactions has been underlined by Marshman & Marshall (1981a) who studied the growth of bacteria on pure proteins adsorbed on clay minerals. Their results were best explained by assuming that protein was bound at two sites; at one, the protein was available to bacteria, at the other it was not. These sites did not appear to coincide with internal and external lattice surfaces. In a second paper, Marshman & Marshall (1981b) suggested that clays such as montmorillonite could affect microbial growth through interactions with the organisms, their substrates, individual enzymes and growth factors. This increases markedly the complexity of the flow of nutrients to organisms. Dashman & Stotzky (1986) have also shown that complexes between different amino acids and peptides and montmorillonite are differentially available to microorganisms. The affinities of permeases for different amino acids are some 100–10 000 times greater than the affinities of clays for the same amino acids, so these are utilised readily; information on peptides was lacking. The differences in affinity of the permeases for different amino acids may account for the differences in the extent of their utilisation. They also postulated that the yield of energy from the intracellular metabolism of some amino acids might be less than the energy required to remove the substrate from clay and transport it into the cell. Thus cysteine was not utilised when bound to montmorillonite or kaolinite whereas proline and arginine were. The requirement for energy to be expended to remove materials from clay and humus may thus be an important consideration when determining the relative amounts of energy available for growth and maintenance.

INPUT AND METABOLISM OF ORGANIC MATTER

As residual organic material sinks through the water column, it becomes increasingly refractory as it is depleted in nitrogen and phosphorus relative to carbon, the C:N:P ratio changing from approximately that of algae (106:16:1) near the surface to 106:3.5:0.11 deeper in the water column (Suess & Müller, 1980). In terrestrial systems, higher plant debris, with a relatively large component of refractory structural polymers such as lignin, cellulose and hemicelluloses which are rich in carbon, is even less suitable as a balanced substrate for microbial growth than the algal detritus in aquatic systems (Table 1).

Table 1. *Element ratios in three major components of forest ecosystems (Melillo & Gosz, 1983)*

	Carbon	Nitrogen	Phosphorus
Vegetation (woody tissue)	1500	10	1
Litter	500	10	1
Soil	120	9.4	1

In both soils and sediments the proportions of nitrogen and phosphorus relative to carbon increases in organic material after deposition. This is a result of the mineralisation of detrital organic carbon and the sequestration of nitrogen and phosphorus into the biomass of the microbial community which degrades the detritus. However, it appears from recent work that despite the C:N:P balance, it is only during the initial stages of detrital decay, when labile soluble carbon is present and available, that microbial degradation of detritus is nitrogen-limited. Thus, the rate of breakdown of wheat straw has been shown to be largely dependent upon the size of an immediately available soluble carbon pool, and of a pool of intermediately available carbon (Reinertson *et al.*, 1984). The addition of available carbon in the form of glucose to wheat straw stimulated its breakdown after 10 days' incubation, but not if added during the initial 10 days when microbial activity was nitrogen-limited (Knapp, Elliott & Campbell, 1983).

Deposited organic detritus is incorporated into the surface layers, the rate of mixing being strongly influenced by the activity of animals which turn these over. O'Brien (1984) has followed downward movement of organic matter in soil by measuring variations in radiocarbon enrichment of organic matter with depth. In pasture soils, earthworms enhanced organic carbon incorporation and transport, in contrast to forest soils with a similar organic content where the smaller animals transported almost no organic matter downwards. Billen (1982) has also shown that bioturbation can influence significantly the vertical transport of organic matter and nutrients in surface layers of sediments.

The organic content of a soil or sediment may be very small in coarse-grained sands or gravels or very large in entirely organic soils. At any one time, it will be the result of a balance between the input of detritus and its removal by mineralisation. Thus, the organic content may be low because the input is low or because, despite a high input, it is being degraded rapidly. In the former case, flux of carbon and energy is small, in the latter case it is large. Much of the organic material may be refractory and unavailable to the microbiota, resulting in the accumulation of undegraded organics, as in the case of peats. Therefore, it is the turnover of the organic matter which determines the total flux of energy through the system, while it is the concentration of *available* organic carbon which determines the amount of energy available to the microflora at any one time.

Energy limitation of growth

Evidence suggests that in the overwhelming majority of soils and sediments the microflora is energy-limited, notwithstanding the large amounts of organic carbon which are sometimes present. Pamatmat *et al.* (1981) suggested that heat outputs from sediments were commensurate with the microbial community being energy-limited and that addition of available electron donors almost invariably stimulated respiration by the soil microflora. Addition of glucose to soil has been shown to stimulate respiration by the soil microflora, to increase soil microbial biomass and soil ATP (Nannipieri, Johnson & Paul, 1978; Sparling, Ord & Vaughan, 1981; Ahmed, Oades & Ladd, 1982) and to stimulate mineralisation of plant detritus (Knapp, Elliot & Campbell, 1983; Reinertsen *et al.*, 1984). Indeed, saturation of the microflora in soil by added glucose permits estimation of microbial biomass, since the maximum initial rate of glucose oxidation is proportional to the biomass present (Anderson & Domsch, 1978).

Following glucose addition, microbial biomass and activity increase but these revert to the original levels as glucose is depleted. Again, addition of hydrogen to slurries of marine sediment stimulates both sulphate reduction and methane formation, reversing the normal inhibition of methane formation by sulphate-reducing bacteria which outcompete methanogenic bacteria for electron donors (Abram & Nedwell, 1978); sulphate reduction in marine sediments is electron donor-limited (Nedwell & Abram, 1979).

The adenylate energy charge is a relative measure of the proportion of energised adenylate, in the form of ATP, compared to the lower-energy ADP and AMP. In metabolically active prokaryotic cells, the energy charge seems to be relatively high (between 0.8 and 0.9), reflecting rapid energy input into the cellular adenylates. In dormant or inactive cells the energy charge falls to values between 0.5 and 0.8, while at energy charges below 0.5 prokaryotic cells die. Eukaryotic cells may survive at these values, however. Energy charges between 0.3 and 0.4 have been measured for soil microorganisms. These charges increased after the addition of glucose but still to a value lower than that obtained for actively growing laboratory cultures (Martens, 1985). These data support the idea of an energy-limited microbial community in soil. However, Brookes, Tate & Jenkinson (1983) found much higher values for soil organisms, as did Tateno (1985) who recorded values of 0.85, reducing to 0.46 when soil was dried. Tateno suggested that inactive soil organisms with low energy charges will decompose quickly, an explanation which seems reasonable judging from the rapidity with which cells in chloroform-fumigated soils are decomposed (Jenkinson, 1966, 1976). Deaney & Gray (1983) have also shown that proteins, nucleic acids and antigens in stained cells decompose quickly in soil. There do not appear to be any measurements of energy charge in aquatic sediments, although their energy-limited nature can be deduced from the stimulation of microbial activity when electron donors are added.

A consequence of the limited energy available in soils and sediments is that growth rates and microbial production rates are slow. Calculations of approximate microbial growth rates, based on rates of carbon production compared to biomass carbon in the microbial standing crop, suggest slow growth rates with generation times of about 3.3 days in forest soil (Chapman & Gray, 1986) and about 2.6 days in marine sediments from the Kiel Fjord and Bight (Meyer-Reil et al., 1980). However, estimates of generation times are very

Table 2. *Estimates of generation times of microorganisms in soils and sediments based on productivity:biomass ratios (PB); thymidine incorporation (THY); incorporation of adenine (ADEN); or turnover of ATP pool. Estimates based upon frequency of dividing cells have been omitted as the method over-estimates rates in sediments (Fallon, Newell and Hopkinson, 1983)*

Environment	Generation time (h)	Method	Reference
Soils			
Tundra	93	PB	Parinkina, 1974
Temperate	26–67	PB	Parinkina, 1974
Peat	39	PB	Clarholm & Rosswall, 1980
Humus	66	PB	Clarholm & Rosswall, 1980
Mineral soil	55	PB	Clarholm & Rosswall, 1980
Clay soil	3024	PB	Jenkinson & Ladd, 1981
Broadbalk	15 168	PB	Jenkinson & Ladd, 1981
Deciduous wood-land	79	PB	Chapman & Gray, 1986
Fungi in soil	161–936	PB	Kjøller & Struwe, 1982
Sediments			
Sands, Kiel Fjord	63	PB	Meyer-Reil *et al.*, 1980
Zostera bed	3.8–125	THY	Moriarty & Pollard, 1981
Zostera bed	144–1008	THY	Moriarty & Pollard, 1982
Coastal muds	118–3049	THY	Fallon *et al.*, 1983
Six marine sites	4–166	ADEN	Craven & Karl, 1984
	212–350	ATP	Craven & Karl, 1984

variable and are very dependent upon the method used and the assumptions made (Chapman & Gray, 1986). A selection of data which illustrate this point is given in Table 2. Very long doubling times are obtained if biomass estimates are based upon direct microscopic or fumigation techniques. Indeed, some of the doubling times are so long that it is difficult to believe that they have any biological meaning. Nevertheless, whatever method is used, the generation times are long compared with those encountered in laboratory cultures. Part of the explanation for these extended generation times must also be that a large proportion of the microbial community in most natural environments is inactive (Gray & Williams, 1971b; Stevenson, 1978). The concept that only a small part of the community is active at any one time has attracted increasing support following the development of a number of techniques for *in situ* differentiation of active and inactive biomass and direct measurements of growth rates, e.g. autoradiography (Meyer-Reil, 1978),

Table 3. *Percentage of the viable biomass thought to be active in various natural environments*

Organisms	Habitat	Technique	% active	Reference
Bacteria	Forest soil			
	A1 horizon	FDA	34	Lundgren & Söder-
	A2 horizon		54	ström, 1983
	B horizon		52	
	Agricultural			
	Barley field	ETS	15	Macdonald, 1980
	Manured soil		25	
	Turfed soil		11	
	Compost		23	
	Vegetable soil		23	
	Field soil		31	
	Aquatic			
	Water over			
	sand	Tritiated glucose	2.3–56.2 average 31.3	Meyer-Reil, 1978
Fungi	Pine forest			
	A1 horizon	FDA	2.4	Söderström, 1979
	A2 horizon		4.3	
	B horizon		2.6	

FDA = hydrolysis of fluorescein diacetate
ETS = electron transport system activity

incorporation of $[^{14}C]$-thymidine (Fuhrman & Azam, 1982), the hydrolysis of fluoroscein diacetate (Söderström, 1979) and the reduction of dyes such as iodonitrotetrazolium by respiration (Iturriaga & Rheinheimer, 1975; Macdonald, 1980). Table 3 shows representative data from these investigations.

Soils or aquatic sediments therefore represent environments in which microorganisms, despite the possible presence of large amounts of organic matter, are energy- or nutrient-limited. A large proportion of recalcitrant organic matter and even some potentially labile material is not immediately available. A number of studies have shown that in marine sediments the chemical analysis of *in situ* concentrations of potential microbial substrates such as amino acids and short-chain fatty acids do not reflect the amounts available to bacteria (Christensen & Blackburn, 1980, 1982; Balba & Nedwell, 1982; Parkes, Taylor & Jørck-Ramberg, 1984). Thus, less than 0.01% of the measured alanine (Christensen & Blackburn, 1980) and 14% of the measured acetate (Christensen & Blackburn, 1982) in sediments from the Limfjord in Denmark were readily available to the microflora. This reduction in availability seems to be due to the complexation of substrate molecules with high-molecular-weight organic molecules or metals (Christensen & Blackburn, 1982; Madsen & Alexander, 1985; Thompson & Nedwell, 1985) which, as discussed earlier, may or may not be adsorbed on particle surfaces.

It must be concluded therefore that soils and sediments are generally energy-limited environments in which microbial species adapted to efficient uptake and utilisation of nutrients will be at an advantage. However, the soil and sediment environments are heterogeneous with respect to time in that an occasional input of organic material creates a temporarily large supply of organic substrates, and seasonal inputs of detritus have similar temporary effects. Thus populations of microbes may persist because they are constantly but minimally active at extremely low available nutrient concentrations, or because they grow rapidly during short periods of nutrient abundance and survive in an inactive state during the long intervening times. These two strategies were reflected in the use of the terms autochthonous and zymogenous by Winogradsky (1924) and the more recent suggestion that both oligotrophic and copiotrophic organisms can coexist in nature (Poindexter, 1981). It is not necessary to suppose that these strategies are adopted by different organisms, however, for facultative oligotrophs are the organisms that might be best suited to a fluctuating environment. Thus, some copiotrophic bacteria respond to starvation by producing dwarf cells which have slower metabolic rates than the larger forms from which they are derived (Novitsky & Morita, 1977; Morita, 1982; Amy & Morita, 1983). Epifluorescence microscopy has been used to show that such cells comprise a significant proportion of the microflora of soils and sediments. Dow, Whittenbury & Carr (1983) have also shown that the swarmer cells of prosthecate bacteria have low metabolic rates and may thus represent an oligotrophic phase of an otherwise copiotrophic organism.

In a soil or sediment, a microbial cell will only be removed by predation or limited leaching, unlike the situation in a chemostat where the major loss of cells is usually due to wash-out. In a chemostat, a species with a greater affinity for a growth-rate-limiting nutrient may outcompete a second species which will then be removed from the system. In a soil or sediment, such rapid disappearance will not occur but the population will decline slowly as it attracts a diminishing proportion of the growth-limiting resource. Eventually, its continued survival will depend upon its ability to meet its maintenance energy requirements and any mechanisms which enhance energy scavenging or reduce maintenance requirements will increase survival. Chapman & Gray (1981) have examined the maintenance energy requirements of the soil bacterium *Arthrobacter globiformis* which is able to survive starvation for many weeks in culture. They showed that as growth rate and temperature decreased, the amount of energy used for maintenance fell. Following starvation, the specific maintenance rate fell even further which could have been due to a shutting down of cell processes in the whole population, or part of the population of cells as previously described for *Escherichia coli* (Koch & Coffman, 1970). Chapman & Gray (1981) also showed that *A. globiformis* had a high true growth yield and under nitrogen-limiting conditions was capable of sequestering the carbon supply and converting it into larger quantities of glycogen which served as a reserve material.

Since dead cells are not washed out of the soil to any appreciable extent, it follows that they may in turn act as a source of energy for the growth of other cells, a phenomenon termed cryptic growth. Evidence from soils fumigated with chloroform demonstrates that dead microbial cells are quickly recycled, and Chapman & Gray (1986) have demonstrated that this could have an appreciable effect on microbial growth rates in soil. They cite an example where the average generation time of microorganisms in a deciduous woodland soil could be reduced from 8.3 days to 3.3 days. In this particular soil, they also showed that if the specific maintenance rate of the population rose above $0.006\,h^{-1}$, then all of the energy input would be used for maintenance, leaving none for growth. Unfortunately, values for the substrate inputs and microbial biomass in soil and the specific maintenance rate and true growth yield of the populations in these soils cannot be measured accurately and so calculations remain very imprecise.

It has also been suggested that energy and carbon limitation will promote adhesion of microbial cells to surfaces (Brown, Ellwood & Hunter, 1977) and it has been known for some time that organisms adsorbed onto surfaces such as glass beads can remain active at nutrient concentrations in the surrounding liquid which would not support growth in the absence of a surface (Jannasch & Pritchard, 1972). Both permanently and reversibly attached cells are better able to scavenge and metabolise organic molecules adsorbed on the surfaces of solid particles than are unattached cells (Kefford, Kjelleberg

& Marshall, 1982; Hermansson & Marshall, 1985) and thus adhesion may lead to the more rapid metabolism of substrates. Paerl & Merkel (1982) have shown that inorganic nutrients such as phosphate are also taken up more rapidly by bacteria attached to surfaces. Evidence that the metabolism of bacteria is increased at surfaces has also come from studies by Marshall, Stout & Mitchell (1971) who showed that bacteria adsorbed to glass were initially dwarf cells which reverted to normal sizes within 12–24 h. The ATP content per unit biovolume of three marine bacteria is known to be greater following attachment to glass (Kjelleberg & Dahlbeck, 1984) and uptake of labelled glucose and glutamate also increases (Kirschman & Mitchell, 1982). However, Jeffrey & Paul (1986) have shown that while the activity of attached cells of *Vibrio proteolytica* under energy-limiting conditions is greater than that of detached cells, the reverse is true under non-limiting conditions. In addition, attached cells are less sensitive to changes in nutrient concentration, their activity being independent of nutrient concentration below a critical threshold. The fact that attached cells might be buffered against changes in nutrient supply could be of even greater significance in soils and sediments where surfaces are so abundant.

REFERENCES

ABRAM, J. W. & NEDWELL, D. B. (1978). Inhibition of methanogenesis by sulphate reducing bacteria competing for transferred hydrogen. *Archives for Microbiology*, **117**, 89–92.

AHMED, M., OADES, J. M. & LADD, J. N. (1982). Determination of ATP in soils: effect of soil treatments. *Soil Biology and Biochemistry*, **14**, 273–9.

ALLISON, F. E. (1968). Soil aggregation – some facts and fallacies as seen by a microbiologist. *Soil Science*, **106**, 136–43.

AMY, P S. & MORITA, R. Y. (1983). Starvation-survival patterns of sixteen freshly isolated open-sea bacteria. *Applied and Environmental Microbiology*, **45**, 1109–15.

ANDERSON, J. P. E. & DOMSCH, K. H. (1978). A physiological method for the quantitative measurement of microbial biomass in soil. *Soil Biology and Biochemistry*, **10**, 215–21.

BALBA, M. T. & NEDWELL, D. B. (1982). Microbial metabolism of acetate, propionate and butyrate in anoxic sediments from the Colne Point salt marsh, Essex, UK. *Journal of General Microbiology*, **128**, 1415–22.

BILLEN, G. (1982). Modelling the processes of organic matter degradation and nutrient recycling in sedimentary systems. In *Sediment Microbiology*, ed. D. B. Nedwell & C. M. Brown, pp. 15–52. London, Academic Press.

BOSATTA, E. & AGREN, G. I. (1985). Theoretical analysis of decomposition of heterogeneous substrates. *Soil Biology and Biochemistry*, **17**, 601–10.

BOSATTA, E. & BERENDSE, F. (1984). Energy or nutrient regulation of decomposition: implications for the mineralization–immobilization response to perturbations. *Soil Biology and Biochemistry*, **16**, 63–7.

BOSATTA, E. & STAAF, H. (1982). The control of nitrogen turnover in forest litter. *Oikos*, **39**, 143–51.

BROOKES, P. C., TATE, K. R. & JENKINSON, D. S. (1983). The adenylate energy change of the soil microbial biomass. *Soil Biology and Biochemistry*, **15**, 9–16.

BROWN, C. M., ELLWOOD, D. C. & HUNTER, J. R. (1977). Growth of bacteria at surfaces: influence of nutrient concentration. *FEMS Microbiology Letters*, **1**, 163–6.

BURNS, R. G. (1983). Extracellular enzyme–substrate interactions in soil. In *Microbes in their Natural Environments*, ed. J. H. Slater, R. Whittenbury & J. W. T. Wimpenny, pp. 249–98. Cambridge, Cambridge University Press.

CHAPMAN, S. J. & GRAY, T. R. G. (1981). Endogenous metabolism and macromolecular composition of *Arthrobacter globiformis*. *Soil Biology and Biochemistry*, **13**, 11–18.

CHAPMAN, S. J. & GRAY, T. R. G. (1986). Importance of cryptic growth, yield factors and maintenance energy in models of microbial growth in soil. *Soil Biology and Biochemistry*, **18**, 1–4.

CHARD, J. M., GRAY, T. R. G. & FRANKLAND, J. C. (1983). Antigenicity of *Mycena galopus*. *Transactions of the British Mycological Society*, **81**, 503–11.

CHARD, J. M., GRAY, T. R. G. & FRANKLAND, J. C. (1985*a*). Purification of an antigen characteristic of *Mycena galopus*. *Transactions of the British Mycological Society*, **84**, 235–41.

CHARD, J. M., GRAY, T. R. G. & FRANKLAND, J. C. (1985*b*). Use of an anti-*Mycena galopus* serum as an immuofluorescence agent. *Transactions of the British Mycological Society*, **84**, 243–9.

CHRISTENSEN, D. & BLACKBURN, T. H. (1980). Turnover of tracer (^{14}C, ^{3}H-labelled) alanine in inshore marine sediments. *Marine Biology*, **58**, 97–103.

CHRISTENSEN, D. & BLACKBURN, T. H. (1982). Turnover of ^{14}C-labelled acetate in marine sediments. *Marine Biology*, **71**, 113–19.

CHUANG, T. Y. & KO, W. H. (1983). Propagule size: its relation to longevity and reproductive capacity. *Soil Biology and Biochemistry*, **15**, 269–74.

CLARHOLM, M. & ROSSWALL, T. (1980). Biomass and turnover of bacteria in a forest soil and peat. *Soil Biology and Biochemistry*, **12**, 49–57.

COOKE, R. C. & RAYNER, A. D. M. (1984). *Ecology of Saprotrophic Fungi*. London, Longman.

CRAVEN, D. B. & KARL, D. M. (1984). Microbial RNA and DNA synthesis in marine sediments. *Marine Biology*, **83**, 129–39.

DARBYSHIRE, J. F., ROBERTSON, L. & MACKIE, L. A. (1985). A comparison of two methods of estimating the soil pore network available to protozoa. *Soil Biology and Biochemistry*, **17**, 619–24.

DASHMAN, T. & STOTZKY, G. (1986). Microbial utilization of amino acids and a peptide bound on homoionic montmorillonite and kaolinite. *Soil Biology and Biochemistry*, **18**, 5–14.

DEANEY, N. B. & GRAY, T. R. G. (1983). Use of fluorochromes to determine viability of soil biomass. *Abstracts, Third International Symposium on Microbial Ecology*, 1983, 63.

DIX, N. J. (1984). Minimum water potentials for the growth of some litter decomposing agarics and other basiodiomycetes. *Transactions of the British Mycological Society*, **83**, 152–3.

DOW, C. S., WHITTENBURY, R. & CARR, N. G. (1983). The 'shut down' or 'growth precursor' cell – an adaptation for survival in a potentially hostile environment. In *Microbes in their Natural Environments*, ed. J. H. Slater, R. Whittenbury and J. W. T. Wimpenny, pp. 187–247. Cambridge, Cambridge University Press.

DOWSON, C. G., RAYNER, A. D. M. & BODDY, L. (1986). Outgrowth patterns of mycelial cord-forming basidiomycetes from and between woody resource units in soil. *Journal of General Microbiology*, **132**, 203–11.

FALLON, R. D., NEWELL, S. Y. & HOPKINSON, C. S. (1983). Bacterial production in marine sediments: will cell-specific measures agree with whole system metabolism? *Marine Ecology Progress Series*, **11**, 117–19.

FITZPATRICK, E. A. (1971). *Pedology: A Systematic Approach to Soil Science.* Edinburgh, Oliver and Boyd.

FLETCHER, M. & MARSHALL, K. C. (1982). Are solid surfaces of ecological significance to aquatic bacteria? *Advances in Microbial Ecology*, **6**, 199–236.

FOSTER, R. C. & ROVIRA, A. D. (1976). Ultrastructure of the wheat rhizosphere. *New Phytologist*, **76**, 343–52.

FOTH, H. D. (1984). *Fundamentals of Soil Science*, 7th edn. New York, Wiley.

FRADKIN, A. & PATRICK, Z. A. (1982). Fluorescence microscopy to study colonization of conidia and hyphae of *Cochliobolus sativus* by soil microorganisms. *Soil Biology and Biochemistry*, **14**, 543–8.

FRANKLAND, J. C. (1984). Autecology and the mycelium of a woodland litter decomposer. In *The Ecology and Physiology of the Fungal Mycelium*, ed. D. H. Jennings & A. D. M. Rayner, pp. 241–60. London, Cambridge University Press.

FRANKLAND, J. C., BAILEY, A. D., HOLLAND, A. A. & GRAY, T. R. G. (1981). Development of an immunological technique for estimating mycelial biomass of *Mycena galopus* in leaf litter. *Soil Biology and Biochemistry*, **13**, 87–92.

FUHRMAN, J. A. & AZAM, F. (1982). Thymidine incorporation as a measure of heterotrophic bacterioplankton production in marine surface waters: evaluation and field results. *Marine Biology*, **66**, 109–20.

GRAY, T. R. G., HISSETT, R. & DUXBURY, T. (1974). Bacterial populations of litter and soil in a deciduous woodland, II. – Numbers, biomass and growth rates. *Revue d'Ecologie et Biologie du Sol*, **10**, 15–26.

GRAY, T. R. G. & WILLIAMS, S. T. (1971*a*). *Soil Microorganisms*. London, Longman.

GRAY, T. R. G. & WILLIAMS, S. T. (1971*b*). Microbial productivity in soil. In *Microbes and Biological Productivity*, ed. D. E. Hughes and A. H. Rose, pp. 255–86. Cambridge, Cambridge University Press.

GREENLAND, D. J. (1971). Interactions between humic and fulvic acids and clays. *Soil Science*, **111**, 34.

GREENWOOD, D. J. (1961). The effect of oxygen concentration on the decomposition of organic materials in soil. *Plant and Soil*, **14**, 360–76.

GRIFFIN, D. M. (1972). *Ecology of Soil Fungi.* London, Chapman & Hall.

GRIFFIN, D. M. & LUARD, E. J. (1979). Water stress and microbial ecology. In *Strategies of Microbial Life in Extreme Environments*, ed. M. Shilo, pp. 49–63. Weinheim, Verlag Chemie.

HERMANSSON, M. & MARSHALL, K. C. (1985). Utilization of surface-localized substrate by non-adhesive marine bacteria. *Microbial Ecology*, **11**, 91–105.

HOFMANN, J. V. (1917). Natural reproduction from seed stored in the forest floor. *Journal of Agricultural Research*, **11**, 1–26.

ITURRIAGA, R. & RHEINHEIMER, G. (1975). A simple method for counting bacteria with active electron transport systems in water and sediment samples. *Kieler Meeresforschungen Sonderheft*, **31**, 83–6.

JANNASCH, H. W. & PRITCHARD, P. H. (1972). The role of inert particulate matter in the activity of aquatic microorganisms. *Memoire Istitut Italiano di Idrobiologia*, **29**, Supplement, 289–303.

JEFFREY, W. D. & PAUL, J. H. (1986). Activity of an attached and free-living *Vibrio* sp. as measured by thymidine incorporation, *p*-iodonitrotetrazolium reduction, and ATP/DNA ratios. *Applied and Environmental Microbiology*, **51**, 150–6.

JENKINSON, D. S. (1966). Studies on the decomposition of plant material in soil. II. Partial sterilization of soil and the soil biomass. *Journal of Soil Science*, **17**, 280–302.

JENKINSON, D. S. (1976). The effects of biocidal treatments on metabolisms in soil – IV. The decomposition of fumigated organisms in soil. *Soil Biology and Biochemistry*, **8**, 203–8.

JENKINSON, D. S. & LADD, J. N. (1981). Microbial biomass in soil: measurement and turnover. In *Soil Biochemistry*, vol. 5, ed. E. A. Paul & J. N. Ladd, pp. 415–71. New York, Dekker.

JONES, J. G. & SIMON, B. M. (1980). Decomposition processes in the profundal region of Blelham Tarn and the Lund tubes. *Journal of Ecology*, **68**, 493–512.

JØRGENSEN, B. B. (1977). Bacterial sulphate reduction within reduced microniches of oxidized marine sediments. *Marine Biology*, **41**, 7–17.

JØRGENSEN, B. B. (1982). Mineralisation of organic matter in the sea bed – the role of sulphate reduction. *Nature (London)*, **296**, 643–5.

JØRGENSEN, B. B. (1983). Processes at the sediment–water interface. In *SCOPE 21. The Major Biogeochemical Cycles and their Interactions*, ed. B. Bolin & R. B. Cook, pp. 477–509. Chichester, John Wiley.

KEFFORD, B., KJELLEBERG, S. & MARSHALL, K. C. (1982). Bacterial scavenging: utilization of fatty acids localized at a solid–liquid interface. *Applied and Environmental Microbiology*, **133**, 257–60.

KIRSCHMAN, D. & MITCHELL, R. (1982). Contribution of particle-bound bacteria to total microheterotrophic activity in five ponds and two marshes. *Applied and Environmental Microbiology*, **43**, 200–9.

KJELLEBERG, S. & DAHLBECK, B. (1984). ATP level of a starving surface-bound, and a free-living marine *Vibrio* sp. *FEMS Microbiology Letters*, **24**, 93–6.

KJØLLER, A. & STRUWE, S. (1982). Microfungi in ecosystems: fungal occurrence and activity in litter and soil. *Oikos*, **39**, 389–422.

KNAPP, E. B., ELLIOTT, L. F. & CAMPBELL, G. S. (1983). Microbial respiration and growth during the decomposition of wheat straw. *Soil Biology and Biochemistry*, **15**, 319–23.

KOCH, A. L. & COFFMAN, R. L. (1970). Diffusion, permeation, or enzyme limitation: a probe for kinetics of enzyme induction. *Biotechnology and Bioengineering*, **12**, 651–77.

LANYI, J. K. (1979). Life at low water activities: group report. in *Strategies of Microbial Life in Extreme Environments*, ed. M. Shilo, pp. 125–35. Weinheim, Verlag Chemie.

LUNDGREN, B. (1984). Size classification of soil bacteria: effects on microscopically estimated biovolumes. *Soil Biology and Biochemistry*, **16**, 283–4.

LUNDGREN, B. & SÖDERSTRÖM, B. (1983). Bacterial numbers in a pine forest soil in relation to environmental factors. *Soil Biology and Biochemistry*, **15**, 625–30.

MCCLAUGHERTY, C. A., PASTOR, J., ABER, J. D. & MELILLO, J. M. (1985). Forest litter decomposition in relation to soil nitrogen dynamics and litter quality. *Ecology*, **66**, 266–75.

MACDONALD, R. M. (1980). Cytochemical demonstration of catabolism in soil microorganisms. *Soil Biology and Biochemistry*, **12**, 419–23.

MADSEN, E. L. & ALEXANDER, M. (1985). Effects of chemical speciation on the mineralization of organic compounds by microorganisms. *Applied and Environmental Microbiology*, **50**, 342–9.

MARSHALL, K. C. (1976). *Interfaces in Microbial Ecology*. Cambridge, Massachusetts, Harvard University Press.

MARSHALL, K. C., STOUT, R. & MITCHELL, R. (1971). Selective sorption of bacteria from seawater. *Canadian Journal of Microbiology*, **17**, 1413–16.

MARSHMAN, N. A. & MARSHALL, K. C. (1981a). Bacterial growth on proteins in the presence of clay minerals. *Soil Biology and Biochemistry*, **13**, 127–34.

MARSHMAN, N. A. & MARSHALL, K. C. (1981b). Some effects of montmorillonite on the growth of mixed microbial cultures. *Soil Biology and Biochemistry*, **13**, 135–41.

MARTENS, R. (1985). Estimation of the adenylate energy charge in unamended and amended agricultural soil. *Soil Biology and Biochemistry*, **17**, 765–72.

MELILLO, J. M. & GOSZ, J. R. (1983). Interactions of biogeochemical cycles in forest ecosystems. In *SCOPE 21. The Major Biogeochemical Cycles and their Interactions*, ed. B. Bolin & R. B. Cook, pp. 177–222. Chichester, John Wiley.

MEYER-REIL, L. A. (1978). Autoradiography and epifluorescence microscopy combined for the determination of number and spectrum of actively metabolising bacteria in natural waters. *Applied and Environmental Microbiology*, **36**, 506–12.

MEYER-REIL, L. A., BOLTER, M., DAWSON, R., LIEBEZEIT, G., SZWERINSKI, H. & WOLTER, K. (1980). Interrelationships between microbiological and chemical parameters of sandy beach sediment, a summer aspect. *Applied and Environmental Microbiology*, **39**, 797–802.

MORIARTY, D. J. W. & POLLARD, P. C. (1981). DNA synthesis as a measure of bacterial productivity in seagrass sediments. *Marine Ecology Progress Series*, **5**, 151–6.

MORIARTY, D. J. W. & POLLARD, P. C. (1982). Diel variation of bacterial productivity in seagrass (*Zostera capricornia*) beds measured by the rate of thymidine incorporation into DNA. *Marine Biology*, **72**, 165–73.

MORITA, R. Y. (1982). Starvation-survival of heterotrophs in the marine environment. *Advances in Microbial Ecology*, **6**, 171–98.

MORTLAND, M. M. (1970). Clay–organic complexes and interactions. *Advances in Agronomy*, **22**, 75.

NANNIPIERI, P., CECCANTI, B., CERVELLI, S. & SEQUI, P. (1978). Stability and kinetic properties of humus–urease complexes. *Soil Biology and Biochemistry*, **10**, 143–7.

NANNIPIERI, P., JOHNSON, R. L. & PAUL, E. A. (1978). Criteria for measurement of microbial growth and activity in soil. *Soil Biology and Biochemistry*, **10**, 223–9.

NEDWELL, D. B. & ABRAM, J. W. (1979). Relative influence of temperature and electron donor and electron acceptor concentrations on bacterial sulphate reduction in saltmarsh sediment. *Microbial Ecology*, **5**, 67–72.

NEWELL, K. (1984). Interaction between two decomposer basidiomycetes and a collembolan under Sitka Spruce: grazing and its potential effects on fungal distribution and litter decomposition. *Soil Biology and Biochemistry*, **16**, 235–40.

NOVITSKY, J. A. & MORITA, R. Y. (1977). Survival of a psychrophilic marine vibrio under long term nutrient starvation. *Applied and Environmental Microbiology*, **33**, 635–41.

O'BRIEN B. J. (1984). Soil organic carbon fluxes and turnover rates estimated from radiocarbon enrichment. *Soil Biology and Biochemistry*, **16**, 115–20.

PAERL, H. W. & MERKEL, S. M. (1982). Differential phosphorus assimilation in attached versus unattached microorganisms. *Archives for Hydrobiology*, **93**, 125–34.

PAMATMAT, H. M., GROF, G., BENTSSON, W. & NOVAK, C. S. (1981). Heat production, ATP concentration and electron transport activity of marine sediments. *Marine Ecology Progress Series*, **4**, 135–43.

PARINKINA, O. H. (1974). Bacterial production in tundra soils. In *Soil Organisms and Decomposition in Tundra*, ed. A. J. Holding, O. W. Heal, S. F. Maclean, Jr and P. W. Flanagan, pp. 65–77. Stockholm, Tundra Biome Steering Committee.

PARKES, R. J., TAYLOR, J. & JØRCK-RAMBERG, D. (1984). Demonstration, using *Desulfobacter* sp., of two pools of acetate with different biological availabilities in marine pore water. *Marine Biology*, **83**, 271–6.

POINDEXTER, J. S. (1981). Oligotrophy: fast and famine existence. *Advances in Microbial Ecology*, **5**, 63–89.

REINERTSEN, S. A., ELLIOTT, L. F., COCHRAN, V. L. & CAMPBELL, G. S. (1984). Role of available carbon and nitrogen in determining the rate of wheat straw decomposition. *Soil Biology and Biochemistry*, **16**, 459–64.

ROGERS, H. J. (1979). Adhesion of microorganisms to surfaces: some general considerations of the role of the envelope. In *Adhesion of Microorganisms to Surfaces*, ed. D. C. Ellwood, J. Melling and P. Rutter, pp. 29–55. London, Academic Press.

SERBAN, A. & NISSENBAUM, A. (1986). Humic acid association with peroxidase and catalase. *Soil Biology and Biochemistry*, **18**, 41–4.

SIALA, A. & GRAY, T. R. G. (1974). Growth of *Bacillus subtilis* and spore germination in an acid forest soil observed by a fluorescent antibody technique. *Journal of General Microbiology*, **81**, 191–8.

SIALA, A., HILL, I. R. & GRAY, T. R. G. (1974). Populations of spore-forming bacteria in an acid forest soil, with special reference to *Bacillus subtilis*. *Journal of General Microbiology*, **81**, 183–90.

SÖDERSTRÖM, B. A. (1979). Seasonal fluctuations of active fungal biomass in horizons of a podsolised pine-forest soil. *Soil Biology and Biochemistry*, **11**, 149–54.

SPARLING, G. P., ORD, B. G. & VAUGHAN, D. (1981). Microbial biomass and activity in soils amended with glucose. *Soil Biology and Biochemistry*, **13**, 99–104.

SPOSITO, G. (1984). *The Surface Chemistry of Soils*. Oxford, Oxford University Press.

STEVENSON, F. J. (1982). *Humus Chemistry*. New York, Wiley.

STEVENSON, L. H. (1978). A case for dormancy in aquatic systems. *Microbial Ecology*, **4**, 127–33.

STILLINGER, F. H. (1980). Water revisited. *Science, New York*, **209**, 451.

SUESS, E. (1980). Particulate organic carbon flux in the oceans – surface productivity and oxygen utilization. *Nature (London)*, **288**, 260–3.

SUESS, E. & MÜLLER, P. J. (1980). Productivity, sedimentation rate and sedimentary organic matter in the oceans. In *Biogéochemie de la Matière Organique a l'Interface Eau–Sediment Marin*, ed. R. Daumas, pp. 17–26. Paris, CNRS.

SWIFT, M. J., HEAL, O . W. & ANDERSON, J. (1979). *Decomposition in Terrestrial Ecosystems. Studies in Ecology*, vol. 5. Oxford, Blackwell.

TATENO, M. (1985). Adenylate energy charge in glucose-amended soil. *Soil Biology and Biochemistry*, **17**, 387–8.

THENG, B. K. G. (1979). *Formation and Properties of Clay–Polymer Complexes*. Amsterdam, Elsevier.

THOMPSON, L. A. & NEDWELL, D. B. (1985). Existence of different pools of fatty acids in anaerobic model ecosystems and their availability to microbial metabolism. *FEMS Microbiology Ecology*, **31**, 141–6.

VIGNEAUX, M., DUMON, J. C., FAUGERES, J. C., GROUSSET, F., JOUANNEAU, J. M., LATOUCHE, C., POUTIERS, J. & PUJOL, C. (1980). Matières organiques et sedimentation en milieu marin. In *Biogéochemie de la Matière Organique a l'Interface Eau–Sediment Marin*, ed. R. Daumas, pp. 113–28. Paris, CNRS.

WINOGRADSKY, S. (1924). Sur la microflore autochthone de la terre arable. *Comptes Rendus Hebdomadaires des Séances de l'Academie de Sciences, Paris*, **178**, 1236–9.

BIOLOGICAL CONTROL WITHIN MICROBIAL COMMUNITIES OF THE RHIZOSPHERE

J. M. LYNCH

GCRI, Worthing Road, Littlehampton, West Sussex BN17 6LP, UK

INTRODUCTION

The ectorhizosphere surrounds plant roots, the endorhizosphere is within the root and the root surface is the rhizoplane (Lynch, 1982). In order to consider interactions between micro-organisms in the rhizosphere, the availability of substrates on which members of the community depend will be described. Whereas micro-organisms within the root can derive substrates directly from the plant, those exterior to the root are dependent on rhizodeposition products (Whipps & Lynch, 1985). The latter include exudates, lysates, sloughed-off cells and mucilage, and the total carbon released including respired CO_2 can amount to 40% w/w of the total photosynthate produced by an actively growing plant. Energy budgets of microbial populations on roots have been attempted by counting viable cells (proportional to biomass) in relation to the carbon flow from roots and it appears that a greater microbial biomass is formed than can be accounted for in the substrate energy provided by roots. However there is difficulty in determining the biomass of a microbial cell *in vivo*. Bacterial cells can appear much smaller on roots than they do in luxuriant growth media; this may be at least in part because they grow slowly and they have little opportunity to deposit any energy reserves. Even in plant solution culture the growth rate of bacteria in the rhizosphere of barley may be only $0.029\,h^{-1}$ for the first 4 days and $0.007\,h^{-1}$ at 7–16 days (Barber & Lynch, 1977). Furthermore, at such low rates of substrate input from roots, any oligotrophic growth would take on more significance, especially for plants growing experimentally or commercially in flowing nutrient media where forced aeration of the root system could introduce substantial quantities of carbon compounds from the gas stream. These ambiguities need resolution to assist the analyses of the influence of the rhizosphere on crop productivity (Whipps & Lynch, 1986). The purpose of the present review, however, is to assess only one facet of that influence, the potential control of pathogenic species in the rhizosphere, by considering specifically a few examples of bacterial and fungal antagonists of plant pathogens. Until recently, much of the analysis of antagonism has been concerned primarily with the pathogen, with little attention paid to the nature and quantification of the interaction (Cook & Baker, 1983). However, many exciting developments are now taking place where microbial ecologists, physiologists and geneticists can play a complementary role to the plant pathologist. The examples of bacterial/fungal and fungal/fungal antagonisms are selected for this review with an emphasis on the mode(s) of action, particularly where chemical metabolites are involved. However, it should be recognized that as yet few of the metabolites identified can be proved to have ecological significance (Lynch, 1976), in the sense that they are biologically active in the form and in the concentration at which they are actually present in the soil/plant system.

COMMUNITY STRUCTURE

Definitive studies on the bacterial population of the rhizosphere were carried out in Canada between 1938 and 1957 (Lochhead, 1959). The counterparts to this work for saprotrophic fungi employed a technique of root washing which removed fungal spores from roots and allowed the subsequent culture on agar and identification of hyphal fungi (Harley & Waid, 1955). However, these approaches were not focussed on symbionts. *Rhizobium* is the bacterial mutualistic symbiont that has received greatest study and indeed the original definition of the rhizosphere by Hiltner (1904) primarily concerned *Rhizobium*. The fungal mutualistic symbionts receiving the greatest attention have been the ectotrophic and endotrophic (vesicular-arbuscular) mycorrhiza (Harley & Smith, 1983). Additionally, bacterial and fungal antagonistic symbionts (pathogens) populate the rhizosphere; physiologically these can be biotrophic (utilizing materials derived from living host tissues) or necrotrophic (living on dead host tissues) and the complexity of this physiological interaction has recently been reviewed (Whipps & Lynch, 1986). In addition to fungi (including yeasts) and bacteria, algae, protozoa and soil animals can also inhabit the rhizosphere, and thus the community structure is very complex. Attempts have been made to study the communities in chemostats using a glucose feed to mimic rhizodeposition substrates (Coleman *et al.*, 1978) although polysaccharides might have proved more effective. Whereas this approach has some value it would be more useful to include the root itself in experimental systems because not only are natural substrates provided but also microbial activity is greater on the rhizoplane. Indeed a recent analysis of microbial adhesion to the root surface has been undertaken (James, Suslow & Steinback, 1985). Adherence was determined by first incubating roots of intact radish seedlings with bacteria, then washing and homogenizing the roots, followed by dilution-plate counting. The binding of bacteria to the roots was irreversible, reaching half its maximum level within 5 min, and it was concentration-dependent in the range 10^6 to 10^8 CFU ml^{-1}. This was little affected by the source or type of bacterium, there being no correlation with the hydrophobicity of the bacterial cell surface. Electrostatic factors appear to be involved; binding in one strain tested was prompted by divalent cations (Ca^{2+} and Mg^{2+}) but unaffected by monovalent cations (Na^+ and K^+). Such a study is potentially useful in quantifying the likely effectiveness of both pathogens and antagonists in colonizing the root.

Plant pathologists have often used the term 'inoculum potential' of a pathogen, defined as the fungal energy for growth per unit area of host surface (Garrett, 1970), but the term is conceptual and non-quantifiable. A quantitative term is 'infection efficiency', which is the proportion of the soil inoculum capable of causing infection of hosts, if favourably placed to do so (Gilligan, 1983). For rhizosphere communities I prefer the term 'colonization potential', which is the number of bacterial cells, length of fungal hyphae or biomass which colonizes a unit length or weight of root (Bennett & Lynch, 1981*b*). In gnotobiotic culture, this was similar for three rhizosphere bacteria and independent of inoculum size in the range 10^3–10^7 cells per mg dry weight of root. However, when the bacteria were grown together the colonization potential of a *Pseudomonas* sp. on a wheat root was sustained, that of a *Mycoplana* sp. was decreased and a *Curtobacterium* sp. disappeared from the population (Bennett & Lynch, 1981*a*). This concept is potentially useful in quantifying pathogen/antagonist interactions and has been investigated for seven antibiotically marked *Pseudomonas* strains in the rhizosphere of potato (*Solanum tuberosum* L.) plants growing in field soil under controlled environmental conditions (Loper, Haack & Schroth, 1985). For two strains, the colonization potential was related to initial seed piece inoculum level; this points to the necessity of testing concepts developed under gnotobiotic conditions in natural soil. The study also showed that colonization potential of the entire root system correlated with *in vitro* osmotolerance of the seven strains and was maximal at low soil temperature (12 °C) even though bacterial

growth on root tip-proximal segments was greater at higher temperatures, thus indicating no relation between bacterial growth rate on
root-segments and colonization potential of the entire root.

The concept of colonization potential was also investigated at an
extra level of complexity by monitoring *field* populations of *Pseudomonas fluorescens*, antagonistic to the pathogen take-all, on seminal
roots of winter wheat (Weller, 1984). Maximum populations of 10^3
to 10^6 CFU cm^{-1} root were found on root sections near the seed
and the root tip. The introduced bacteria (resistant to rifampicin
and nalidixic acid as markers) compete well with indigenous bacteria,
comprising at least 25% of the fluorescent pseudomonads detected
by plate counts 48 days after planting.

The modes of interaction between plant pathogens and their antagonists in the rhizosphere can usually be considered as parasitism
(hyperparasitism) or predation of one organism by another, antibiosis, or competition where demand exceeds immediate supply of
materials or space. An indirect mode of action in induced resistance
or cross-protection where a micro-organism induces the plant to produce the antagonistic action (Kuc, 1982). These various interactions
have been discussed by Whipps (1986). Hyperparasitism will not
be further discussed in the present article but it has recently been
reviewed in detail (Whipps, Lewis & Cooke, in press). The modes
of action are not necessarily exclusive and even simple *in vitro* tests
can be difficult to classify. For example, an inhibition zone created
by one organism against another is usually considered to be due
to antibiotic production by the antagonist but it could also be the
consequence of the secretion of a cell wall-degrading extracellular
enzyme (Fig. 1).

Natural rhizospheres are often inimical to pathogens because they
have antagonists as part of the rhizosphere community. The soil
of such rhizospheres is regarded as 'suppressive' and provides a useful
source of biocontrol agents.

BACTERIAL ANTAGONISTS

Agrobacterium rhizogenes *strain 84*

The only bacterial biocontrol agent which is available commercially
is *Agrobacterium rhizogenes* strain 84, which is used to control crown
gall caused by *Agrobacterium tumefaciens*, *A. rhizogenes* and *A.*

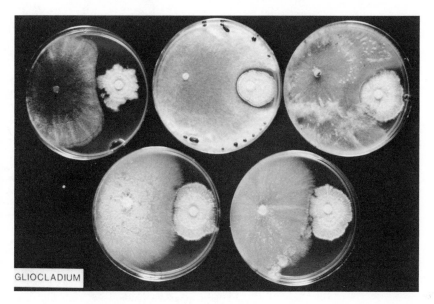

Fig. 1. *In vitro* antagonism of *Gliocladium roseum* to some plant pathogens. The antagonist is placed on the right and the pathogen on the left of each plate. The pathogens are (left to right) top row: *Phomopsis* sp., *Sclerotinia sclerotiorum*, *Botrytis cinerea*; bottom row: *Fusarium oxysporum* f. sp. *dianthi*, *F. oxysporum* f. sp. *narcissi*.

rubi (all biotypes of *A. radiobacter*) on fruit trees (almond, peach, rose, cherry). The virulence genes which cause the pathogen to induce tumours are located on a large (200 kb) plasmid (the Ti-plasmid); the T-DNA (24 kb) is transferred to the plant cell (Kerr & Tate, 1984). The antagonist most commonly used (strain 84) is a non-pathogenic strain of the pathogen and the action depends on reducing the pathogen:non-pathogen ratio to less than one. The antagonist produces an antibiotic, agrocin, which is coded for on a plasmid and is taken up by the pathogen (Kerr & Tate, 1984). Other antagonistic strains produce other agrocins (specifically low-molecular-weight antibiotics produced by *Agrobacterium* which act against other agrobacteria). Strains carrying the Ti-plasmid induce the synthesis by the plant of the opines octopine, nopaline and agropine, which can be catabolized by the inciting bacteria but not by the plant or other micro-organisms. Strain 84 only controls pathogens with the nopaline-producing Ti-plasmid, but these are the pathogens responsible for most economic damage. Strain 84 kills pathogens with agrocin 84 because it is actively transported into the pathogens via a permease which also transfers other opines (agrocinopines) into its own cells. Thus the genes for sensitivity to agrocin 84 which

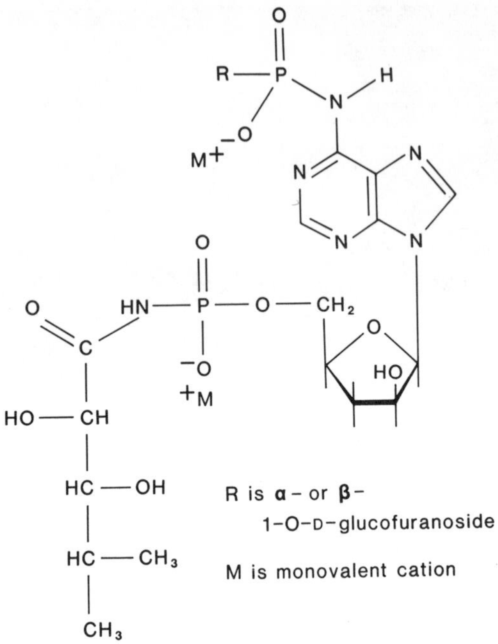

R is α- or β-
1-O-D-glucofuranoside

M is monovalent cation

Agrocin 84

are located in the nopaline Ti-plasmid cannot be regarded as 'suicide genes' because they also promote the uptake of a food base.

However, the genes on plasmid pAg 396 controlling agrocin 84 production (and resistance) can be transferred from the antagonist to a pathogenic recipient in the presence of nopaline. Thus, the reason for failure of the antagonist in some situations is that large populations of the pathogen in soil have given rise to high concentrations of nopaline. However, unlike many biocontrol systems, the mechanism of antagonist action and reasons for breakdown and hence variability are known. This allows optimism for the future because perhaps, for example, an antagonist strain with a defective plasmid transfer system could be developed, such that the capacity for antibiotic production is retained solely by the antagonist.

An interesting recent development in *Agrobacterium* studies is that plasmid RP4 from *Escherichia coli* has been used to mobilize the agrocin 84-encoding plasmid pAg 396 from *A. tumefaciens* strain 396 to *A. tumefaciens* strains C58 and C5831 and to *Rhizobium meliloti* (Hendson & Thomson, 1986). *Rhizobium* spp. can often colonize the rhizosphere of plants which are hosts to *A. tumefaciens* and the possibility therefore arises that the modified *R. meliloti* strain might be used in biological control of crown gall. This option would need to be investigated in field trials.

Pseudomonas fluorescens/Bacillus mycoides

One of the major steps in our understanding of take-all disease of cereals and grasses (caused by *Gaeumannomyces graminis*) was the observation that the incidence of disease declines naturally over a period of about 4 years during wheat monoculture. The soil is then 'suppressive', in contrast to 'conducive' when the disease is rampant. The natural decline can be induced by transferring suppressive soil into conducive soil (Shipton, Cook & Sitton, 1973). It is now generally considered that bacteria are one of the active agents in inducing the decline, particularly *Pseudomonas fluorescens* which is a common rhizosphere colonist (Weller & Cook, 1983), although several other interpretations of decline have been put forward (Asher & Shipton, 1981). When strains of these bacteria isolated from such soils were applied to wheat seeds, they suppressed take-all in both greenhouse- and field-grown winter and spring wheat (Fig. 2). These bacteria

Fig. 2. Biological control of take-all disease (caused by *Gaeumannomyces graminis* var. *tritici*) by a fluorescent *Pseudomonas* sp. introduced on the seed. The plot area had been treated with methyl bromide to eliminate naturally occurring root pathogens of wheat and to make the soil conducive to take-all. The healthy plants on the left received no pathogen or antagonist inoculum. The disease was introduced to the other plants but the plants in the centre had antagonist inoculum as well (Weller & Cook, 1983).

Table 1. *Biological control of take-all* (Gaeumannomyces graminis) *of wheat by a soil drench of* Bacillus pumilus *(based on the data of Capper & Campbell, 1986)*

Inoculation of antagonist	Shoot weight (g)	Percent root-infection at 13 wk	Yield (g m^{-2})	1000 grain weight (g)
Uninoculated control	3.9	44	66	21.5
Inoculated				
At sowing	5.2	27	104[a]	24.4
6 weeks post sowing	4.7	35	93[b]	23.6
At sowing and 6 week post sowing	6.0	27	134	26.1
Double strength at sowing and 6 week post sowing	6.9	20	134	27.1
Double strength at sowing	5.9	26	144	26.2

All results were significantly different from the control at $P = 0.01$ except [a] ($P = 0.05$) and [b] ($P = 0.1$). The inoculant was produced in tryptic soy broth and used as a soil drench on the plots (10^8 cells or spores per ml).

improve crop growth by disease suppression only and not by promoting growth. Antibiotic-resistant inoculants exceed 10^7 CFU g^{-1} root tissue after 3 weeks but the mechanism of action of the antagonism is not yet reported. There have been variable responses from the inoculants and it appears unlikely that this will be explained until the action is elucidated.

Control of take-all has similarly been achieved with *Bacillus mycoides* and *Bacillus pumilus* (Capper & Campbell, 1986) but again no mode of action has been suggested. The results of the most successful trial reported are shown in Table 1. However, it must be emphasized that this is one trial only and for the concept of biological control to be useful as an agricultural practice as opposed to an ecological observation, consistency of response must be obtained or at least the situations where an antagonist will not be effective must be established. In this series of trials disease control was not obtained when there was a low natural incidence of disease on the trial sites or especially dry soil conditions. It can generally be expected that bacteria would be less effective as antagonists in the latter conditions because they are usually more sensitive to low water potentials than are fungi (Lynch, 1983; see also Nedwell & Gray, this volume).

In contrast to the antagonists of take-all, the mode of action of strain Pf-5 of *Pseudomonas fluorescens*, which is effective in suppressing the damping-off of cotton seedlings by *Pythium ultimum*, has been studied (Howell & Stipanovic, 1980). This strain produces an antibiotic, pyoluteorin, and treating cotton seed with the bacterium

Pyoluteorin

or its antibiotic at the time of planting in pathogen-infested soil increased seedling survival from 33 to 65% and from 28 to 71%, respectively. As with most antibiotics it could be inactivated by soil colloids and was therefore only effective as a seed treatment. This makes the damping-off diseases (seed and seedling blights) a good potential target for biological control. However, damping-off diseases are usually caused by a complex of pathogens and in this respect it was disappointing that the antibiotic was not active against other fungi associated with cotton seedling diseases (*Alternaria* sp., *Fusarium* sp., *Thielaviopsis basicola*, *Rhizoctonia solani* and *Verticillium dahliae*). A strain of *Pseudomonas fluorescens* inhibiting growth of *V. dahliae*, *T. basicola* and *Alternaria* spp. *in vitro* produced an antibiotic, pyrrolnitrin, which, when used as a seed treatment, gave

Pyrrolnitrin

increased survival of cotton seedlings in soil infested with *R. solani* (Howell & Stipanovic, 1979).

Enterobacter cloacae

Enterobacter cloacae has proved effective against the damping-off of pea and cucumber and the discovery of its activity demonstrates a useful ecological selection concept (Hadar *et al.*, 1983). Cucumber and pea seeds were pre-germinated in aerated water until radicle emergence and then planted into soil infested with *Pythium* spp. Pre-germination greatly reduced disease incidence relative to non-germinated seeds. The pre-germination caused rapid development of bacteria on the seed (10^7 CFU per seed) and these seeds were less susceptible to *Pythium* than seeds germinated aseptically. The disease resistance could be achieved by treating dry seeds with the total bacterial population from germinated seeds but one of the bacteria, *E. cloacae*, isolated from the total microbiota was particularly effective. It could form bacterial sheaths around *Pythium ultimum* with resulting hyphal lysis.

E. cloacae inoculated onto seeds colonized plant roots in soil-free moist chambers and when growing in sterilized soil it colonized the entire rhizosphere; this contrasted with *Trichoderma* spp. (Chao *et al.*, 1986). However, the colonization was poor in untreated soil. In autoclaved soil with fungi added, *E. cloacae* colonized well, inhibiting *Trichoderma harzianum* but when bacteria were added, *E. cloacae* was inhibited and *T. harzianum* grew well. Clearly these complex interactions could limit the exploitation of *E. cloacae* but irrigation may help to distribute the inoculum over the root system. The possibility of dual inoculation also arises, however, because *T. harzianum* is a biocontrol agent itself.

E. cloacae does not appear to produce antibiotics, toxins or cell wall-degrading enzymes but it binds firmly to the hyphae of *Pythium ultimum*, probably mediated by a lectin-type interaction (Nelson *et al.*, 1986) (Fig. 3). *In vitro* studies showed that the sugar composition of the growth medium determined the inhibitory effect of *E. cloacae* on *P. ultimum* and that growth inhibition was linked to binding of the bacteria to the hyphae. On seeds, the interaction was greatest when small amounts of carbohydrate were exuded by the seed. Treating cucumber seeds with sugars (D-glucose, D-galactose, sucrose, *N*-acetyl-D-glucosamine or β-methyl-D-glycoside) that interfered with binding eliminated the biological control activity. Further analysis of lectin-like activity in this and other interactions may lead to more reliable formulations of biocontrol agents.

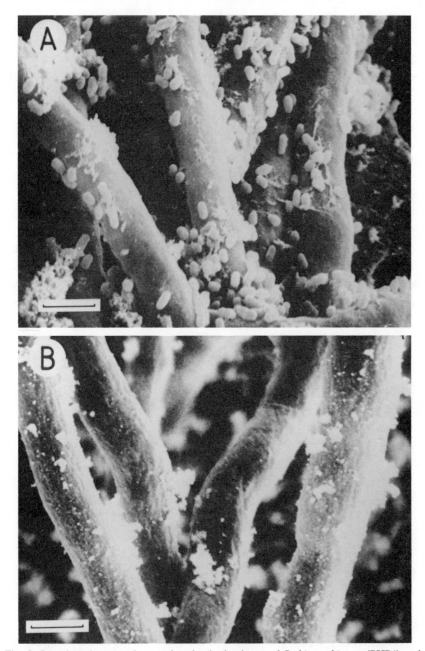

Fig. 3. Scanning electron micrographs of paired cultures of *Pythium ultimum* (PHP4) and *Enterobacter cloacae* (NRRLB-14095). A, No sugar added to paired cultures; after 24 h, non-adhering bacteria were removed by rinsing paired cultures with distilled water. B, *P. ultimum* grown in the presence of 10 mM sucrose; after 24 h, non-adhering bacteria were removed by rinsing paired cultures with a sucrose solution (10 mM). Bars = 5 μm. (From Nelson *et al.*, 1986, with permission.)

It is possible that such bacteria could have a direct effect on seed and seedling vigour, affecting their resistance to disease. In a study using *Azotobacter chroococcum*, barley seedling root extension was sometimes stimulated following bacterial inoculation but when nitrate was present, extension was always inhibited because the bacterium could compete with the seed for available oxygen (Harper & Lynch, 1979). The possibilities of such variable interactions occurring should be considered in developing any seed inoculation procedures.

Plant Growth-Promoting Rhizobacteria (PGPR)

A milestone in rhizosphere and biological control studies was the recognition by Kloepper *et al* (1980) that specific strains of the *Pseudomonas fluorescens-putida* group could rapidly colonize potato roots and produce significant yield increases (up to 144%). The bacteria produced an extracellular siderophore (microbial iron transport agent), pseudobactin, (Teintze & Leong, 1981; Teintze *et al.*, 1981)

Pseudobactin

which was claimed to chelate iron and make it unavailable to pathogens and sub-clinical pathogens or deleterious rhizobacteria. Uptake is critical above pH 5 because the Fe^{3+} is highly insoluble and not

easily available to plants or bacteria; this is when effective siderophore production by an organism places it at a competitive advantage. Primary evidence for this action was that whereas the bacteria
and pseudobactin were equally effective in promoting potato growth
in greenhouse assay, Fe(III)EDTA (with or without the bacteria)
and Fe(III) pseudobactin were ineffective. Furthermore, the PGPR
exhibited antibiosis to *Escherichia coli* K12–194, which does not produce a siderophore, but were inactive against *E. coli* K-12 AN193
which produces the siderophore enterbactin. It was recognized that
pseudobactin was an antibiotic but in a series of succeeding papers
by the authors, they have never established the relative importance
of chelation and antibiosis in the mode of action. Certainly several
investigators worldwide have observed positive growth responses
from plants to these and similar bacteria, but the response has usually
varied between experiments, which is a common hallmark of bacterial inoculation trials (Lynch, 1983). In a large series of seedling
assays, plant growth responses of bacterial inoculants in the absence
of pathogens varied between large positive and negative values
(Elliott & Lynch, 1985), an effect which may be due to stability
of bacterial plasmids which are present in root-colonizing pseudomonads (R. Wheatcroft, L. F. Elliott & J. M. Lynch, unpublished).

Irrespective of the considerations on variability of response, it
should be considered whether plants avoid iron deficiency in such
a system by having an uptake mechanism from the bacterial cells
or an effective siderophore production system of their own. Recent
evidence (Becker, Hedges & Messens, 1985) indicates that purified
pseudobactin in concentrations as low as $0.1\mu M$ and the siderophore
from *Pseudomonas aureofaciens* ATCC 15926 inhibit the uptake of
ferric iron from $^{59}FeCl_3$ by the roots of pea and maize plants sufficiently to reduce the synthesis of chlorophyll and this has been interpreted as competitive binding (Fig. 4). In the same series of
experiments aerobactin from *Klebsiella pseumoniae* had no effect
on the uptake of ^{59}Fe by maize roots. These observations demonstrated that iron uptake by roots from the bacterial siderophores
does not occur but did not establish whether the roots produce siderophores themselves. In contrast, agrobactin (Ong, Peterson & Neilands, 1979), a linear catechol siderophore from *Agrobacterium
tumefaciens* strain B6, enhances the uptake of ferric iron into young
pea and bean plants with a consequent increase in the synthesis
of chlorophyll (Becker, Messens & Hedges, 1985). The genes involved
in the biosynthesis of pseudobactin have been cloned (Moores *et*

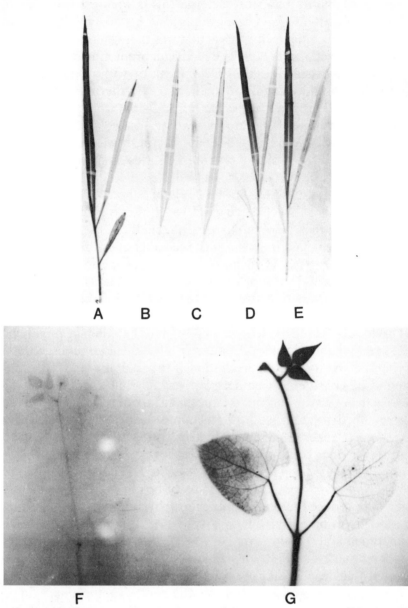

Fig. 4. Autoradiograms of plants grown in a low-iron medium containing ^{59}Fe. A–E are maize plants (from Becker, Hedges & Messens, 1985, with permission) and F–G are bean plants (from Becker, Messens & Hedges, 1985, with permission). Plant A grew in medium free of exogenous siderophores, plants B and C grew in the presence of 10 μM pseudobactin and the siderophore from *Pseudomonas aureofaciens* ATCC 15926, respectively, and plants D and E grew in the presence of the same chelators at 100-fold lower concentrations. Plant F grew in medium free of exogenous siderophores, whilst plant G grew in medium containing 5 μM agrobactin.

Agrobactin

al., 1984). However, it remains to be demonstrated *in vivo* that siderophore production is the critical gene product in PGPR. Indeed, siderophore-mediated iron assimilation systems are widely distributed in bacteria and fungi and the genes for enterobactin and aerobactin production have been identified at the molecular level (Neilands, 1984). Most siderophores are classed chemically as hydroxamic acids or catechols, providing ferric ion-specific ligands. The role of iron in rhizosphere ecology in soil as opposed to *in vitro* systems is very complex and affects, for example, the fungal production of the growth regulator ethylene (Lynch, 1972). Similarly, excessive uptake of Fe^{2+} is toxic to plants (Benckiser *et al.*, 1984). It seems to me that if pseudobactin is the critical product of the PGPR then its antibiotic properties may be more important.

Another complication that should be considered in siderophore action is the influence of other nutrients. The cations Mg^{2+}, Mn^{2+}, Ca^{2+}, Co^{2+}, Cu^{2+}, Zn^{2+}, Al^{3+}, K^+, Na^+, NH_4^+ or Ni^+ compete with Fe^{3+} on binding sites of the siderophore *in vitro*, but Fe^{3+} has a greater affinity than the other cations (Elad & Baker, 1985a). Added cations delay the multiplication of *Pseudomonas* spp. in soil. The siderophore itself inhibits the germination of *Fusarium oxysporum* f.sp. *cucumerinum* chlamydospores, an effect nullified by excess iron in the soil. Siderophore production is associated with carbon substrate availability (Elad & Baker, 1985b). Disease suppressiveness correlates with siderophore production in liquid culture by pseudomonads when *Fusarium solani* is the pathogen but not *F. oxysporum* formae speciales. Chlamydospores of *F. solani* f. sp. *phaseoli* have greater weights, volume and iron content per propagule than those of *F. oxysporum* f. sp. *cucumerinum*. It thus appears that the larger chlamydospores of *F. solani* may not require exogenous iron and energy for complete germination and successful infection. On the basis of the evidence presented it can be predicted that carbon and mineral content of soil will affect siderophore production in a complex manner, and the action may thus depend on the carbon and mineral reserves of the pathogen propagules.

In some respects it might be considered that PGPR is a misnomer because no evidence has ever been presented on their direct effect on plant growth. In a review of seed and root bacterization, Brown (1974) concluded that bacteria could stimulate plant growth by the production of growth regulators although the evidence for this came largely from the observation that bacteria in pure culture can produce supernatants with growth regulator-like activity in bioassays. However, she also recognized the potential antagonistic activity of the bacteria against plant pathogens. More recently Okon (1985) has concluded that whereas *Azospirillum* inoculants are unlikely to contribute significantly to plant nutrition by N_2-fixation, some of the observed increases in yield following inoculation could result from greater nutrient uptake as the bacteria increased nitrogen, phosphorus and potassium uptake from mineral solutions, an effect which might be mediated by the production of growth regulators from the bacteria. While those regulators need not be those which are currently recognized by plant physiologists, caution should certainly be exercised in implicating them in the bacterial mediation of plant growth because typically such substances show stimulatory effects at low concentrations and inhibitory effects at higher concentrations. Indeed an increase in growth regulator pool of the plant can induce disease-like symptoms (Pegg, 1984). The PGPR concept is potentially important in agriculture and the scientific understanding of plant/micro-organism interactions. However, to gain a clearer understanding of this it is essential that more experiments be carried out at both plant and microbial physiological levels.

FUNGAL ANTAGONISTS

Biological control procedures have most commonly been studied by plant pathologists/mycologists and the greatest interest has been on fungal/fungal antagonism (Cook & Baker, 1983). In a recent review, the potential of *Coniothyrium minitans* and *Gliocladium roseum* as antagonists of sclerotial-forming pathogens was outlined and it was demonstrated that these antagonists could attack different stages of the life cycle of *Sclerotinia sclerotiorum* of lettuce (Lynch & Ebben, in press). *C. minitans* is particularly effective at invading sclerotia mycoparasitically, an important feature in disease carry-over between crops, whereas *G. roseum* can attack hyphae preferentially.

Here I will mention two further groups of fungal/fungal interactions, studied in my institute at the mode of action level.

Penicillium patulum, *Aspergillus clavatus* and *A. terreus*

In studies first reported in 1945, when there was a substantial interest in soil as a source of antibiotics, the fungi *Penicillium patulum*, *Aspergillus clavatus* and *A. terreus* were tested as potential biocontrol agents because they all produced the antibiotic patulin (Grossbard, 1952). It was recognized that glucose as an energy source was necess-

Patulin

ary to elevate the antibiotic titre, but wheat straw, lawn mowings, bracken, timothy grass and sugar-beet pulp could also provide this energy. The concept of substrate supply to the biocontrol agent for metabolism is critical and it is amazing how many recent investigations have failed to recognize this.

The importance of Grossbard's work was the demonstration of antibiotic activity *per se* rather than the demonstration of biocontrol potential. Indeed, in common with so many other studies, the control of damping-off disease of tomato by these fungal antagonists was variable between experiments.

Trichoderma and *Gliocladium* spp.

Trichoderma is probably the most extensively studied fungal antagonist. The species studied most are *T. hamatum*, *T. harzianum* and *T. viride*. *T. viride* has been exploited commercially on impregnated wood dowels placed in drilled holes of the trunk and major limbs of plum, peach or nectarine trees (Dubos & Ricard, 1974) to control silver leaf disease caused by *Stereum (Chondrostereum) purpureum*. Currently, the major interest is activity against soil-borne diseases and patents have been filed concerning selection of new strains. Together with *Gliocladium*, which also has potential for biocontrol, the biology and ecology of *Trichoderma* species has been reviewed by Papavizas (1985) and their biochemical potential has been reviewed by Eveleigh (1985).

The question of ecological success of *Trichoderma* has been somewhat equivocal. It produces a large amount of extracellular cellulases and xylanases which makes it a particularly effective degrader of lignocelluloses in soil such as straw (Harper & Lynch, 1985). As such it can be considered to be a zymogenous organism according to Winogradsky's terminology or an *r*-strategist in modern ecological terminology. Thus, while there is an abundant supply of substrate present it can be expected to be at a competitive advantage. When the substrate supply is exhausted, it will survive as spores but will not necessarily be at a competitive advantage over other species. If it has been active in decomposition, the spore numbers will likely be greater than those of other species in the vicinity of decayed substrates so that entry of fresh substrate into the soil will again place it at a competitive advantage. Indeed the whole concept of biological control largely depends on relative propagule numbers of pathogen to antagonist. *Fusarium* spp. which are common pathogens, also have powerful extracellular cellulases and xylanases and the hyphal (colonial) growth rate on agar media is similar to that of *Trichoderma* spp. (J. M. Lynch, unpublished). Where the pathogen has extensively colonized crop residues, one strategy of biological control with *Trichoderma* would be to elevate the antagonist population over the pathogen by inoculation.

Competition for substrates is not the only attribute that is sought in a useful antagonist; indeed the more ecological strategies used by the antagonist against the pathogen, the more likely it will be to succeed. It is well recognized that *Trichoderma* spp. are mycoparasitic in that their hyphae coil around hyphae of other fungi (Fig. 5). When the fungus is grown on a nutrient-rich medium there is no obvious direct interaction between the fungal hyphae but when water agar is the growth substrate, penetration of the pathogen cell walls by the antagonist is evident (C. J. Ridout, J. R. Coley-Smith & J. M. Lynch, unpublished). Such penetration almost certainly

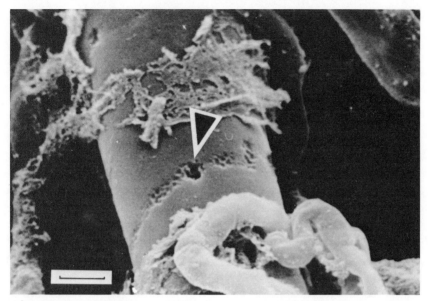

Fig. 5. Coiling of *Trichoderma viride* around *Rhizoctonia solani*. Note points of cell wall degradation and penetration (arrowed) where the antagonist has become detached from the pathogen. Bar = 2 μm.

involves the production of cell wall-degrading enzymes by the antagonist. *Trichoderma* produces $\beta 1 \rightarrow 3$ glucanase and chitinase and these enzymes have been claimed to be agents inducing cell wall lysis (Elad, Chet & Henis, 1982). The extracellular enzymes of *Trichoderma* have been characterized in detail (Ridout, Coley-Smith & Lynch, 1986, and unpublished), with chitinase (molecular weight 77 600), $\beta 1 \rightarrow 3$ glucanase (molecular weights 28 183 – major, 10 000 – minor) and proteinase (molecular weights 64 565, 25 118 and 10 000) as the major components. Indeed the proteinase is present in greatest amounts and it is possible that this enzyme could be involved in the degradation of pathogen hyphal walls.

Early studies by Dennis & Webster (1971) indicated the activity of *Trichoderma* spp. against *Rhizoctonia solani* and other test fungi by a volatile antibiotic. In these early studies, there was no full chemical characterization but the activity in *Trichoderma harzianum* did appear to be associated with a coconut smell. Recently, N. Claydon & J. R. Hanson (unpublished) have characterized the smell and antibiotic activity as 6n-pentyl-2H-pyran-2-one, with the pentenyl pyrone as a minor component. This material is volatile and raises some interesting possibilities ecologically because volatile materials

6n-pentyl-2H-pyran-2-one

are ideal as soil fungicides. At present where soil has to be sterilized the fumigants used can be toxic to workers and can also leave harmful residues in soil. Further work by Claydon & Hanson has also characterized a non-volatile fungicide.

One of the useful attributes of *Trichoderma* is that it has a broad host range (Table 2). We have found a strain of *T. viride* which controlled *Rhizoctonia solani* on lettuce grown in glasshouse soils on six occasions during the year (C. J. Ridout, J. R. Coley-Smith & J. M. Lynch, unpublished). It has already been indicated that lack of reproducibility has been a common feature of biocontrol systems but may be reduced with protected crops which are grown with some control of environment.

Both the strain and delivery system of the antagonist is critical in the effect of *Trichoderma* and *Gliocladium* (Lewis & Papavizas, 1985). Eight of 14 isolates of *Trichoderma* spp. and *Gliocladium virens* reduced the survival of *Rhizoctonia solani* by at least 50% in pathogen-infested beet seed in soil amended with a sand/cornmeal inoculum of the pathogen, but mycelial preparations were always more effective than conidia or free mycelia (Table 3). In these experiments isolates of *T, hamatum* and *G. virens* were more effective than those of *T. harzianum* and *T. viride* but this cannot always be expected. The mycelial preparations were effective in preventing damping-off of cotton, sugar beet and radish seedlings in the glasshouse. There was no correlation between populations density of antagonist and pathogen survival or damping-off and therefore it seems that the physiological state of the antagonist is likely to be more critical than its biomass. There was a highly significant negative correlation between plant stand (% of plants surviving) and pathogen survival in soil. Damping-off diseases would seem to be a good target for biological control because the interaction takes place over a short time period (weeks). There would probably be more scope for sustaining the antagonist population level during this short period when

Table 2. Trichoderma *spp. as biological control agents*

in vitro		in vivo	
Pathogen	Antagonist	Pathogen	Antagonist
Botrytis cinerea	T. harzianum[a,c]	Ceratocystis ulmi[c]	T. viride
	T. pseudokoningii[c]	Hymenomycetes[c]	T. viride
Fusarium culmorum	T. harzianum[a]		T. polysporum
F. oxysporum	T. harzianum[a,b]	Rhizoctonia solani[a,c]	T. harzianum
	T. koningii[b]		T. viride
	T. viride[b]	Sclerotium rolfsii[c]	T. harzianum
F. solani	T. harzianum[a]	Sclerotinia	
		sclerotiorum[a]	T. harzianum
Fomes annosus	T. koningii[b]	Stereum purpureum[c]	T. harzianum
	T. longi-		
	brachiatum[b]		T. polysporum
	T. viride[b]		T. viride
		Verticillium fungicola[c]	T. polysporum
Pythium ultimum	T. koningii[b]		
	T. viride[b]		
Pyrenochaeta			
lycopersici	T. harzianum[a]		
Rhizoctonia solani	T. hamatum[a,b]		
	T. harzianum[a]		
	T. koningii[b]		
	T. longi-		
	brachiatum[b]		
	T. viride[a,b]		
Sclerotinia			
sclerotiorum	T. harzianum[a]		
S. trifolium	T. harzianum[c]		

[a] Personal observations
[b] Dennis & Webster (1971)
[c] Cited by Eveleigh (1985)

the plant is susceptible to the pathogen than with diseases which affect later stages of crop growth. The damping-off diseases may also provide useful 'bioassays' for other diseases thereby enabling more rapid screening of antagonists. Such screens would advance the study of biological control more rapidly. In this context it is interesting that a high correlation has been found between the ability of isolates of *Trichoderma* to prevent bean disease caused by *Sclerotium rolfsii* in the glasshouse and to prevent germination of pathogen sclerotia on soil plates in the laboratory (Henis, Lewis & Papavizas, 1984).

Table 3. *Effect of Trichoderma hamatum on survival and saprophytic growth of* Rhizoctonia solani, *on antagonist proliferation in soil and on seedling damping-off (after Lewis & Papavizas, 1985)*

Antagonist and preparation	Pathogen survival in infested beet seed (%)	Pathogen saprophytic growth (% beet seed colonization in soil)	Antagonist population (CFU g^{-1} soil)	Plant stands in glasshouse soil at 3 weeks (%)		
				Cotton	Sugar beet	Radish
Tm-23						
Conidia[a]	97	92	5×10^2			
Conidial preparation[b]	96	95	1×10^3			
Mycelium[c]	94	83	4×10^3			
Mycelial preparation[d]	33	0	2×10^8	90	69	73
TRI-4						
Conidia[a]	95	94	2×10^3			
Conidial preparation[b]	98	90	6×10^2			
Mycelium[c]	88	78	2×10^5			
Mycelial preparation[d]	2	0	4×10^9	80[e]	58[e]	72
Control (non-infected)	97	89	3×10^3	94[e]	46[e]	78[e]

[a] Conidia were added directly to soil at 5×10^3 g^{-1} soil; [b] Conidial preparations were conidia on bran mixed immediately with soil (1:200, w/v) to give antagonist density of 5×10^3 conidia g^{-1} soil; [c] Mycelium was grown on potato-dextrose broth for 2 days and added to soil at 5×10^3 propagules g^{-1} soil; [d] Mycelial preparations were 3-day-old cultures of antagonist on bran added to soil (1:200 w/v) to give antagonist density of 5×10^3 propagules g^{-1} soil; [e] Means of two experiments.

In glasshouse cropping trials of lettuce, *Gliocladium roseum* inoculants have been useful in suppressing *Sclerotinia sclerotiorum*, although its inability to penetrate the sclerotia make it less attractive than *Coniothyrium minitans* as a biocontrol agent (Lynch & Ebben, in press).

Production of the antibiotic gliovirin by *Gliocladium virens* against *Pythium ultimum*, the cause of damping-off disease of cotton,

Gliovirin

appears to be important in the antagonist/pathogen interaction (Howell, 1982; Howell & Stipanovic, 1983). Supporting evidence has been provided by the observation that a mutant with enhanced gliovirin production was more inhibitory to *P. ultimum* in culture than the parent isolate and showed similar efficacy as a seedling disease-suppressant, even though its growth rate was reduced when compared to the parent isolate.

CONCLUSION – FUTURE DIRECTIONS

There is great scope and impetus for manipulation of rhizosphere communities by enhancing the activity of beneficial organisms and by inoculation of pathogen antagonists. The major problem in studies thus far is variability or response. Significant advances will probably only be made when the modes of action of antagonist activity are known. The control of crown gall is an excellent example because here sensible science has indicated how plasmid transfer can limit the effectiveness in some situations. The time is now right for modifying the antagonist genetically perhaps to reduce plasmid transfer. With many other agents, there is little opportunity immediately for the geneticist and molecular biologist to become involved with the microbial ecologist/plant pathologist because the gene products have not yet been identified, although with some this could be imminent.

Defining the gene products will aid in developing screening and selection processes for better isolates and also in producing larger-scale production systems.

The *in vitro* construction of biological control agents as a means to limit inconsistency of effects has been discussed by Lindow (1986). One route is to use altered pathogens with their deleterious traits removed by genetic manipulation. Some of the genes which determine virulence, production of plant cell and wall-degrading enzymes, plant growth inhibitors and ice nucleation have been identified. Removal of these may enable their use as biocontrol agents in the rhizosphere if they can be satisfactorily introduced into the environment. More directly, genes that might increase the effectiveness of antagonists, such as those conferring the production of bacteriocins, antibiotics and chitinase, have also been identified. It would also be useful to construct strains with increased osmotolerance, nutritional diversity or biocide resistance. However, a danger inherent in this approach is that the modification of the strains will result in a loss of ecological competitiveness; this can sometimes result from the induction of antibiotic resistance in strains or even by strain maintenance on laboratory media as opposed to soil. Furthermore, whereas it seems most unlikely that there will be any serious environmental hazard in the use of genetically engineered strains, it must be admitted that few studies have been undertaken on gene flow in the soil ecosystem and the potential of plasmid transfer is real (Kerr & Tate, 1984). There is, therefore, still much to be gained from the identification and isolation of beneficial organisms from natural rhizosphere communities and attempting to elevate their contribution to natural populations by inoculation in suitable formulations which will promote their activity against deleterious organisms, particularly where there is already a chemical or physical stress in the environment. It is also possible that this approach might proceed more satisfactorily by using mixed inocula with communities containing members which are mutually dependent on each other such as have been observed in laboratory studies of straw decomposition (Lynch and Harper, 1985). However, whereas this latter approach is interesting ecologically, the economics of mixed culture inoculations may make it an unrealistic commercial prospect. Inoculation is not the only route to the control of rhizosphere microbial communities; modification of farming practices is an alternative. For example, foot-rots of wheat caused by *Fusarium culmorum* in the Pacific northwest of the USA are controlled by a wheat/fallow cycle which increases the availability of water to plants and makes them more resistant to invasion by the pathogen (Cook, 1980). However, the cost incurred in a fallow might make the introduction of a biocontrol agent to allow continuous cropping economically attractive.

ACKNOWLEDGEMENTS

I am grateful to Professor G. E. Harman and Dr E. B. Nelson for providing a photograph and experimental results in advance of publication, to Dr R. W. Hedges and Mr C. J. Ridout for photographs, to Dr R. Campbell for comments on Table 1, and to Dr J. M. Whipps for critical review of the manuscript.

REFERENCES

ASHER, M. J. C. & SHIPTON, P. J. (1981). *Biology and Control of Take-all*. London, Academic Press.

BARBER, D. A. & LYNCH, J. M. (1977). Microbial growth in the rhizosphere. *Soil Biology and Biochemistry*, **9**, 305–8.

BECKER, J. O., HEDGES R. W. & MESSENS, E. (1985). Inhibitory effect of pseudobactin on the uptake of iron by higher plants. *Applied Environmental Microbiology*, **49**, 1090–5.

BECKER, J. O., MESSENS, E. & HEDGES, R. W. (1985). The influence of agrobactin on the uptake of ferric iron by plants. *FEMS Microbiology Ecology*, **31**, 171–5.

BENCKISER, G. SANTIAGO, S., NEUE, H. U., WATANABE, I. & OTTOW, J. C. G. (1984). Effect of fertilization on exudation, dehydrogenase activity, iron-reducing populations and Fe^{++} formation in the rhizosphere of rice (*Oryza sativa* L.) in relation to iron toxicity. *Plant and Soil*, **79**, 305–16.

BENNETT, R. A. & LYNCH, J. M. (1981a). Bacterial growth and development in the rhizosphere of gnotobiotic cereal plants. *Journal of General Microbiology*, **125**, 95–102.

BENNETT, R. A. & LYNCH, J. M. (1981b). Colonization potential of rhizosphere bacteria. *Current Microbiology*, **6**, 137–8.

BROWN, M. E. (1974). Seed and root bacterization. *Annual Review of Phytopathology*, **12**, 181–97.

CAPPER, A. L. & CAMPBELL, R. (1986). The effect of artificially inoculated antagonistic bacteria on the prevalence of take-all disease of wheat in field experiments. *Journal of Applied Bacteriology*, **60**, 155–60.

CHAO, W. L., NELSON, E. B., HARMAN, G. E. & HOCH, H. C. (1986). Colonization of the rhizosphere by biological control agents applied to seeds. *Phytopathology*, **76**, 60–5.

COLEMAN, D. C., COLE, C. V., HUNT, H. W. & KLEIN, D. A. (1978). Trophic interactions in soils as they affect energy and nutrient dynamics. I. Introduction. *Microbial Ecology*, **4**, 345–9.

COOK, R. J. (1980). *Fusarium* foot rot of wheat and its control in the Pacific Northwest. *Plant Disease*, **64**, 1061–6.

COOK, R. J. & BAKER, K. F. (1983). *The Nature and Practice of Biological Control of Plant Pathogens*. St Paul, American Phytopathological Society.

DENNIS, C. & WEBSTER, J. (1971). Antagonistic properties of species groups of *Trichoderma*. II. Production of volatile antibiotics. *Transactions of the British Mycological Society*, **57**, 363–9.

DUBOS, B. & RICARD, J. L. (1974). Curative treatment of peach trees against silver leaf disease (*Stereum purpureum*) with *Trichoderma viride* preparations. *Plant Disease Reporter*, **58**, 147–50.

ELAD, Y. & BAKER, R. (1985a). Influence of trace amounts of cations and sidero-phore-producing pseudomonads on chlamydospore germination of *Fusarium oxysporum*. *Phytopathology*, **75**, 1047–52.

ELAD, Y. & BAKER, R. (1985b). The role of competition for iron and carbon in suppression of chlamydospore germination of *Fusarium* spp. by *Pseudomonas* spp. *Phytopathology*, **75**, 1053–9.

ELAD, Y., CHET, I. & HENIS, Y. (1982). Degradation of plant pathogenic fungi by *Trichoderma harzianum* in carnation. *Plant Disease*, **65**, 675–7.

ELLIOTT, L. F. & LYNCH, J. M. (1985). Plant growth inhibitory pseudomonads colonizing winter wheat (*Triticum aestivum* L.) roots. *Plant and Soil*, **84**, 57–65.

EVELEIGH, D. E. (1985). *Trichoderma*. In *Biology of Industrial Micro-organisms*, ed. A. L. Demain and N. A. Solomon, pp. 489–509. Menlo Park, Benjamin Cummings Publishing Co.

GARRETT, S. D. (1970). *Pathogenic Root-infecting Fungi*. Cambridge, Cambridge University Press.

GILLIGAN, C. A. (1983). Modeling of soilborne pathogens. *Annual Reviews of Phyto-pathology*, **21**, 45–64.

GROSSBARD, E. (1952). Antibiotic production by fungi on organic manures and in soil. *Journal of General Microbiology*, **6**, 295–310.

HADAR, Y., HARMAN, G. E., TAYLOR, A. G. & NORTON, J. M. (1983). Effects of pregermination of pea and cucumber seeds and of seed treatment with *Enterobacter cloacae* on rots caused by *Pythium* sp. *Phytopathology*, **71**, 569–72.

HARLEY, J. L. & SMITH, S. E. (1983). *Mycorrhizal Symbiosis*. London, Academic Press.

HARLEY, J. L. & WAID, J. S. (1955). A method of studying active mycelia on living roots and other surfaces in the soil. *Transactions of the British Mycological Society*, **38**, 104–18.

HARPER, S. H. T. & LYNCH, J. M. (1979). Effects of *Azotobacter chroococcum* on barley seed germination and seedling development. *Journal of General Microbiology*, **112**, 45–51.

HARPER, S. H. T. & LYNCH, J. M. (1985). Colonisation and decomposition of straw by fungi. *Transactions of the British Mycological Society*, **85**, 655–61.

HENDSON, M. & THOMSON, J. A. (1986). Expression of an agrocin-encoding plasmid of *Agrobacterium tumefaciens* in *Rhizobium meliloti*. *Journal of Applied Bacteriology*, **60**, 147–54.

HENIS, Y., LEWIS, J. A. & PAPAVIZAS, G. C. (1984). Interaction between *Sclerotium rolfsii* and *Trichoderma* spp.: relationship between antagonism and disease control. *Soil Biology and Biochemistry*, **16**, 391–5.

HILTNER, L. (1904). Uber neuere Erfahrungen und Probleme auf dem Gebiet der Bodenbakteriologie und unter besonderer Beruchsichtigung der Frundungung und Brache. *Arbeiten der Deutschen Landwirtschaftsgesellschaft Berlin*, **98**, 59–78.

HOWELL, C. R. (1982). Effect of *Gliocladium virens* on *Pythium ultimum*, *Rhizoctonia solani* and damping off of cotton seedlings. *Phytopathology*, **72**, 496–8.

HOWELL, C. R. & STIPANOVIC, R. D. (1979) Control of *Rhizoctonia solani* and cotton seedlings by *Pseudomonas fluorescens* and an antibiotic produced by the bacterium. *Phytopathology*, **69**, 480–2.

HOWELL, C. R. & STIPANOVIC, R. D. (1980). Suppression of *Pythium ultimum*-induced damping-off of cotton seedlings by *Pseudomonas fluorescens* and its antibiotic, pyoluteorin. *Phytopathology*, **70**, 712–5.

HOWELL, C. R. & STIPANOVIC, R. D. (1983). Gliovirin, a new antibiotic from *Gliocladium virens*, and its role in the biological control of *Pythium ultimum*. *Canadian Journal of Microbiology*, **29**, 321–4.

JAMES, D. W., SUSLOW, T. V. & STEINBACK, K. E. (1985). Relationship between rapid, firm adhesion and long-term colonization of roots by bacteria. *Applied and Environmental Microbiology*, **50**, 392–7.

KERR, A. & TATE, M. E. (1984). Agrocins and the biological control of crown gall. *Microbiological Sciences*, **1**, 1–4.

KLOEPPER, J. W., LEONG, J., TEINTZE, M. & SCHROTH, M. N. (1980). Enhanced plant growth by siderophores produced by plant-growth-promoting rhizobacteria. *Nature, London*, **286**, 885–6.

KUC, J. (1982). Induced immunity to plant disease. *Bioscience*, **32**, 854–60.

LEWIS, J. A. & PAPAVIZAS, G. C. (1985). Effect of mycelial preparations of *Trichoderma* and *Gliocladium* on populations of *Rhizoctonia solani* and the incidence of damping-off. *Phytopathology*, **75**, 812–17.

LINDOW, S. E. (1986). *In vitro* construction of biological control agents. In *Biotechnology and Crop Improvement and Protection*, ed. P. R. Day, pp. 185–98. Croydon, British Crop Protection Council Monograph No. 34.

LOCHHEAD, A. G. (1959). *Qualitative Studies of Soil Micro-organisms. I–XV (1938–1957)*. Ottawa, Canada Department of Agriculture.

LOPER, J. E., HAACK, C. & SCHROTH, M. N. (1985). Population dynamics of soil pseudomonads in the rhizosphere of potato (*Solanum tuberosum* L.). *Applied and Environmental Microbiology*, **49**, 416–22.

LYNCH, J. M. (1972). Mode of ethylene formation by *Mucor hiemalis*. *Journal of General Microbiology*, **83**, 407–11.

LYNCH, J. M. (1976) Products of soil micro-organisms in relation to plant growth. *CRC Critical Reviews in Microbiology*, **5**, 67–107.

LYNCH, J. M. (1982). Interactions between bacteria and plants in the root environment. In *Bacteria and Plants*, ed. M. E. Rhodes-Roberts & F. A. Skinner, pp. 1–23. London, Academic Press.

LYNCH, J. M. (1983). *Soil Biotechnology. Microbiological Factors in Crop Productivity*. Oxford, Blackwell Scientific Publications.

LYNCH, J. M. & EBBEN, M. H. E. (in press). The use of micro-organisms to control plant disease. *Journal of Applied Bacteriology*, symposium supplement.

LYNCH, J. M. & HARPER, S. H. T. (1985). The microbial upgrading of straw for agricultural use. *Philosophical Transactions of the Royal Society of London, Series B*, **310**, 221–6.

MOORES, J. C., MAGAZIN, M., DITTA, G. S. & LEONG, J. (1984) Cloning of genes involved in the biosynthesis of pseudobactin, a high affinity iron transport agent of a plant growth-promoting *Pseudomonas* strain. *Journal of Bacteriology*, **163**, 55–8.

NEILANDS, J. B. (1984). Siderophores of bacteria and fungi. *Microbiological Sciences*, **1**, 9–14.

NELSON, E. B., CHAO, W.-L., NORTON, J. M., NASH, G. T. & HARMAN, G. E. (1986). Attachment of *Enterobacter cloacae* to *Pythium ultimum* hyphae: possible role in the biological control of *Pythium* pre-emergence damping-off. *Phytopathology*, **76**, 327–35.

OKON, Y. (1985). *Azospirillum* as a potential inoculent for agriculture. *Trends in Biotechnology*, **3**, 223–8.

ONG, S. A., PETERSON, T. & NEILANDS, J. B. (1979). Agrobactin, a siderophore from *Agrobacterium tumefaciens*. *Journal of Biological Chemistry*, **254**, 1860–5.

PAPAVIZAS,, G. C. (1985). *Trichoderma* and *Gliocladium*: biology, ecology and potential for biocontrol. *Annual Review of Phytopathology*, **23**, 23–54.

PEGG, G. F. (1984). The role of growth regulators in plant disease. In *Plant Diseases: Infection, Damage and Loss*, ed. R. K. S. Wood and G. J. Jellis, pp. 29–48. Oxford, Blackwell Scientific Publications.

RIDOUT, C. J., COLEY-SMITH, J. R. & LYNCH, J. M. (1986). Enzyme activity and electrophoretic profile of extracellular protein induced in *Trichoderma* spp. by cell walls of *Rhizoctonia solani*. *Journal of General Microbiology*, **132**, 2345–52.

RUPPEL, E. G., BAKER, R., HARMAN, G. E., HUBBARD, J. P., HECKER, R. J. & CHET, I. (1983). Field tests of *Trichoderma harzianum* Rifai aggr. as a biocontrol agent of seedling disease in several crops and *Rhizoctonia* root rot of sugar beet. *Crop Protection*, **2**, 399–408.

SHIPTON, P. J., COOK, R. J. & SITTON, J. W. (1973). Occurrence and transfer of a biological factor in soil that suppresses take-all of wheat in eastern Washington. *Phytopathology*, **63**, 511–17.

TEINTZE, M., HOSSAIN, M. B., BARNES, C. L., LEONG, J. & VAN DER HELM, D. (1981). Structure of ferric pseudobactin, a siderophore from a plant growth promoting *Pseudomonas*. *Biochemistry*, **20**, 6446–57.

TEINTZE, M. & LEONG, J. (1981) Structure of pseudobactin A, a second siderophore from plant growth promoting *Pseudomonas* B10. *Biochemistry*, **20**, 6457–62.

WELLER, D. M. (1984). Distribution of a take-all suppressive strain of *Pseudomonas fluorescens* on seminal roots of winter wheat. *Applied and Environmental Microbiology*, **48**, 897–9.

WELLER, D. M. & COOK, R. J. (1983). Suppression of take-all of wheat by seed treatments with fluorescent pseudomonads. *Phytopathology*, **73**, 463–9.

WHIPPS, J. M. (in press). Use of micro-organisms for biological control of vegetable diseases. In *Aspects of Applied Biology*. Wellesbourne, Association of Applied Biologists.

WHIPPS, J. M., LEWIS, K. & COOKE, R. C. (in press). Mycoparasitism. In *Mycoparasitism and Plant Disease Control*, ed. M. N. Burge. Manchester, Manchester University Press.

WHIPPS, J. M. & LYNCH, J. M. (1985). Energy losses by the plant in rhizodeposition. *Annual Proceedings of the Phytochemical Society of Europe*, **26**, 59–71.

WHIPPS, J. M. & LYNCH, J. M. (1986). The influence of the rhizosphere on crop productivity. *Advances in Microbial Ecology*, **9**, 187–244.

GENETIC INTERACTIONS AND DEVELOPMENTAL VERSATILITY DURING ESTABLISHMENT OF DECOMPOSER BASIDIOMYCETES IN WOOD AND TREE LITTER

A. D. M. RAYNER*, LYNNE BODDY† and C. G. DOWSON*

*School of Biological Sciences, University of Bath, Claverton Down, Bath, BA2 7AY, UK
†Department of Microbiology, University College, Newport Road, Cardiff, CF2 1TA, UK

INTRODUCTION

Crucial as they are in community biology, the processes by which decomposer basidiomycetes (i.e. excluding biotrophs such as the Ustilaginales and Uredinales) establish themselves within natural habitats have been neglected. This neglect has probably been partly due to the naive view of establishment as simply involving the arrival of propagules at or below the resource surface, followed, under suitable conditions, by mycelial outgrowth. However, the issues are obviously far more complex because establishment processes necessarily involve a dynamic interplay between the activities of the basidiomycete thallus and both abiotic and biotic components of its microenvironment, the biotic components including other genotypes of the same species. In consequence, the thallus is exposed, often sequentially, to selection pressures favouring different and even opposite attributes. For example, attributes favouring rapid ramification in a domain may not be appropriate to defence or effective exploitation of resources in that domain.

An impoverished view of the nature and properties of basidiomycete mycelium, as something resembling animated cotton wool (Wood, 1985), has further clouded understanding of the manner in which these changing selection pressures can be coped with. For example, it was once widely considered that sufficient genetic flexibility could be achieved by hyphal fusion (anastomosis) between different mycelia of the same species to form a complex, physiologically unified genetic mosaic. From this mosaic, genetic components best fitted to the changing ecological settings faced by the mycelium as colonization proceeded could be selected sequentially (Burnett, 1965). Such a pattern of behaviour would be unique amongst eukaryotes and at variance with modern ideas about the action of natural selection on individuals rather than groups (Williams, 1971; Carlile, in press).

Growing awareness of two fundamental attributes places these issues in a different perspective, suggesting that the mycelium of basidiomycetes is a body form which resembles that of fully-fledged multicellular organisms. First, being composed of an intercommunicating system of apically extending, branching tubes (hyphae) filled with protoplasm, the mycelium is like an indeterminate embryo, with different parts able to select from a range of alternative modes of development to suit distinctive functional requirements (Rayner & Coates, in press). Thus a series of superimposable switch mechanisms control the outcome of contrasting patterns of morphogenesis such as cells and hyphae, branching and extension, diffuse and coherent growth, and juvenility and senescence. It is this facility, rather than genetic heterogeneity, which confers on the individual the developmental versatility enabling it to span the heterogeneous, changing and often discontinuous niches it occupies naturally.

The second important attribute is that basidiomycete hyphae possess powerful recognition responses which condition the occurrence and outcome of fusion with other hyphae, or propagules, of the same or different genotype (Rayner, 1986a). A long-range signalling system determines the primary ability of hyphae to grow towards receptive sites prior to fusion with other hyphae, conidia or basidiospores, and generally seems to operate within a species, or between closely related species, regardless of genotype. Fusion and subsequent reactions are determined by contact stimuli, which do depend on genotype.

Self-fusions, within the same thallus, or between thalli of identical genotype, occur readily and convert the mycelium from a radiate communication system to a network. Protoplasmic destruction and erosion of septa associated with nuclear migration are not found, although in strictly monokaryotic and dikaryotic mycelia a remarkable process of nuclear replacement occurs in recipient compartments (Aylmore & Todd, 1984; Todd & Aylmore, 1985; Ainsworth & Rayner, 1986).

Non-self interactions between thalli differing in genotype at typically multiallelic or polygenic recognition loci can be of three distinctive types. Sexual acceptance allows entry of donor nuclei into recipient hyphae, usually followed by nuclear migration associated with erosion of septa. It is conditioned by the presence of complementary mating alleles between interacting homokaryons, or homokaryons and heterokaryons. Rejection responses, associated with somatic incompatibility which obviates formation of unified mosaics (see above), involve what appears to be a programmed cycle of protoplasmic vacuolation and destruction in contiguous compartments. These responses occur between homokaryons which are not mating-competent and generally in interactions involving heterokaryons, where their presence is related to the non-receptiveness of the mycelia formed by mating to nuclear migration. Parasitic interactions are often initiated by similar signalling systems, but they do not usually result in protoplasmic continuity: entwining and penetration of recipient ('host') hyphae are the usual outcome and rejection responses are not elicited – at least initially.

The importance of these two major attributes of basidiomycete mycelia in determining patterns of establishment in wood and litter will now be examined. First, the nature of the problems which successful colonists must solve before establishment in these habitats will be considered, followed by an account of the wide variety of colonization strategies used to solve these problems. Developmental versatility and recognition responses can then be understood within the context of the distinctive requirements of different colonization strategies. It should be noted that whilst this discussion is mostly limited to basidiomycetes, many of the arguments will apply to other fungal groups, especially ascomycetes. Nonetheless, those basidiomycetes which degrade refractory lignocellulosic substrates, operate over years, decades or even centuries and this underlies the diversity of their colonization strategies and their developmental versatility, making them an excellent example for discussion.

PROBLEMS OF ESTABLISHMENT IN WOOD AND LITTER

For the purposes of this chapter, 'wood' will be taken to include all the major durable components, $\geqslant 1$ cm diameter which are not shed from trees on a regular seasonal or annual basis. Litter includes all the fallen debris, excluding wood, which tends to accumulate at ground level below trees. As such, wood and litter provide exceedingly heterogeneous habitats and a wide variety of problems for potential basidiomycete colonists (Fig. 1). Accordingly, four principal determinants of patterns of establishment can be identified: the

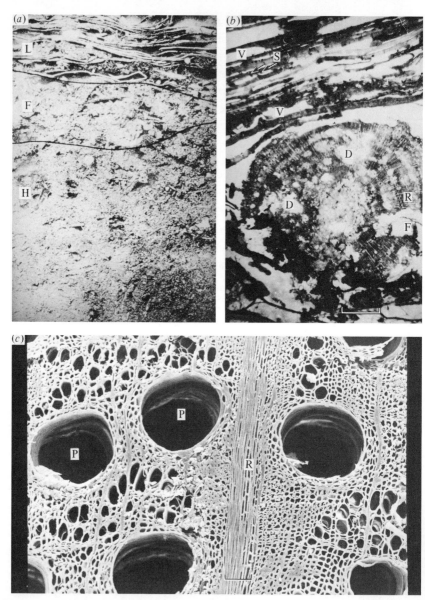

Fig. 1. Spatial heterogeneity of wood and litter microhabitats. (a) Detail of a gelatine-embedded vertical section through the top 5 cm of a well developed organic soil in a *Castanea sativa* (sweet chestnut) woodland showing the litter (L), fermentation (F) and humus (H) sub-horizons. (From Swift, Heal & Anderson, 1979). (b) Detail of a section through the litter layer and a rotten twig in the same site as (a), illustrating voids between leaves (V), voids resulting from decomposition (D), anatomical voids of the vessels and medullary rays (R), fungal hyphae (F) and dark fungal stroma (S). Scale bar = 1 cm. (From Swift, Heal & Anderson, 1979.) (c) Scanning electron micrograph of *Quercus* sapwood showing different sized vessels (P) and medullary ray cells (R). Scale bar = 100 μm. (Courtesy of M. Hale, unpublished.)

spatiotemporal distribution of resources, the quality of these resources, microclimatic conditions, and the extent to which a resident microflora has already become established.

Discontinuous versus continuous distribution of resources

As is implicit in the above definitions, critical differences occur between wood and litter with respect to the times and places at which they become available for colonization by basidiomycetes. These differences may be the primary reason for the incomplete, but nevertheless clear division of K-selected (i.e. vegetatively persistent – see below) decomposer basidiomycetes into wood- or litter-inhabiting classes.

With respect to wood, colonization is often – perhaps almost invariably in undisturbed woodlands – initiated in the standing tree (Rayner & Boddy, in press). Understanding of the processes of colonization by wood decomposers must therefore begin with consideration of factors governing establishment before fall. Furthermore, both before and after fall, woody resources are distributed discontinuously, so that fungal colonists must possess effective means of migration.

With respect to litter components, their duration on the standing tree is so much less than that of wood that there is little opportunity for establishment of basidiomycetes prior to fall except for a few highly specialized forms and true parasites. Hence the major opportunity for decomposer basidiomycetes to colonize occurs at ground level, where a second fundamental difference between wood and other litter becomes apparent. Whilst individual components of the litter, e.g. individual twigs, petioles and fruits, are discontinuously distributed, in the mass these components form a relatively homogeneous layer which provides a distinctive habitat. A third major difference is that this layer is regularly or continuously replenished by seasonal or continuous litter fall. Seasonal litter fall in particular will alter microenvironmental conditions radically as well as enriching the habitat with new substrates for exploitation by the decomposer community.

Resource quality and microclimate

The microenvironment of basidiomycete thalli growing in wood and litter is determined by the interaction between *resource quality*, that is intrinsic physicochemical properties such as structure and chemical composition, and *microclimate*, which encompasses extrinsic conditions, notably of moisture, temperature and aeration. A detailed discussion of the microenvironments occurring within wood and litter systems has been provided by Boddy (1984). Here the most salient features determining the direction and extent of basidiomycete growth will be discussed in outline. These are considered to be first the presence of constitutive or induced barriers limiting access to nutrient resources within the plant tissues, secondly the occurrence of unfavourable moisture and gaseous regimes, and thirdly the presence of allelopathic substances.

Wood

In wood, resource quality and microclimate are influenced by the primary nature of this tissue as a device for carrying water and minerals from an underground source through the aerial environment to a distant photosynthesizing canopy. Trees have evolved a structure which provides for conduction of water in long columns under considerable tension whilst conserving this commodity by limitation of access of gases to the functioning conduits through cavitation or evaporation. The resulting plumbing systems, combined with the mechanisms for their repair and maintenance, coincidentally create a spatiotemporally highly heterogeneous set of microenvironmental conditions for establishment of fungal growth (Rayner, 1986*b*).

The predominantly axially oriented systems of conduits, together with the life support system of radially aligned medullary rays (Fig. 1c), provide potential routes of access for fungal hyphae, which in the absence of other constraints will result in wedge-shaped, columnar colonization zones. However, in the living tree, the presence of an intact bark layer impedes access to these passageways. Moreover, functionally intact sapwood filled with water will be inimical to mycelial growth, although access of air or gas to these passageways, resulting from injury or stress to the tree, cavitation or heartwood formation, will counteract this. However, alleviation of one problem for the fungus commonly introduces others. These include the formation of sealant zones (by suberization and production of gums, resins or tyloses) and infusion of the tissues with allelopathic chemicals which are absent from functionally intact sapwood. Furthermore, although the gaseous phase is often prominent in older wood, it is often relatively anoxic and enriched with carbon dioxide (Rayner, 1986*b*).

In dead wood, barriers to fungal growth will invariably be constitutive rather than induced. Microenvironmental conditions of moisture and temperature will fluctuate in response to the external environment whilst gaseous regimes will be related to both external conditions and the activity of the decomposer community. Furthermore, the bark which formerly had been a barrier may now shelter the wood from external extremes and provide readily available nutrients from the cambial layer.

Litter

Microenvironmental conditions in litter need to be considered at two levels – within the litter system as a whole, and within individual litter components. With respect to the latter, distinctive conditions result from the presence of an array of allelopathic chemicals and anatomical features. However, the relatively small size of litter components and their greater surface area to volume ratios allows them to equilibrate more readily with ambient conditions. Induced barriers are of lesser importance, given that basidiomycete colonization occurs after fall.

Regarding litter systems as a whole, microenvironmental conditions and routes of access for fungi will, as in wood, be governed by systems of fluid-filled voids, and there will be marked stratification below the surface layer (Fig. 1a, b). However, by comparison with wood, the void systems will be more heterogeneous and discontinuously distributed. Also the relative permeability of the outer layer will allow much more rapid gaseous exchange with the external atmosphere. Hence, gaseous composition will often, except in waterlogged situations, approximate to atmospheric air. Diffusion paths will be short and the desiccating or fluctuating moisture conditions in surface layers will give way to a more constant environment further down. Finally, the mobility of macro- and microfauna within the litter system provides a further point of departure from wood, at least during early stages of decomposition (see Anderson, this volume).

Extent of prior colonization

Of obvious significance to a colonist is whether or not, when it arrives, its potential habitat is already occupied, and if it is, by what. If the habitat is occupied, then successful establishment may depend on mechanisms allowing replacement of the previous resident(s).

If the habitat is unoccupied, then success may be determined by the ability to exploit this situation as rapidly as possible, particularly by utilization of easily assimilable substrates before only the more refractory lignocellulosic ones remain. Unless deliberately cut or broken off by wind or storm, wood will generally be substantially decayed by the time it becomes 'available' to non-pathogenic decomposer basidiomycetes at ground level. By contrast, litter, although colonized by phylloplane organisms, pathogens and endophytes, will not contain a firmly established community of decomposer basidiomycetes at the time of fall.

COLONIZATION STRATEGIES

The foregoing account has summarized the widely varying biotic and abiotic environmental factors which limit the establishment of decomposer basidiomycetes in wood or litter. In order to understand how these lead to diverse modes of establishment it is necessary to identify the behavioural attributes required to overcome the various constraints. This can be done by considering first the general ecological strategies adopted by organisms in response to the primary determinants of their natural distribution and then the strategies specific to colonization wood and litter.

General concepts: r and K-selection; ruderal, combative and stress-tolerant strategies

During evolution, different types of selection pressure have resulted in polarization between two sorts of organisms with respect to life span and reproductive commitment. K-selected organisms characteristically have a long individual life span and a slow or intermittent commitment to reproduction. The converse applies in r-selected organisms (Harper & Ogden, 1970). Analysis of the selection pressures involved reveals that they are of three types: environmental stress (S-selection), competitive stress (C-selection) and disturbance (R-selection). According to one scheme favoured currently by some plant ecologists (Grime, 1979) and fungal ecologists (Pugh, 1980; Cooke & Rayner, 1984) these selection pressures result in stress tolerance strategies (S-selected), competitive or combative strategies (C-selected; Cooke & Rayner regarded 'combative' as being a more appropriate term for fungi) and ruderal strategies (R-selected). C-

and *S*-selection can be considered as forms of *K*-selection, whilst *R*-selection is equivalent to *r*-selection. It is important to note that as the primary strategies are part of a spectrum of *behaviour*, they should *not* be used to classify individual *organisms* which may exhibit combinations of the different strategies either at one and the same time or at different times during their life cycle. This proviso aside, behavioural attributes of decomposer basidiomycetes which can be understood within the general context of the three primary strategies will now briefly be addressed, and their relevance to community development pathways outlined.

Ruderal strategies

Ruderal strategies are promoted by disturbance, which can be defined as any sudden environmental event which, either by *destruction* of resident biomass or *enrichment* of the habitat, provides a virgin resource for colonization. These strategies are based on rapid arrival, capture and conversion to fungal biomass of easily assimilable growth substrates, and rapid commitment to reproduction before competitors become established.

Combative strategies

Combative strategies are promoted in undisturbed habitats, relatively free from stress, and in which there is consequently a high potential incidence of competitors. Success therefore depends on the ability to defend domain which has been occupied using the process of 'primary resource capture' or to sequester domain from previous residents via 'secondary resource capture' (Cooke & Rayner, 1984; Rayner & Webber, 1984).

Stress-tolerant strategies

These strategies depend on tolerance of an environmental stress, defined as any more or less continuously imposed feature other than competition, which limits the production of biomass by the majority of organisms under consideration, e.g. extremes of temperature, humidity, aeration and grazing pressure. A special form of stress tolerance, latent invasion, allows sparse development in a habitat where stress prevents colonization by potential competitors, followed by rapid capitalization, i.e. luxuriant mycelial outgrowth from the previously established inoculum, after a lessening of the stressful conditions. This gives a decisive advantage to the latent invader in primary resource capture.

Strategies and community development pathways

Understanding of successional changes in fungal communities has been obscured by overemphasis of floristic changes and lack of appreciation of the wide variety of factors which can result in replacement of one individual, population or community by another (Rayner & Todd, 1979; Cooke & Rayner, 1984; Rayner & Webber, 1984). One approach to rationalizing the complex sequences of events is to consider the available pathways along which community development may be channelled by four major determinants of community change: stress-aggravation, stress-alleviation, intensification of combat, and disturbance. A simplified scheme based on the behavioural strategies which will predominate in communities developing under a range of circumstances is presented in Fig. 2.

Colonization strategies of wood decay basidiomycetes in standing trees

To the wood-decaying basidiomycete, the standing tree represents an enormous reserve of food. However, access to this reserve is impeded by all those physical and chemical factors which ensure proper functioning of the tree. Five distinctive colonization strategies can be identified whereby these factors are overcome, circumvented or tolerated. These are: unspecialized opportunism, active pathogenesis, specialized opportunism, heartrot, and desiccation tolerance. All are based on the unsuitability of functionally intact sapwood as a habitat for mycelial growth (Rayner, 1986b; Rayner & Boddy, in press). Before outlining the essential features of these colonization strategies, the same proviso made with respect to general strategies must be made, to the effect that they represent identifiable nodal points in a continuum of behaviour.

Unspecialized opportunism

This is exhibited when normally inaccessible sapwood is made suddenly available for colonization by injury or rapid death of the bark. Damage represents both a disturbance and an elimination of a major

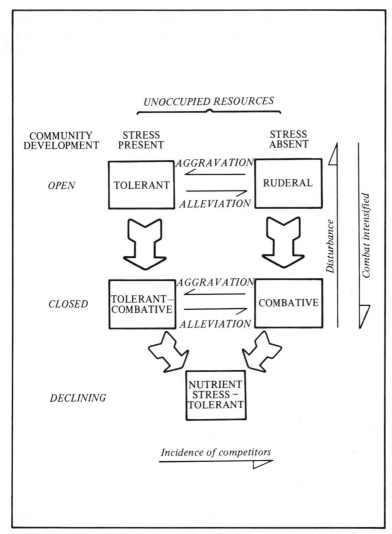

Fig. 2. Diagram of possible community development pathways from colonization of a totally unoccupied resource, through an open community stage with still unoccupied resources available for primary capture, to a closed community with all primary capture of domain completed. The pathways culminate in a declining stage characterized by severe nutrient stress. In the absence of competitors, developing tolerant communities may progress directly to declining tolerant communities without an intermediate combative stage. (From Rayner & Webber, 1984.)

stress barrier. Hence a pattern of community development can be expected in which there is initial selection of organisms with ruderal attributes followed by establishment of combative communities. This is consistent with numerous observations of the colonization of wounded angiospermous trees by pioneer communities dominated by non-basidiomycetous fungi and bacteria causing discoloration but not decay, or by basidiomycetes such as *Chondrostereum purpureum* which cause little decay, reproduce rapidly and are readily replaced (Mercer, 1982). True decay fungi, such as species of *Coriolus*, *Bjerkandera* and *Stereum*, which do not exhibit strong selectivity for particular types of tree, attain dominance at a later stage. However, whilst such colonization sequences resemble those depicted in Fig. 2, they have also been interpreted in terms of progressive breakdown of the defences of living tree tissues (Shigo, 1979).

Limitation of colonization following wounding has been ascribed to active host defence (Shigo, 1979, 1984). However, the view favoured here is that repair mechanisms which seal off damaged from functionally intact tissue are really responsible (Boddy & Rayner, 1983a; Rayner, 1986b). The immediate effect of injury or bark death is exposure of sapwood tissues to air and consequent drying due to cavitation and evaporation. The penetration of air depends on the location, timing and severity of damage, and defines exactly those tissues which become available for colonization by opportunists. Penetration by air is, in turn, dependent on the distribution of void space resulting from wood anatomy, on whether water columns are under tension or pressure, and on the rate of sealing damage by production of gums, resins, tyloses etc. For example, deep wounds of the trunk commonly give rise to colonization zones in the form of two inverted wedge-shaped cones which extend mostly upwards rather than downwards, in autumn rather than in spring, and in older rather than in recent sapwood (Coutts, 1976; Leben, 1985; Rayner, 1986b). The sealant zones delimiting the colonized regions are of two types: 'barrier' zones between wood extant at the time of damage and wood formed subsequently, and 'reaction' zones within tissues extant at the time of damage (Shain, 1979).

Active pathogenesis

The distinction between active pathogens and opportunists is that the latter rely on other agents to alleviate unfavourable conditions in functionally intact sapwood while the former do so themselves. They achieve this by killing living tissues, notably in the cambium and medullary rays, and by destroying pit membranes, both of which help to generate and extend regions affected by cavitation. Unlike parasites of non-woody plants or plant parts, the pathogenic activities of these fungi can be seen as a means of preventing maintenance of the hostile microenvironmental conditions which bar their way to already dead xylem, rather than as a mode of nutrition in itself.

Critical to the success of active pathogens is the establishment of a sufficient inoculum base from which an attack can be made on the host tissues before the defence and/or repair mechanisms of the latter deny access to the fungus. Establishment of this base may be achieved by a variety of means, including initial exploitation of other colonization strategies, notably heartrot, wound colonization and specialized opportunism (see below). However, the primary involvement of an active pathogenic mechanism is at least certain in one major case, the ectotrophic infection of living roots. As its name implies, this habit involves the superficial spread of a mycelial front over or within the bark, in advance, sometimes by 1 m or more, of occupation of the wood cylinder itself. It is the hallmark of several economically important tree pathogens including *Armillaria* spp., *Heterobasidion annosum*, *Phellinus noxius*, *P. weirii*, *Rigidoporus lignosus*, *Inonotus tomentosus* and several *Ganoderma* spp. Whilst Garrett (1970) regarded the ectrotrophic habit as a mechanism for 'diluting out' host resistance, it is better seen as a way of establishing an effective, extensible inoculum from which outer host cells, including cambium, are killed, hence opening the way to the wood cylinder. It is also clear that this mechanism can only operate in environments such as soil or litter where the mycelial front can be supplied by connections with already colonized material (see below).

Specialized opportunism
The basis for this strategy is that particular version of stress tolerance referred to earlier as latent invasion. Accordingly, fungi using this strategy capitalize on alleviation of microenvironmental stress brought about by factors other than their own activities, in which sense they are opportunistic, whilst being in a position so to capitalize by first becoming established under stressful conditions, hence being specialized. The involvement of this strategy in establishment of decay in trees has not yet been demonstrated unequivocally, but there is strong circumstantial evidence for it, especially in relation to the occupation of domain by individual genotypes (see below). Basidiomycetes believed to possess this strategy include a wide range of species whose pathogenicity has never been established directly, but which often exhibit considerable preference (selectivity) for trees of a particular type. Examples include *Piptoporus betulinus* on *Betula*, *Oudemansiella mucida* on *Fagus*, *Peniophora limitata* on

Fraxinus, and *Peniophora quercina* and *Stereum gauspatum* on *Fagus*. Host-selective xylariaceous ascomycetes in *Hypoxylon* and related genera exhibit similar behaviour (Rayner & Boddy, 1986 and in press).

A general mechanism underlying specialized opportunism may be limitation of water supply to functionally intact sapwood arising from internally or externally imposed stresses: that is, stress to the tree results in stress-alleviation for fungi and predisposes the tree to 'infection' by specialized opportunists. Drought stress is probably a particularly frequent factor allowing establishment of specialized opportunists, which often occur in trunks or branches lacking obvious major wounds which serve as colonization foci for air-borne fungi with other strategies.

Heartrots

Once erroneously regarded as the primary cause of decay in standing trees, the heartrot fungi circumvent the problems of colonizing sapwood by growing in the heartwood, where living cells are absent or rare, and in which there is often a relatively extensive gaseous phase. Nonetheless, conditions in the heartwood are commonly highly stressful, gaseous conditions being far from atmospheric and the tissues commonly being suffused with inhibitory chemicals (extractives), generally phenolics. Accordingly, heartrot fungi frequently exhibit a high degree of host selectivity, grow slowly, persist for many years and lack combative ability against less specialized decay fungi. They are amongst the best examples of pure stress-tolerant strategies in the fungal world.

Desiccation tolerance

When bark or sapwood function is entirely lost from standing trunks or attached branches, the underlying wood experiences fluctuations in moisture content and may dry out. Probably quite a large number of stress-tolerant decay fungi can exploit this situation but at present very little detailed ecological information is available about them. However, their principal adaptations may be expected to include production of mycelia and reproductive bodies capable, for example, by production of chlamydospores or mucilage, of dormant survival in the dry state and rapid resumption of function when wetted. Amongst homobasidiomycetes, *Rhodotus palmatus*, *Peniophora cinerea*, *P. lycii* and *Schizopora paradoxa* are good examples, whereas amongst heterobasidiomycetes, jelly fungi such as *Exidia*, *Tremella*,

Auricularia and *Dacrymyces* are probably able to tolerate fluctuations in moisture availability. Xylariaceous ascomycetes and gelatinous discomycetes, e.g. species of *Bulgaria*, *Neobulgaria* and *Ascocoryne*, probably possess at least partial desiccation tolerance.

Integration of colonization strategies during decay community development in a standing tree
From the foregoing account, a complex picture emerges of the processes by which communities of fungi, including basidiomycetes, combine to bring about the eventual demise and decomposition of the standing tree. At first the emphasis will be primarily on the direct interaction between the fungi and tree, and the variety of ways in which the selectively hostile stress conditions imposed actively or passively by the tree can be countered, alleviated, overruled or by-passed. Invading fungi will hence largely be separated from one another by their differing colonization strategies, so that interfungal interactions will largely be confined to those within or between populations with shared or overlapping strategies.

As time proceeds, and the tree begins to decline, so the emphasis will change. Pathogens in the very act of directly alleviating conditions for themselves will predispose sapwood to colonization by specialized and unspecialized opportunists. Unspecialized opportunists previously confined to regions within the vicinity of damage will begin to extend their domain into previously uncolonized territory. Death and colonization will spread to branches of progressively higher order. Heartrot fungi, previously confined within the central wood cylinder, may begin to encroach outwards, only to be met by newly established residents in sapwood. Dead, exposed limbs may lose so much water that fungal activity within them can only be sustained by species with desiccation-tolerant strategies.

In many parts of the tree the stress conditions which first dictated colonization are progressively alleviated and intensification of combative conditions occurs. The ensuing battles will first involve the resident community of pioneers, with the opportunists perhaps having the edge over the pathogens and heartrot fungi. Later, truly combative fungi may establish themselves and begin to replace the pioneers. From the air these may include, in angiosperms, such species as *Phlebia radiata* and *Coriolus versicolor*. From the soil, combative cord-formers such as *Hypholoma fasciculare* and *Phanerochaete velutina* will invade (see below). The intense decay caused by some of these fungi will, in addition to that caused by active pathogens, hasten the fall of limbs and the windthrow of trunks. The stage is then set for new cycles of invasion and interactions on the woodland floor.

Strategies in felled or fallen wood and litter

To reiterate, apart from roots, neither wood nor litter will normally arrive at the woodland floor in an uncolonized state, although the nature of the resident microbial communities in each may be very different. Consequently, basidiomycetes which actively decompose wood and litter in this location exhibit the capacity either to defend domain which they have captured before or soon after fall, or to gain access to domain by secondary resource capture. The difference between defensive and attacking strategies is often reflected in a further major distinction in behaviour between resource-unit restriction and non-restriction (Cooke & Rayner, 1984; Rayner, Watling & Frankland, 1985). Mycelia of unit-restricted fungi are confined to individual resource-units, be these petioles, fruits, twigs, or branches. Mycelia of non-unit-restricted fungi are not confined to individual units of wood or components of litter, but can migrate between these units.

Defensive strategies and unit restriction

With respect to wood, it is clear that some decay fungi can persist for many years, having first colonized the tree while it was still standing, and without growing out to colonize new domain. This applies best to certain xylariaceous ascomycetes, such as *Daldinia concentrica* in *Fraxinus*, so that whilst some species of *Pleurotus, Ganoderma, Piptoporus* and *Phellinus* can persist, many of the basidiomycetes colonizing before fall appear to be rather readily replaced thereafter. Certain species of *Coriolus* and *Stereum*, e.g. *C. versicolor* and *S. hirsutum*, which colonize cut or broken wood primarily by air-borne spores, although capable of replacing pioneers appear to be primarily defensive thereafter, and may persist for several years.

With regard to litter, good examples of unit-restricted basidiomycetes are provided by the genera *Marasmius* and *Mycena sensu lato*. Many exhibit a high degree of selectivity, e.g. *Marasmius buxi* on *Buxus* leaves, *Marasmius hudsonii* on *Ilex* leaves and *Mycena strobilina* on *Pinus* cones (Rayner, Watling & Frankland, 1985). How they become established is unclear, but it may be that some sort of latent invasion mechanism, allowing establishment before fall, is involved: otherwise they must arrive, presumably as spores, and be able to establish by replacing the previously resident microflora. The remarkable tropical agaric *Crinipellis perniciosa* serves as a useful illustration of principles: basidiospores infect developing cocoa (*Theobroma cacao*) tissues, eliciting formation of abnormal growths (brooms) which become ramified by a non-culturable mycelium; following death of the broom, a culturable mycelium develops which produces crops of fruit bodies for up to several years (Hedger, 1985; Wheeler, 1985).

Attacking strategies and non-unit restriction
The life of unit-restricted fungi on the woodland floor, both in wood and litter components, is ultimately often limited by the presence there of highly combative non-unit-restricted fungi. The role of the latter is to scavenge the woodland floor, consuming all appropriate resource units that fall within their path, then to move on inexorably until they meet another equally or more combative non-unit-restricted individual of the same or a different species. However, relating to the distinction between wood and litter described earlier, the form of the mobile mycelial units differs markedly according to habitat type. Wood inhabitants ramify the woodland floor in the form of linear mycelial aggregates, mycelial cords and rhizomorphs, which interconnect between spatially discontinuous woody resource units. Litter inhabitants produce diffuse mycelia which ramify the litter layer as a whole: these are the fairy-ring formers, notably species of *Clitocybe, Collybia, Marasmius* and *Mycena* (Cooke & Rayner, 1984; Thompson, 1984). Intermediate between the wood and litter non-restricted fungi just described are various species of *Marasmius, Marasmiellus* and *Crinipellis* which migrate between litter components by rhizomorphic growth, and in humid environments such as tropical rain forest even become epiphytic in intact tree leaves and shoots whilst still attached (Hedger, 1985).

Ruderal strategies
Ruderal strategies on the woodland floor can only be pursued if some form of disturbance occurs. In the case of wood, this is very common, especially in managed woodland or forest, since cutting or felling of trees is an effective means of disturbance. Sequences of colonization along the lines depicted in Fig. 2 are thus promoted,

beginning with predominance of ruderal communities relatively inactive in decomposition, encompassing such basidiomycetes as *Schizophyllum commune, Chondrostereum purpureum, Corticium evolvens, Stereum sanguinolentum* and *Flammulina velutipes*. Combative, active decay species then become dominant including primarily defensive species of *Stereum* and *Coriolus* and non-restricted cordforming species such as *Hypholoma fasciculare, Phallus impudicus, Tricholomopsis platyphylla* and *Phanerochaete velutina* (Coates & Rayner, 1985*a*, *b*, *c*).

Animal activity and fire represent other causes of disturbance. *Coprinus* spp. probably include some of the best examples of ruderal litter-inhabiting basidiomycetes.

Stress-tolerant strategies
Replacement of previous mycofloras, or components thereof, need not always imply superior combative ability of succeeding individuals. As indicated in Fig. 2, stress aggravation, e.g. resulting from nutrient depletion or desiccation, can also be important. However, unit and non-unit restriction have been emphasized in this section because of their implications for the relation between developmental regulation and colonization strategies.

DEVELOPMENTAL REGULATION AND COLONIZATION
STRATEGIES

It should be clear from the foregoing that a thallus form suitable for one colonization strategy may be quite unsuitable for another. Moreover, the whole process of establishment of domain may involve sequential stages such as arrival, establishment of an inoculum base, primary resource capture, resource exploitation, defence, and secondary resource capture, all in an environmental setting which may be constantly changing via stress alleviation, stress aggravation, disturbance and intensification of combat. It would indeed be a challenge to account for all this on the basis of a view of the basidiomycete mycelium as little more than an assemblage of protoplasm-filled duplicating units (hyphal tips) between which there is little or no intercommunication or functional differentiation. On the other hand, the variability of behaviour ('mutability') of mycelia has always been recognized by mycologists, but regarded as mysterious, if not treacherous and frustrating, by experimentalists.

When set within its true, ecological context, the intrinsic variability of mycelial growth becomes not frustrating but the explanation for the versatility in development which is required for establishment in spatiotemporally heterogeneous environments. A new picture emerges of the vegetative thalli of basidiomycetes, and indeed other fungi, as entities which can adopt a variety of alternative forms ('modes' as Gregory (1984) called them) conferring different functional properties. Hence individuals with different colonization strategies can adopt suitably distinctive developmental patterns, whilst the same individual may be able to switch from one pattern to the other as circumstances dictate. Moreover, fungi with narrow ecological niches may become fixed into particular developmental patterns, whilst greater plasticity will be retained by those with broader niches.

In trying to bring some order to the chaos of observed variability in development of fungal thalli, Rayner & Coates (in press) postulated the existence of at least five distinctive sets of alternative modes of morphogenesis. They suggested further that these alternatives are modulated genetically by a series of superimposable switch mechanisms which are *cued* by a wide variety of endogenous and exogenous stimuli. The nature of the exogenous stimuli is presumably dependent on the environmental signals which would normally be encountered by the fungi concerned. Rayner and Coates detailed possible criteria aiding distinction of simple mutations and direct environmental effects on metabolic functioning from the proposed switch mechanisms. Therefore, rather than elaborating on these mechanisms, our approach here will be to focus on the ecological implications of the alternative states which Rayner & Coates proposed.

Determinate/cellular – interdeterminate/filamentous transitions

Conversion from determinate to indeterminate morphogenesis occurs at spore germination and in reverse at sporogenesis. A growing number of fungi, including many basidiomycetes, are also known to be able to switch between unicellular (yeast-like) and mycelial forms. This capacity has naturally aroused much interest, but the accompanying neglect of other dimorphisms is epitomized by the frequent use of the term 'dimorphism' to cover only the specific case of mycelial–yeast dimorphism (Stewart & Rogers, 1983).

Discussion of cellular–filamentous transitions in an ecological context can thus apply to two distinct issues: the relative merits of spore production *versus* mycelial spread in dissemination, and of yeasts *versus* hyphae in primary resource capture. In both cases, the relative merits of the cellular and filamentous modes can be approximated, respectively, to the demands of *r*- and *K*-selection.

The differing characteristics of spores and mycelium as means of arrival at resource surfaces by basidiomycetes has been discussed in some detail by Rayner, Watling & Frankland (1985) who pointed out that the greater the reliance on spores, the greater will be the tendency for resource unit restriction. To summarize, spores can be produced in large numbers but, unless associated with a vector, they lack means of ensuring their arrival at suitable resources other than by random selection. Furthermore, colonization from spores will be effected from localized foci, opportunities for input of water and nutrients will be limited, and buffering against hostile influences at the resource surface will be absent. Hence spores lack inoculum potential (cf. Garrett, 1970).

Arrival by mycelium is a particular feature of those non-unit-restricted basidiomycetes colonizing wood or litter on the ground or in humid aerial environments. Much greater inoculum potential can be brought to bear by the mycelium, enhancing the ability to replace residents or overcome stress barriers. Colonization is not limited to localized foci, nutrients and water may be imported from an already established food base and, by formation of compacted structures such as cords (see also later), buffering against hostile abiotic conditions can be achieved.

Because the hyphae grow from their tips, mycelial growth can be highly polarized – to a degree depending on branch angle and frequency, which are controlled by separate switch mechanisms (see below). As a result, considerable directionality can be achieved during growth between discontinuously distributed nutrient depots, so conserving energy by reducing wastage of biomass due to growth in an unproductive direction. This is illustrated by the behaviour of *Hypholoma fasciculare* during growth between woody resource units in soil (Fig. 3*a–h*). This figure demonstrates powerfully the remarkable ability of the mycelium to behave as a co-ordinated unit (Dowson, Rayner & Boddy, 1986): as an army it sends out scouting parties, establishes lines of communication, brings in reinforcements by redirecting the movements of its troops, conquers domain, establishes bases, and moves on.

The conservation of growth polarity illustrated in Fig. 3 for a mycelial system which is compacted into mycelial cords (see also below) also has considerable bearing on those diffuse mycelial systems of litter-inhabiting basidiomycetes which form fairy rings. The annular shape of fairy rings has always been difficult to explain, since although build-up of allelopaths and nutrient limitation have been suggested as reasons for the absence of mycelium from the central regions, leaching and enrichment by litter fall should alleviate these stresses. Cooke & Rayner (1984) suggested that the annulus was due to the establishment of a source–sink relation conserving polarity between the trailing edge and mycelial front, and this now seems a satisfactory explanation. In field experiments with *Clitocybe* rings, we excised segments of the mycelium and reorientated them within, outside or behind the annulus. Within the annulus the reoriented mycelial front ceased growth; outside or behind the annulus, growth continued with conserved polarity (Dowson, Rayner & Boddy, unpublished).

Once arrival has been effected, propagation of unicells and mycelial proliferation have complementary advantages and disadvantages. Hence unicells can be well equipped for rapid capture and conversion to biomass of easily assimilable substrates, for dispersion in mobile media, and for tolerance of adverse water potential, aeration and nutrient limitation (see also Nedwell & Gray, this volume). On the other hand mycelia are well equipped for conquest of fixed spatial domain, breakdown of refractory substrates via extracellular enzyme action, functional compartmentalization, and penetration or obviation of physicochemical barriers.

Just as in certain fungal pathogens of insects and vertebrates, the facility to switch between unicellular and mycelial morphogenesis may be expected to be critical to successful establishment of many wood- and litter-inhabiting basidiomycetes. However, the full extent to which this applies is not yet clear. It is established that production of a unicellular stage is important in the transmission of wood-decaying species of *Stereum* and *Amylostereum* by wood wasps, and the chlamydosporic mycelia of *Rhodotus palmatus* and *Schizopora paradoxa* seem likely to be important in their desiccation-tolerant strategies (Rayner & Boddy, 1986). But the greatest opportunities for deployment of a unicellular stage in establishment would seem to be within the sapstream of standing trees; this is after all the mechanism by which wilt fungi such as *Ceratocystis ulmi* and *Verticillium* spp. gain access to the vascular tissues. Such a mechanism would readily provide the basis for a specialized opportunist strategy but unfortunately it has yet to be demonstrated, for certain, in practice.

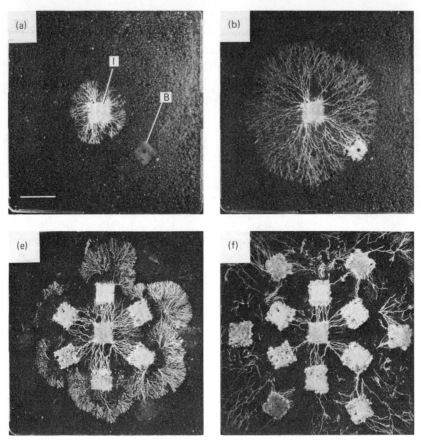

Fig. 3. Growth of non-restricted, cord-forming mycelium of *Hypholoma fasciculare* through non-sterile soil between woody resource units. (*a–d*) Growth from an inoculum wood block (I) towards an uninoculated 'bait' wood block (B). (Scale bar = 4 cm) (*e, f*) Growth from a central inoculum to an hexagonal array of baits, showing symmetrical outgrowth patterns. (*g, h*) As (*e, f*) but with three baits adjacent to the central inoculum removed after initial contact. Note the cessation of growth where baits have been removed, and asymmetric outgrowth from residual baits to colonize the outermost wood blocks. (Photographs from Springham, 1986).

Wood-inhabiting heterobasidiomycetes, particularly in the Tremel-lales, are well known to produce yeast phases as part of their life cycle (see below), but the colonization strategies of most of these fungi remain obscure, except for some which parasitize homobasidio-mycetes. In homobasidiomycetes, by contrast, where latent invasion mechanisms are strongly implicated, production of yeast phases has yet to be confirmed, although production of conidia, especially by haploid homokaryotic mycelia, is undisputed. However, evidence is beginning to accrue that homobasidiomycetes, including wood-

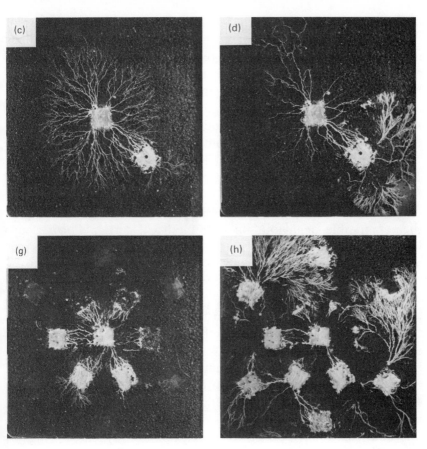

decaying forms possessing putative latent invasion strategies, may after all be able to produce yeast phases (Rayner & Coates, in press and unpublished; Prillinger, 1984, 1986 and in press), and that phenolics produced by the tree in response to wounding or stress could provide cues for reversion to mycelium (Dowson & Rayner, unpublished). If verified, these observations would have far-reaching implications, but for now they must be considered unsubstantiated.

Alterations in internode length and branch-angle: 'gear shifts'

The switch between unicellular and filamentous growth is in fact just the primary example of how growth resources can be reapportioned for greater or lesser polarity by morphogenetic controls in fungi. Subservient to this is a further system analogous to the gearbox of a forward-driven motor vehicle which enables selection of different degrees of conversion of engine torque to forward motion. Thus mycelia can adjust their degree of polarity in two main ways: by altering internode length (associated with hyphal diameter and branching frequency), and by varying branch-angle to provide different degrees of alignment of marginal hyphae (Rayner *et al.*, 1985; Rayner & Coates, in press). An equivalent system appears to control polarity of yeast pseudomycelia (unpublished observations).

The existence of gear shifts in basidiomycete mycelia is often evident from patterns of mycelial outgrowth from a germinating spore (Fig. 4*a*) and can become 'fixed' in striking 'slow dense/fast effuse' dimorphisms such as that illustrated in Fig. 4*b*. They also encompass the remarkable behaviour on agar media of certain *Phlebia* spp. in which the mycelial margin is composed of rapidly extending, appressed, coenocytic hyphae, and is followed from behind by a consolidatory phase of septate, aerial hyphae (Fig. 4*c*).

The facility for 'changing gear' has immense ecological implications. For example, a high (fast forward) gear will facilitate exploration and coverage of domain, together with rapid extraction of easily assimilable nutrients (cf. *r*-selection), whilst a low gear may aid in initial establishment of an inoculum base, exploitation of refractory resources, consolidation of territorial gains and stress tolerance (cf. *K*-selection). These principles are illustrated by the alternations between exploratory and consolidatory growth evident during the migration between food bases of *Hypholoma fasciculare* (Fig. 3) as well as the growth form just mentioned in *Phlebia*, where, in addition, the coenocytic hyphae have been found to lack the recognition responses necessary for effective combat (Boddy & Rayner, 1983*b*; *cf.* below).

Aerial versus appressed or submerged growth

Production of mycelium which is not in intimate contact with the substratum (referred to here as 'aerial mycelium') acts both as a drain on resources from trophic mycelium, and as a means of freeing growing hyphae from physicochemical constraints within the substratum. Aerial and non-aerial growth may, in different basidiomycetes, be closely coupled, partially uncoupled (resulting in endogenously or exogenously regulated rhythmicity), or largely uncoupled, resulting in distinctive growth phases, dimorphisms or polymorphisms.

Amongst Basidiomycotina, striking aerial–non-aerial dimor-phisms occur in the wood decaying members of the Hymenochaeta-ceae. In *Hymenochaete corrugata*, the two colony forms have similar extension rates, but the appressed form is yellow-brown whilst the aerial form is white (Fig. 4*d*, *e*). Both forms develop on the natural substratum, the appressed type being associated with more-decayed wood (Sharland, Burton & Rayner, 1986), a feature which is of particular interest because only this type possesses the tyrosinase and laccase activities which are associated with ligninolytic activity and secondary metabolism (P. R. Sharland, personal communica-tion; *cf.* Ander & Eriksson, 1978; Kirk & Fenn, 1982). A very similar dimorphism occurs in *Phellinus tremulae* except that here the appressed pigmented form has a slower extension rate (i.e. there may be co-expression of the slow dense switch) and can grow over a greater range of temperatures than the aerial form (Hiorth, 1965; Niemelä, 1977). Similar behaviour also occurs in *Rigidoporus ligno-sus*, the root pathogen of rubber, with the interesting implication that the superficial ectotrophic mycelium may be in a different func-tional mode from the laccase-producing mycelium within the wood cylinder (Boisson, 1968; Geiger, Nandris & Goujon, 1976). This places the ectotrophic infection habit in an entirely new perspective, and indeed may have general implications regarding the switch from mycelial arrival and establishment to exploitative growth in lignocel-lulosic basidiomycetes.

Compacted versus diffuse morphogenesis

At critical times during mycelial development in basidiomycetes, the initial divergent growth of hyphae is superseded by convergent growth, perhaps mediated by positive autotropisms (Ainsworth & Rayner, 1986) resulting in hyphal fusion (*cf.* Fig. 4*c*) and aggregation. By such means are generated the compacted, tissue-like structures referred to generally as plectenchymatous, e.g. fruit bodies, sclero-tia, stromata, pseudosclerotia, rhizomorphs and mycelial cords. The exact form of these plectenchymatous structures is probably dictated principally by how localized or generalized is the compaction process, and by the mycelial form on which it operates (Rayner & Coates, in press). Hence, generalized expression of this phenomenon within a plane of diffuse mycelium probably accounts, together with pro-liferation of branching, for the formation of the crust-like plates which, when formed within substrata, delimit what have been termed

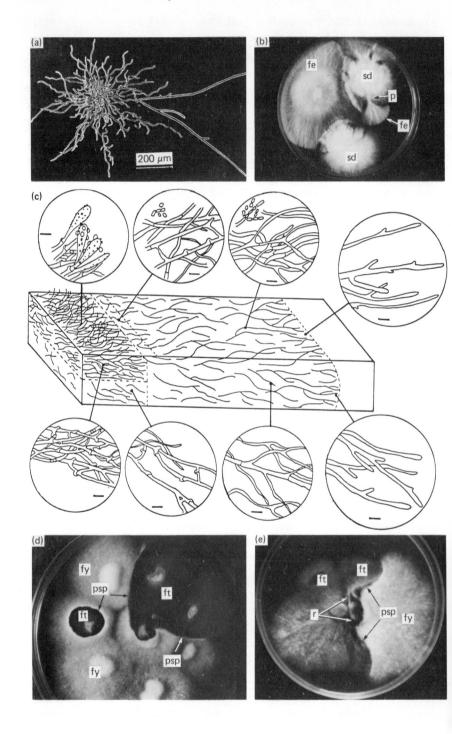

pseudosclerotia (Campbell, 1933). Interestingly, juxtaposition of the aerial and non-aerial forms in *Hymenochaete corrugata* results in formation of such a pseudosclerotial plate (Fig. 4*d*, *e*), whose primary ecological function seems to be defensive against potential combatants and protective against adverse abiotic environmental factors, especially desiccation.

Compaction of outwardly extending, collaterally aligned hyphal systems results, by contrast, in formation of linear organs such as mycelial cords and rhizomorphs. Depending on the degree of polarity of these structures, a spectrum of types can be formed, from the apically dominant true rhizomorphs such as those of *Armillaria* spp., to apically diffuse forms (Fig. 5). The extreme polarity of *Armillaria* rhizomorphs is implicit in the fact that they exhibit an extension rate an order of magnitude greater than that of diffusely growing hyphae (Rishbeth, 1968). The primary function of these linear organs is as connectives, allowing supply from a food base to either a sporophore or an actively growing mycelial front (*cf.* Fig. 3), whence they commonly contain an internal system of wide 'vessel' hyphae (see Rayner *et al.*, 1985). Although often produced in culture as a result of combative interactions, they are probably not normally directly involved in combat in nature, since systems of different individuals can often interdigitate significantly in soil. However, once at a food base the increased inoculum potential they confer may improve combative ability significantly. Where systems of different individuals or species come into contact, reactions equivalent to somatic incompatibility and hyphal interference often occur.

In *Steccherinum fimbriatum*, cord systems have been found which exhibit a slow dense/fast effuse switch, demonstrating further how mechanisms of recognition and developmental regulation characteristic of diffuse hyphal systems can be recapitulated at the level of hyphal aggregates.

Fig. 4. (*a*) Development of a colony of *Coniophora puteana* from a single basidiospore, showing progressive shifts in extension rate of marginal hyphae. (After Kemper, 1937.) (*b*) Outgrowth of mycelia from fruit body tissue of *Hypholoma fasciculare* onto 2% malt agar, showing slow dense (sd)/fast effuse (fe) dimorphism and origin of fast-effuse sectors by 'point growth' (p). Compare with outgrowth patterns in soil shown in Fig. 3. (*c*) Colony characteristics of a dikaryotic culture of *Phlebia radiata* growing through 2% malt agar, illustrating the marginal zone of coenocytic, non-anastomosing hyphae which is superseded by septate, anastomosed hyphal growth with clamp-connections. Scale bars = 10 μm. (After Boddy & Rayner, 1983*b*.) (*d*) Subcultures from a single colony of *Hymenochaete corrugata* which have grown out in 'flat' (ft) and 'fluffy' (fy) non-aerial and aerial forms, the two colony types interacting to produce a pseudosclerotial plate (psp). (From Sharland, Burton & Rayner, 1986.) (*e*) Rejection reaction (r) between two different heterokaryons of *H. corrugata* and associated changes in morphogenesis (*cf.* Fig. 4*d*). (From Sharland, Burton & Rayner, 1986.)

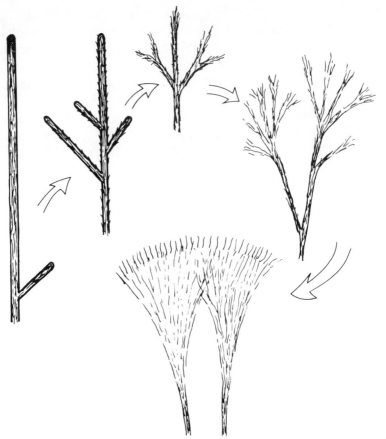

Fig. 5. Diagram illustrating the spectrum of compacted mycelial outgrowth patterns resulting in production of linear organs. Progression from strongly rhizomorphic outgrowth (far left) to diffuse outgrowth followed by consolidation (bottom centre) is associated with loss of apical control over extension of marginal hyphae resulting in increased branching and loss of apical coherence. (From Rayner *et al.*, 1985.)

Juvenility and senescence

In addition to possessing a 'gearbox', basidiomycete mycelia also appear to possess a braking system, enabling their potentially infinite capacity for growth to be brought to a halt. Evidence for this system can be seen in Fig. 3 where contact with a 'bait' results in the entire margin of initial exploratory growth being brought to a halt. Activation of the braking system may generally be a prequisite for redirection of growth resources prior to adoption of a new morphogenetic mode. Indeed it has been suggested that the periodic fruiting of fairy-ring fungi may be conditioned by an endogenous switch of this sort: since these non-unit-restricted fungi do not encounter a boundary to the resources they occupy, some such mechanism is required to cue the transition from vegetative to reproductive growth (Lysek, 1984).

So what is the basis for this braking system? There is increasing evidence that this may involve elicitation of programmed 'senescence' pathways associated with activation of phenoloxidase systems and melanization; in certain ascomycetes there is further evidence for involvement of cytoplasmic determinants under the control of nuclear genes (Daboussi-Bareyre, 1980; Esser *et al.*, 1984).

The association with phenoloxidase activity and melanization recalls the behaviour of certain aerial–non-aerial dimorphic forms discussed earlier, notably *Hymenochaete corrugata*, where the formation of a melanized pseudosclerotial plate (Fig. 4*d*, *e*) now assumes a special significance. A common mechanism relating damage, phenoloxidase systems, senescence, melanization, induction of plectenchyma formation, sporogenesis and rejection responses (see below) now seems to be emerging (Rayner & Coates, in press; see also Ross, 1985). This would revolutionize our understanding of the behaviour of basidiomycete thalli in nature.

RELATION OF RECOGNITION SYSTEMS TO COLONIZATION STRATEGIES AND MYCELIAL DOMAINS

As indicated earlier, powerful recognition and response systems occur in basidiomycetes which condition acquiescence to, or rejection or acceptance of, self- and non-self within species. The existence of these mechanisms may further relate to the outcome and mechanisms of combative interspecific interactions (Rayner, 1986*a*). The rejection mechanisms are now much more widely known than a decade ago, and are commonly referred to as somatic or vegetative incompatibility. They are proving a valuable tool in analysing population structure (Brasier, 1984; Rayner & Boddy, 1986) because of the often easily observed demarcation zones between different genotypes both in culture (see Fig. 4*e*) and in nature (Fig. 6*a*). Evidence is also accumulating that both rejection and acceptance responses involve modulation of the developmental switches described in the previous section, the rejection responses for example involving activation of the senescence pathways (Rayner & Coates, in press; see

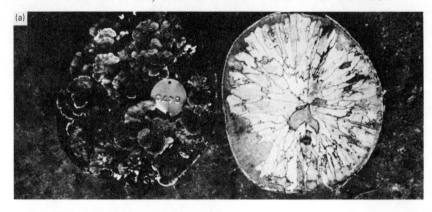

Fig. 6. Typical colonization patterns of upper (*a*) and lower (*b*) cut surfaces of beech (*Fagus sylvatica*) logs placed upright with their bases buried in the litter of a deciduous woodland. Numerous genets of *Coriolus versicolor*, delimited by narrow dark interaction zone lines in the decayed wood, are present near the upper cut surface, whilst a single genet of the cord-forming *Tricholomopsis platyphylla* has virtually sole occupancy of the wood adjacent to the lower surface. (From Coates & Rayner, 1985*c*.)

Fig. 4*e*). However, here only the ecological significance of recognition responses in relation to colonization strategies will be considered in any depth.

Outcrossing versus non-outcrossing strategies

Outcrossing, that is sexual conjugation followed by diploidization and meiosis between genetically different haploid homokaryotic lines, results in the production of variable basidiospore progeny from a single basidiocarp. Hence the degree of genetic variability within decomposer basidiomycete populations will be determined by the degree to which outcrossing occurs. This in turn can be regulated in two ways: (1) by the extent to which sexual or asexual mechanisms are primarily responsible for propagation of the population; and (2) by interconversion between essentially apomictic and heteromictic, i.e. non-outcrossing and outcrossing, life cycles.

Sexual versus asexual propagation
Where a mating (= homogenic incompatibility = outcrossing = heterothallic = heteromictic) system is functional within a population, then the balance between basidiosporogenesis and other (asexual) modes of propagation will be decisive in ultimately determining the dynamics and structure of that population. Hence, asexual propagation will result in proliferation of particular discrete genotypes as either spatially discontinuous clones or spatially contiguous individual mycelia (henceforth the term 'genet' will be used to describe such population sub-units, following the recommendation of Brasier & Rayner, in press). Conversely, predominance of sexual reproduction will result in a spatiotemporally heterogeneous population, comprising many different genets.

Asexual propagation can involve either mitotic production of discrete propagules – spores, microsclerotia, sclerotia, etc. – or mycelial spread. Relatively little is known about the role of asexual propagules in the population dynamics of decomposer basidiomycetes. Although many species do produce conidia, these are by no means as prolific as in many ascomycetes, and they are commonly haploid – consequently any homokaryotic mycelia established from them are likely to become converted to heterokaryons by mating (see below). Chlamydospores and sclerotial bodies, by contrast, may more commonly be heterokaryotic but they would principally help to ensure survival of a genet in a particular location, rather than facilitating its spread to others. However, the distribution of individual genets of *Athelia rolfsii* on Californian golf greens is indicative of dissemination of sclerotia during cultivation (Punja & Grogan, 1983).

By contrast with asexual popagules, there is little doubt that mycelial spread is, and has been, fundamental in the proliferation of individual genets of non-unit-restricted wood and litter-decomposing basidiomycetes over considerable areas of ground. Hence genets occupying more than a hectare of ground have been detected in rhizomorphic or cord-forming wood decomposers, including species of *Armillaria* (Korhonen, 1978; Anderson *et al.*, 1979; Thompson & Boddy, 1983), *Tricholomopsis platyphylla* (Thompson & Rayner,

1982), and *Phanerochaete velutina* (Thompson & Boddy, 1983). Somewhat smaller genets have been detected in the root pathogens *Heterobasidion annosum* (Chase & Ullrich, 1983; Stenlid, 1985) and *Phellinus weirii* (Childs, 1963) which depend more on root–root contacts for establishment of their ectotrophic infections than on migratory mycelium. On the basis of current evidence, fairy rings appear generally to represent extensive, stable, individual genets, at least in *Marasmius oreades* (Burnett & Evans, 1966; R. C. Aylmore, personal communication), *Clitocybe nebularis* and *C. flaccida* (Dowson, Rayner & Boddy, unpublished). In *C. nebularis*, isolates of the same somatic compatibility type were obtained from different fruit bodies in the same ring some 50 m apart along the circumference.

With respect to establishment from mycelium or basidiospores in individual woody resource units, Fig. 6(*a*, *b*), showing colonization of cut beech (*Fagus sylvatica*) logs, is instructive: mycelial colonization is associated with formation of a spatially extensive individual genet, whereas basidiospore colonization is associated with numerous discrete genets. Moreover, Coates & Rayner (1985*a*, *b*) have demonstrated that artificially high basidiospore inoculum loads can restrict the size of the domains of individual genets to such a degree as to restrict markedly the size – and even the production – of fruit bodies, as well as decay rates.

These colonization patterns in cut logs, however, illustrate only the sort of population structure consequent upon enrichment disturbance. There is growing evidence that in the aerial portions of standing trees the structure of decay populations, originating presumably following arrival as propagules and establishment by strategies other than unspecialized opportunism, is often very different. This is to the effect that individual genets, which are specific to a particular tree, commonly have very extensive domains – often having virtually sole occupancy of a trunk or branch. Correspondingly, they tend to produce the sometimes colossal fructifications which could only be supported by a mycelium commanding a very considerable resource pool (Rayner *et al.*, 1984; Rayner & Boddy, 1986).

The reasons behind the formation of such extensive genets probably varies with respect to the different colonization strategies, but all may be due in some way to the selectively stressful conditions which condition successful establishment. Thus, in heartrots the establishment of extensive individual domains probably takes place (under stress conditions militating against potential competitors) over many years or decades by a process of slow mycelial spread from a colonization focus. Selection of positionally or otherwise advantaged genets as colonization proceeds deeper into the wood may further reduce the incidence of intraspecific competition.

In the case of some specialized opportunism strategies, formation of extensive genets (often several metres long) appears, in contrast to heartrots, to occur very rapidly, perhaps within a single growing season. As previously suggested, pre-establishment within the vascular system by means of an easily dispersed inoculum, such as yeast cells, would provide a suitable explanation. An interesting possibility, detected with *Piptoporus betulinus* in *Betula*, is that larger numbers of genets may become established in trees whose root systems are infected by *Armillaria* than in trees predisposed to colonization by other factors (Adams, 1982; Rayner & Boddy, in press). A similar situation has been detected with the ascomycete *Daldinia concentrica* in *Fraxinus*.

'Apomictic' versus 'heteromictic' life cycles

Besides production of asexual propagules such as conidia and vegetative reproduction by means of mycelial spread, another means of achieving dissemination of an individual genet is by eliminating the requirement for conjugation between mating-type-compatible homokaryons from the sexual cycle. This can be effected if field homokaryons acquire the capacity to complete the life cycle, including formation of basidiospores and even meiosis, without themselves becoming heterokaryotized. Such 'non-outcrossing' strategies have been detected in a number of populations of wood-decomposers. Their hallmark is the production of identical homokaryotic basidiospore progeny from naturally collected fruit bodies, which give rise to somatically compatible mycelia. Often non-outcrossing populations become subdivided into several to numerous groups, between which a rejection (somatic incompatibility) response occurs directly, without any intervening heterokaryotic stage. A particular study of non-outcrossing populations has been made in the genus *Stereum* (Ainsworth, in press). In summary, whilst some taxonomic species of *Stereum* have so far been found to be composed of entirely outcrossing populations, e.g. *S. gausapatum* and *S. rugosum*, others contain reproductively isolated outcrossing and non-outcrossing subpopulations, e.g. *S. hirsutum*, *S. sanguinolentum*, and in yet other cases closely related species pairs occur, one outcrossing, the other non-outcrossing, e.g. *S. rameale* and *S. ochraceoflavum*. Preliminary indications, in need of much further substantiation, are that non-outcrossing is characteristic of populations which are on the edge of their range, (e.g. Scandinavian populations of *S. hirsutum*), or which possess colonization strategies with a strong ruderal element, e.g. *S. sanguinolentum* in Europe.

These indications tie in with the general expectation that *K*-selected populations exposed to heterogeneous and changing biotic or abiotic constraints, e.g. those colonizing genetically variable unwounded standing trees, will tend to preserve variability. By contrast, those colonizing a widely available, effectively homogeneous resource, and which commit rapidly to reproduction, will be based on *r*-selection favouring the proliferation of particular well-fitted genets.

Homokaryon–heterokaryon transitions in outcrossing species

An ecologically neglected facet of the life cycles of outcrossing decomposer basidiomycetes is the occurrence of two distinctive phases, the haploid homokaryotic primary thallus, resulting from germination of basidiospores, and the heterokaryotic (common) or diploid (rare) secondary thallus originating from sexual conjugation between primary thalli. This is obviously vital to considerations of establishment, since where basidiospores represent the principal method of arrival, it will be the primary thalli which may predominantly be responsible for establishment.

This situation assumes even greater significance when it is realized that primary and secondary thalli commonly exhibit fundamentally different patterns of morphogenesis – often corresponding directly to the alternative developmental modes itemized in the previous section.

Hence, homokaryons often develop either as budding, yeast-like forms, which is a characteristic feature of some heterobasidiomycetes, or as 'slow dense' mycelia in which the marginal hyphae have short internodes and wide-angled branching. By contrast, heterokaryons often develop as 'fast effuse' mycelia, with marginal hyphae having long internodes and/or acute-angled branches. Heterokaryons also commonly have an enhanced capacity, compared with homokaryons, to produce compact structures such as mycelial cords, rhizomorphs, pseudosclerotia, pseudorhiza and reproductive fruit bodies (Rayner *et al.*, 1985).

These differences in morphogenetic properties between homokaryons and heterokaryons have sometimes been attributed directly to the genes controlling mating, i.e. the mating factors. However, exceptions to the general trends just mentioned do occur and, more importantly, these trends also seem very likely to be related to differences in the ecological function of primary and secondary thalli. Thus, depending on the rapidity with which they are likely to be converted by mating into heterokaryons, homokaryons will be responsible to a greater or lesser extent for primary establishment of a colonization base in previously unoccupied habitats. By contrast, heterokaryons will be responsible for secondary extension of the colonization base, combat with other individuals and elaboration of structures by which to exit from resource units. The homokaryon–heterokaryon transition may thus best be regarded as a very important cue, rather than being directly responsible for the morphogenetic changes which accompany it.

In order to establish the ecological importance of homokaryon–heterokaryon transitions, it is important to have available some data concerning the longevity of homokaryotic thalli and the kinetics of their conversion to heterokaryons under natural conditions. Unfortunately few such data exist, but there are some indications that considerable differences may exist between different fungal populations with regard to rates of access and migration through established homokaryotic thalli. In the case of *Coriolus versicolor* and *Flammulina velutipes*, use of homokaryotic thalli as a means of viable trapping of basidiospores from the atmosphere served to demonstrate how readily such thalli would be heterokaryotized following arrival on their surface of spores of complementary mating type (Adams *et al.*, 1984; Williams, Todd & Rayner, 1984). These experiments also served to demonstrate, in a woodland site, how heterogeneous was the basidiospore rain of *Coriolus versicolor*: no mating type repeats were attained during a year's sampling at monthly intervals. In a different experiment, inoculation of homokaryons of *C. versicolor* into logs led to establishment of on average two to three heterokaryons per homokaryon inoculated, it being evident that heterokaryosis was complete within 3 months (Williams, Todd & Rayner, 1981). A similar experiment in which cut beech logs were exposed to a natural air-borne inoculum revealed that whilst homokaryons of *C. versicolor* and *Bjerkandera adusta* could be isolated from surface wood over a full 2 year period, indicating continual recruitment of spores from the atmosphere, at depth homokaryons were never found more than 6 months after initial exposure (Coates & Rayner, 1985a).

In a recent series of experiments with the cord-forming species *Hypholoma fasciculare* and *Phanerochaete velutina*, wood blocks permeated with homokaryotic mycelia were placed directly into nonsterile soil, and mycelial cord systems allowed to grow out therefrom. The mycelia were sampled after several months. In *H. fasciculare*, which is readily heterokaryotized, often associated with a marked slow dense/fast effuse switch in laboratory culture, all the mycelia were heterokaryotic and often more than one heterokaryon was formed per inoculum. In *P. velutina*, which is much less readily heterokaryotized in laboratory culture, only around 25% of the mycelia sampled were heterokaryotic, the remainder remaining homokaryotic even though forming quite well-defined cord systems (Dowson, Rayner & Boddy, unpublished).

Clearly, there is much further experimental work needed in this field.

Parasitism as a strategy for establishment and domain capture

As already indicated, colonization of cut or broken wood by basidiospores of such fungi as *Coriolus versicolor* and *Bjerkandera adusta* often result in establishment of populations of genets with individually small domains within the wood (*cf.* Fig. 6). Correspondingly, the fruit bodies of such fungi which habitually occupy small domains tend to be of small size. However, two decomposer basidiomycetes, *Pseudotrametes gibbosa* and *Lenzites betulina*, seem to colonize fallen or cut wood rather than standing trees – yet they possess large domains and correspondingly sizeable fruit bodies.

This at first sight enigmatic situation is resolved by realizing that both *P. gibbosa* and *L. betulina* employ a remarkable establishment strategy not unlike that of strangler figs or temporarily parasitic ants (Rayner, Boddy & Dowson, in press). Thus, associated with lack of recognition and activation of rejection responses by their hosts, individual genets of these fungi are able specifically to parasitize and then take over the domain of whole populations of genets of *Coriolus* species (in the case of *L. betulina*) or *Bjerkandera* species (in the case of *P. gibbosa*) which had pioneered the colonization process.

CONCLUSIONS

We hope to have demonstrated during the course of this overview, some of the new ideas which are emerging as a consequence of a deeper understanding of the versatility and responsiveness of the vegetative thalli of decomposer basidiomycetes. There is much to be substantiated and almost unlimited scope for further exploration. Above all, here is a field where the fusion of ecological, plant pathological, developmental, genetic and molecular approaches is not only a desirable but also a realistic prospect. Thereby may be provided an example for biologists in general.

REFERENCES

ADAMS, T. J. H. (1982). 'Piptoporus betulinus, some aspects of population biology. Ph.D. thesis, University of Exeter.
ADAMS, T. J. H., WILLIAMS, E. N. D., TODD, N. K. & RAYNER, A. D. M. (1984). A species-specific method of analysing populations of basidiospores. Transactions of the British Mycological Society, 82, 359–61.
AINSWORTH, A. M. (in press). Mycelial and hyphal interactions in holocoenocytic Basidiomycotina. In Evolutionary Biology of the Fungi, ed. A. D. M. Rayner, C. M. Brasier & D. Moore. Cambridge, Cambridge University Press.
AINSWORTH, A. M. & RAYNER, A. D. M. (1986). Responses of living hyphae associated with self and non-self fusions in the basidiomycete Phanerochaete velutina. Journal of General Microbiology, 132, 191–201.
ANDER, P. & ERIKSSON, K.-E. (1978). Lignin degradation and utilization by microorganisms. Progress in Industrial Microbiology, 14, 1–58.
ANDERSON, J. B., ULLRICH, R. C. ROTH, L. F. & FILIP, G. M. (1979). Genetic identification of clones of Armillaria mellea in coniferous forests in Washington. Phytopathology, 69, 1109–11.
AYLMORE, R. C. & TODD, N. K. (1984). Hyphal fusion in Coriolus versicolor. In The Ecology and Physiology of the Fungal Mycelium, ed. D. H. Jennings & A. D. M. Rayner, pp. 103–25. Cambridge, Cambridge University Press.
BODDY, L. (1984). The micro-environment of basidiomycete mycelia in temperate deciduous woodlands. In The Ecology and Physiology of the Fungal Mycelium, ed. D. H. Jennings & A. D. M. Rayner, pp. 261–89. Cambridge, Cambridge University Press.
BODDY, L. & RAYNER, A. D. M. (1983a). Origins of decay in living deciduous trees: the role of moisture content and a re-appraisal of the expanded concept of tree decay. New Phytologist, 94, 623–41.
BODDY, L. AND RAYNER, A. D. M. (1983b). Mycelial interactions, morphogenesis and ecology of Phlebia radiata and Phlebia rufa from oak. Transactions of the British Mycological Society, 80, 437–48.
BOISSON, C. (1968). Mise en évidence de deux phases mycéliennes successives au cours du développment du Leptoporus lignosus (Kl.) Heim. Comptes Rendus des Séances de l' Academie de Science, Paris, Série D, 266, 1112–15.
BRASIER, C. M. (1984). Inter-mycelial recognition systems in Ceratocystis ulmi: their physiological properties and ecological importance. In The Ecology and Physiology of the Fungal Mycelium, ed. D. H. Jennings & A. D. M. Rayner, pp. 451–97. Cambridge, Cambridge University Press.
BRASIER, C. M. & RAYNER, A. D. M. (in press). Whither terminology below the species level in the fungi? In Evolutionary Biology of the Fungi, ed. A. D. M. Rayner, C. M. Brasier & D. Moore. Cambridge, Cambridge University Press.

BURNETT, J. H. (1965). The natural history of recombination systems. In *Incompatibility in Fungi*, ed. K. Esser & J. R. Raper, pp. 98–113. Berlin, Springer-Verlag.

BURNETT, J. H. & EVANS, E. J. (1966). Genetical homogeneity and the stability of the mating-type factors of 'fairy rings' of *Marasmius oreades*. *Nature, London*, **210**, 1368–9.

CAMPBELL, A. H. (1933). Zone lines in plant tissues. I. The black lines formed by *Xylaria polymorpha* (Pers.) Grev. in hardwoods. *Annals of Applied Biology*, **20**, 123–45.

CARLILE, M. J. (in press). Gene flow and genetic exchange – their promotion and prevention. In *Evolutionary Biology of the Fungi*, ed. A. D. M. Rayner, C. M. Brasier & D. Moore. Cambridge, Cambridge University Press.

CHASE, T. E. & ULLRICH, R. C. (1983). Sexuality, distribution and dispersal of *Heterobasidion annosum* in pine plantations of Vermont. *Mycologia*, **75**, 825–31

CHILDS, T. W. (1963). *Poria weirii* root rot. In *Symposium on Root Diseases of Forest Trees, Corvallis, Oregon, 1962. Phytopathology*, **53**, 1124–7.

COATES, D. & RAYNER, A. D. M. (1985a). Fungal population and community development in cut beech logs. I. Establishment via the aerial and cut surface. *New Phytologist*, **101**, 153–71.

COATES, D. & RAYNER, A. D. M. (1985b). Fungal population and community development in cut beech logs. II. Establishment via the buried cut surface. *New Phytologist*, **101**, 173–81.

COATES, D. & RAYNER, A. D. M. (1985c). Fungal population and community development in cut beech logs. III. Spatial dynamics, interactions and strategies. *New Phytologist*, **101**, 183–98.

COOKE, R. C. & RAYNER, A. D. M. (1984). *Ecology of Saprotrophic Fungi*. London & New York, Longman.

COUTTS, M. P. (1976). The formation of dry zones in the sapwood of conifers. I. Induction of drying in standing trees and logs by *Fomes annosus* and extracts of infected wood. *European Journal of Forest Pathology*, **6**, 372–81.

DABOUSSI-BAREYRE, M. J. (1980). Heterokaryosis in *Nectria haematococca*. *Journal of General Microbiology*, **116**, 425–33.

DOWSON, C. G., RAYNER, A. D. M. & BODDY, L. (1986). Outgrowth patterns of mycelial cord-forming basidiomycetes from and between woody resource units in soil. *Journal of General Microbiology*, **132**, 203–11.

ESSER, K., KUCK, U., STAHL, U. & TUDZYNSKI, P. (1984). Senescence in *Podospora anserina* and its implications for genetic engineering. In *Ecology and Physiology of the Fungal Mycelium*, ed. D. H. Jennings & A. D. M. Rayner, pp. 343–52. Cambridge, Cambridge University Press.

GARRETT, S. D. (1970). *Pathogenic Root-Infecting Fungi*. Cambridge, Cambridge University Press.

GEIGER, J. P., NANDRIS, D. & GOUJON, M. (1976). Activité des laccases et des peroxydases au sein de vacines d'Hévéa attaquées par le pourridié blanc (*Leptoporus lignosus* (K1.) Heim). *Physiologie Végétale*, **14**, 271–82.

GREGORY, P. H. (1984). The fungal mycelium: an historical perspective. *Transactions of the British Mycological Society*, **82**, 1–11.

GRIME, J. P. (1979). *Plant Strategies and Vegetation Processes*. Chichester & New York, John Wiley.

HARPER, J. L. & OGDEN, J. (1970). The reproductive strategy of higher plants. I. The concept of strategy with special reference to *Senecio vulgars* L. *Journal of Ecology*, **58**, 681–9.

HEDGER, J. N. (1958). Tropical agarics: resource relations and fruiting periodicity. In *Developmental Biology of Higher Fungi*, ed. D. Moore, L. A. Casselton,

D. A. Wood & J. C. Frankland, pp. 41–86. Cambridge, Cambridge University Press.

HIORTH, J. (1965). The phenoloxidase and peroxidase activities of two culture types of *Phellinus tremulae* (Bond.) Bond. & Boriss. *Meddelelser Norske Skogforsöksvesen*, **20**, 249–72.

KEMPER, W. (1937). Zur Morphologie und Cytologie der Gattung *Coniophora*, inbesondere des sogenannten Kellerschwammes. *Zentralblatt für Bakteriologie Parasitenkunde Infektionskrankheiten und Hygiene, Abteilung* II, **97**, 100–24.

KIRK, T. K. & FENN, P. (1982). Formation and action of the ligninolytic system in basidiomycetes. In *Decomposer Basidiomycetes: Their Biology and Ecology*, ed. J. C. Frankland, J. N. Hedger & M. J. Swift, pp. 67–90. Cambridge, Cambridge University Press.

KORHONEN, K. (1978). Interfertility and clonal size in the *Armillaria mellea* complex. *Karstenia*, **18**, 31–42.

LEBEN, C. (1985). Wound occlusion and discoloration columns in red maple. *New Phytologist*, **99**, 485–90.

LYSEK, G. (1984). Physiology and ecology of rhythmic growth and sporulation in fungi. In *The Ecology and Physiology of the Fungal Mycelium*, ed. D. H. Jennings & A. D. M. Rayner, pp. 323–42. Cambridge, Cambridge University Press.

MERCER, P. C. (1982). Basidiomycete decay of standing trees. In *Decomposer Basidiomycetes: Their Biology and Ecology*, ed. J. C. Frankland, J. N. Hedger & M. J. Swift, pp. 143–60. Cambridge, Cambridge University Press.

NIEMELÄ, T. (1977). The effects of temperature on the two culture types of *Phellinus tremulae* (Fungi, Hymenochaetaceae). *Annales Botanici Fennici*, **14**, 21–4.

PRILLINGER, H. (1984). Zur Evolution von Mitose, Meiose und Kernphasenwechsel bei Chitinpilzen. *Zeitschrift für Mykologie*, **50**, 267–352.

PRILLINGER, H. (1986). Morphologische Atavismen bei Homobasidiomyceten durch naturliche und kunstliche Inzucht und ihre Bedeutung für die Systematik. *Berichte der Deutschen Botanischen Gesellschaft*, **99**, 31–42.

PRILLINGER, H. (in press). Yeasts and anastomosis: their occurrence and implications for the phylogeny of Eumycota. In *Evolutionary Biology of the Fungi*, ed. A. D. M. Rayner, C. M. Brasier & D. Moore. Cambridge, Cambridge University Press.

PUGH, G. J. F. (1980). Strategies in fungal ecology. *Transactions of the British Mycological Society*, **75**, 1–14.

PUNJA, Z. K. & GROGAN, R. G. (1983). Hyphal interactions and antagonism among field isolates and single-basidiospore strains of *Athelia (Sclerotium) rolfsii*. *Phytopathology*, **73**, 1279–84.

RAYNER, A. D. M. (1986a) Mycelial interactions – genetic aspects. In *Natural Antimicrobial Systems*, ed. M. E. Rhodes-Roberts, G. W. Gould, A. K. Charnley, R. M. Cooper & R. G. Board, pp. 277–96. Bath, Bath University Press.

RAYNER, A. D. M. (1986b). Water and the origins of decay in trees. In *Water, Fungi and Plants*, ed. P. G. Ayres & L. Boddy, pp. 321–41. Cambridge, Cambridge University Press.

RAYNER, A. D. M. & BODDY, L. (1986). Population structure and the infection biology of wood-decay fungi in living trees. *Advances in Plant Pathology*, **5**, 119–60.

RAYNER, A. D. M. & BODDY, L. (in press). *Fungal Decay of Wood: Biology and Ecology*. Chichester & New York, John Wiley.

RAYNER, A. D. M., BODDY, L. & DOWSON, C. G. (in press). Temporary parasitism of *Coriolus* spp. by *Lenzites betulina*: a strategy for domain capture in wood decay fungi. *Microbiology Ecology*.

RAYNER, A. D. M. & COATES, D. (in press). Regulation of mycelial organization and responses. In *Evolutionary Biology of the Fungi*, ed. A. D. M. Rayner, C. M. Brasier & D. Moore. Cambridge, Cambridge University Press.

RAYNER, A. D. M., COATES, D., AINSWORTH, A. M., ADAMS, T. J. H., WILLIAMS, E. N. D. & TODD, N. K. (1984). The biological consequences of the individualistic mycelium. In *The Ecology and Physiology of the Fungal Mycelium*, ed. D. H. Jennings & A. D. M. Rayner, pp. 509–40. Cambridge, Cambridge University Press.

RAYNER, A. D. M., POWELL, K. A., THOMPSON, W. & JENNINGS, D. H. (1985). Morphogenesis of vegetative organs. In *Developmental Biology of Higher Fungi*, ed. D. Moore, L. A. Casselton, D. A. Wood & J. C. Frankland, pp. 249–79. Cambridge, Cambridge University Press.

RAYNER, A. D. M. & TODD, N. K. (1979). Population and community structure and dynamics of fungi in decaying wood. *Advances in Botanical Research*, **7**, 333–420.

RAYNER, A. D. M., WATLING, R. & FRANKLAND, J. C. (1985). Resource relations – an overview. In *Developmental Biology of Higher Fungi*, ed. D. Moore, L. A. Casselton, D. A. Wood & J. C. Frankland, pp. 1–40. Cambridge, Cambridge University Press.

RAYNER, A. D. M. & WEBBER, J. F. (1984). Interspecific mycelial interactions – an overview. In *The Ecology and Physiology of the Fungal Mycelium*, ed. D. H. Jennings & A. D. M. Rayner, pp. 383–417. Cambridge, Cambridge University Press.

RISHBETH, J. (1968). The growth rate of *Armillaria mellea*. *Transactions of the British Mycological Society*, **51**, 575–86.

ROSS, I. K. (1985). Determination of the initial steps in differentiation in *Coprinus congregatus*. In *Developmental Biology of Higher Fungi*, ed. D. Moore, L. A. Casselton, D. A. Wood & J. C. Frankland, pp. 353–73. Cambridge, Cambridge University Press.

SHAIN, L. (1979). Dynamic responses of differentiated sapwood to injury and infection. *Phytopathology*, **69**, 1143–7.

SHARLAND, P. R., BURTON, J. L. & RAYNER, A. D. M. (1986). Mycelial dimorphism, interactions and pseudosclerotial plate formation in *Hymenochaete corrugata*. *Transactions of the British Mycological Society*, **86**, 158–63.

SHIGO, A. L. (1979). Tree decay – an expanded concept. *U.S. Department of Agriculture, Forest Service Information Bulletin*, No. 419.

SHIGO, A. L. (1984). Compartmentalization: a conceptual framework for understanding how trees grow and defend themselves. *Annual Review of Phytopathology*, **22**, 189–214.

SPRINGHAM, P. (1986). 'Resource capture by cord-forming basidiomycetes.' Project report, University of Bath.

STENLID, J. (1985). Population structure of *Heterobasidion annosum* as determined by somatic incompatibility, sexual incompatibility, and isoenzyme patterns. *Canadian Journal of Botany*, **63**, 2268–73.

STEWART, P. R. & ROGERS, P. J. (1983). Fungal dimorphism. In *Fungal Differentiation: A Contemporary Synthesis*, ed. J. E. Smith, pp. 267–313. New York & Basel, Marcel Dekker Inc.

SWIFT, M. J., HEAL, O. W. & ANDERSON, J. M. (1979). *Decomposition in Terrestrial Ecosystems*. Oxford, Blackwell Scientific Publications.

THOMPSON, W. (1984). Distribution, development and functioning of mycelial cord systems of decomposer basidiomycetes of the deciduous woodland floor. In *The Ecology and Physiology of the Fungal Mycelium*, e.d. D. H. Jennings & A. D. M. Rayner, pp. 185–214. Cambridge, Cambridge University Press.

THOMPSON, W. & BODDY, L. (1983). Decomposition of suppressed oak trees in even-aged plantations. II. Colonization of tree roots by cord and rhizomorph producing basidiomycetes. *New Phytologist*, **93**, 277–91.

THOMPSON, W. & RAYNER, A. D. M. (1982). Spatial structure of a population of *Tricholomopsis phatyphylla* in a woodland site. *New Phytologist*, **92**, 103–14.

TODD, N. K. & AYLMORE, R. C. (1985). Cytology of hyphal interactions and reactions in *Schizophyllum commune*. In *Developmental Biology of Higher Fungi*, ed. D. Moore, L. A. Casselton, D. A. Wood & J. C. Frankland, pp. 231–48. Cambridge, Cambridge University Press.

WHEELER, B. (1985). The growth of *Crinipellis perniciosa* in living and dead cocoa tissue. In *Developmental Biology of Higher Fungi*, ed. D. Moore, L. A. Casselton, D. A. Wood & J. C. Frankland, pp. 103–15. Cambridge, Cambridge University Press.

WILLIAMS, E. N. D., TODD, N. K. & RAYNER, A. D. M. (1971). *Spatial development of populations of Coriolus versicolor*. *New Phytologist*, **89**, 307–19.

WILLIAMS, E. N. D., TODD, N. K. & RAYNER, A. D. M. (1984). Characterization of the spore rain of *Coriolus versicolor* and its ecological significance. *Transactions of the British Mycological Society*, **82**, 323–6.

WILLIAMS, G. C. (1971). *Group Selection*. Chicago, Aldine, Atherton.

WOOD, R. K. S. (1985). The ecology and physiology of the fungal mycelium. *Journal of Experimental Botany*, **36**, 858.

INTERACTIONS BETWEEN INVERTEBRATES AND MICROORGANISMS: NOISE OR NECESSITY FOR SOIL PROCESSES?

J. M. ANDERSON

Wolfson Ecology Laboratory, Department of Biological Sciences, University of Exeter, Exeter EX4 4PS, UK

INTRODUCTION

The operation of soil microbial processes at a gross level is well understood and this knowledge continues to be refined and applied in forestry and agricultural management practices. There has, however, been little explicit recognition of the complexity of the processes regulating nutrient fluxes between litter, soil and plant roots (Frissel & van Veen, 1982). Mechanisms of decomposition and nutrient transformation in soils are usually interpreted in terms of a simple sequence of steps mediated by microorganisms and the roles of invertebrates in these processes are rarely invoked (Anderson & Ineson, 1984). In fact, this is a manifestation of a general dichotomy in soil biology (and perhaps in soil biochemistry) between descriptive and process-orientated research. Swift (1984) contends that studies on the role of microorganisms in decomposition processes fall into two main schools. In one approach the decomposer organisms are broadly defined as 'driving variables' promoting the processes of carbon and nutrient mineralization with little explicit recognition of the structure of the microbial community. This contrasts with the second school which has been concerned with the diversity of decomposer species, and their distribution, abundance and activities. Similar trends are evident in soil zoology (Satchell, 1974; Usher, Booth & Sparks, 1982).

The net result of these two approaches is that we have considerable difficulty integrating information relating to specific organisms into the understanding of soil processes operating on less defined spatial and temporal scales. The fundamental question is whether there are significant functional attributes of these diverse systems, or do the complex interactions of the component species in the community represent 'background noise' which is filtered out through the net contribution of microsite processes to ecosystem-level fluxes of carbon or nutrients?

There is very extensive literature on these interactions which has been reviewed comprehensively elsewhere (Anderson, Coleman & Cole, 1981; Satchell, 1983; Anderson, Rayner & Walton, 1984; Seastedt, 1984; Fitter *et al.*, 1985). Emphasis in this paper is placed on their functional importance for decomposition and nutrient cycling through direct and indirect effects on soil processes. The relative contributions of these effects in a system is determined by spatial and temporal scales on which the organisms interact.

SPATIO-TEMPORAL SCALES OF INTERACTIONS

In very general terms animal–microbial interactions may be considered to operate on micro-, meso- and macro-scales in space and time which are a function of the relative body sizes (Fig. 1) of the

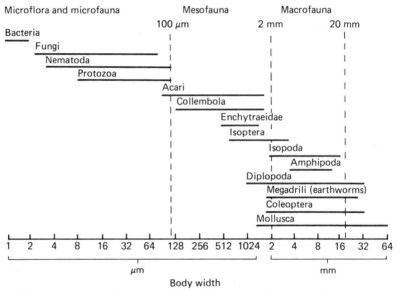

Fig. 1. Functional classification of soil organisms by body width. (After Swift, Heal & Anderson, 1979.)

organisms. The feeding activities of nematodes and protozoans (microfauna) operate within the water-filled pores and surface films of the soil matrix at similar scales to bacteria and fungal hyphae. These microfauna also have comparable generation times, of hours or days, to bacteria and fungi, and are hence able to respond rapidly to temporal changes in microbial populations subject to the constraints placed on their feeding activities by soil pore-size distributions. Collembolans and mites (mesofauna) generally live on litter surfaces and in air-filled macropores in soil. Many species are mycophagous but the proportion of higher plant material ingested increases as a positive function of body size (Anderson, 1975), suggesting that the smaller species are able to feed more selectively on fungi. The generation times of soil mites and collembolans are of the order of weeks or months so that their populations are likely to respond to changes in the quality and quantity of fungal biomass over comparable periods of time. Enchytraeids, fly larvae and some mites have more burrowing habits and in acid woodland soils their faeces may constitute a large and definitive component of the humus fabric (Kubiena, 1953; Rusek, 1985).

The feeding and burrowing activities of the macrofauna transcend most of the microhabitat constraints on the interactions between microorganisms and the smaller groups of soil invertebrates in litter and soil. Hence macrofaunal effects on bacteria and fungi depend upon the frequency of these perturbations to microbial habitats and the phenological responses of the microorganisms.

The indirect physical effects of animals on microbial populations and activities thus increase with increasing body size, but as the body size of invertebrates increases the weight-specific metabolic rates generally decrease within a trophic group. Hence the microfauna, with a small biomass but high densities of individuals and rapid turnover, will have a larger direct metabolic contribution to carbon and nitrogen fluxes through consumption of microorganisms than the same biomass of macrofauna (Petersen & Luxton, 1982).

The direct and indirect effects of invertebrates on microbial populations and nutrient mineralization can be viewed therefore as reciprocal processes with their relative contributions to soil processes determined by the structure of the soil organism community.

DIRECT CONTRIBUTIONS OF INVERTEBRATES TO SOIL PROCESSES

Most of the studies on the nutrient dynamics of bacteriophagy and mycophagy by nematodes and protozoans have been carried out in microcosms set up as analogues of microsite conditions in bulk soils or the rhizosphere. These studies have shown that the microfauna can regulate bacterial and fungal growth dynamics and nutrient uptake by plants. Seedlings of the grass *Bouteloua gracilis* growing with either bacteriophagous protozoans (Elliott, Coleman & Cole, 1979) or mycophagous nematodes (Ingham *et al.*, 1985) showed increased growth and foliar nitrogen concentrations in comparison with controls without animals. Similarly, Clarholm (1985a) has grown wheat plants with and without protozoans in autoclaved arable soil which had been reinoculated with a natural bacterial flora. After 6 weeks the plants grown in soil with protozoans had a 60% larger mass and a higher shoot/root biomass ratio than controls with bacteria only. The mechanism invoked for this effect is the stimulation of carbon-limited bacteria by root exudates as the root invades soil microsites. A third of the nitrogen immobilized by the growth of the bacteria is then released as excretory ammonium ions by protozoans feeding in the rhizosphere. The quantitative importance of these rhizosphere processes has not been demonstrated under field conditions but it is unlikely that they contribute significantly to the mineral nitrogen available to arable crops. Rooting densities greater than $10\,\mathrm{cm}^{-2}$ have rarely been found in spaced crops and only sparsely distributed roots ($<1\,\mathrm{cm}^{-2}$) are needed to exploit nitrate reserves fully in the soil solution (Nye & Tinker, 1977). The difficulty of extrapolating these microsite mechanisms to the field does not negate the potential importance of faunal impacts on microbial processes of nutrient immobilization and release in bulk soil. Reciprocal seasonal trends in protozoan and microbial biomass phosphorus have been demonstrated in a wheat-fallow soil (Fig. 2) and the predator–prey basis of this relationship is supported by calculations for a grassland soil that protozoans (mainly amoebae) can consume four times the mean bacterial biomass per year (Elliott & Coleman, 1977).

More precise estimates of faunal contributions to ecosystem nutrient fluxes are derived from budgets on the feeding, excretion and population dynamics of the component species or functional groups in the community. Some of the most comprehensive studies of this kind have been carried out within the Swedish Ecology of Arable Lands Programme (Rosswall & Paustian, 1984). Detailed nitrogen budgets have been drawn up for four cropping systems: a lucerne ley without fertilizer; a grass ley receiving $200\,\mathrm{kg\,N\,ha^{-1}\,a^{-1}}$; barley receiving $120\,\mathrm{kg\,N\,ha^{-1}\,a^{-1}}$; and barley without fertilizer nitrogen. Results are summarized in Table 1. It was concluded that only the protozoan and nematode faunas excreted significant amounts of mineral nitrogen which was highest in the fertilized grass ley

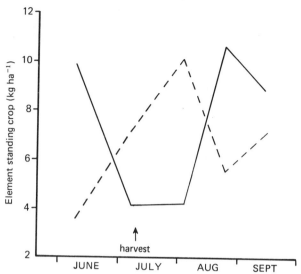

Fig. 2. Dynamics of microbial biomass phosphorus (--) and protozoan carbon (—) in a no-till wheat field (Colorado). Biomass element pools were determined by chloroform fumigation. The values for protozoan carbon have been multiplied by 5. (After Ingham *et al.*, 1985.)

(19% of $25.7\,\mathrm{g\,m^{-2}\,a^{-1}}$, i.e. about $50\,\mathrm{kg\,ha^{-1}\,a^{-1}}$) and lowest in the barley plots (38–$40\,\mathrm{kg\,ha^{-1}\,a^{-1}}$). However, the percentage of total mineralization attributed to faunal consumption of bacteria and fungi was about a third of the total flux in the barley plots compared with a fifth in the leys. A similar contribution of fauna to total nitrogen mineralization was calculated for a Swedish pine forest by Persson (1983). Here the soil fauna biomass was only $1.7\,\mathrm{g\,m^{-2}}$ (in comparison with $120\,\mathrm{g\,m^{-2}}$ fungi and $39\,\mathrm{g\,m^{-2}}$ bacteria) and represented only 4% of heterotroph metabolism. However, the animals consumed 30–60% of microbial production and hence contributed 10–49% of total nitrogen mineralization in the forest soil ($28\,\mathrm{kg\,ha^{-1}\,a^{-1}}$) of which nearly 70% was excretion by nematodes and protozoans.

Earthworm populations were small in both of these Swedish study areas; even in the lucerne ley the mean earthworm biomass was only $50\,\mathrm{kg}$ dry wt $\mathrm{m^{-2}}$ (Lofs-Holmin & Bostrom in Clarholm, 1985*b*). Estimates of the mean earthworm biomass (dry weight) for temperate regions range from about $50\,\mathrm{kg\,ha^{-1}}$ in boreal forests to about $450\,\mathrm{kg\,ha^{-1}}$ in the most densely populated New Zealand pastures. Lee (1983) has calculated from these figures (assuming a production: biomass ratio of 2–5) that the turnover of nitrogen through these

Table 1. *Nitrogen mineralization and total soil organic matter nitrogen balance for four arable cropping systems in Sweden (from Rosswall & Paustian, 1984)*

Cropping system	Fertilizer treatment (kg N ha^{-1} a^{-1})	Estimated net mineralization (g m^{-2} a^{-1})	Relative mineralization (% of soil N)	Mineralization via fauna (% of net mineralization)	Total soil N balance (g N m^{-2} a^{-1})
Lucerne ley	0	21.1	2.8	20	−5.3
Meadow ley	200	25.7	3.4	19	−8.2
Barley	0	10.5	1.4	36	−8.7
Barley	120	12.8	1.8	31	−7.4

populations amounts to 10–$225 \, kg \, N \, ha^{-1} a^{-1}$ with excretion contributing a further 18–$50 \, kg \, N \, ha^{-1}$. The assimilation efficiency of earthworms is low and so these populations are maintained by large throughputs of soil and litter material. With moderate or high populations of earthworms most of the plant litter is consumed each year in temperate regions and cast production ranges from 10–500 tonnes $ha^{-1} a^{-1}$. Most estimates for pastures and grasslands are about 40–50 tonnes $ha^{-1} a^{-1}$, representing a 3–4 mm annual increment on the soil surface (Lee, 1983). Considerable casting occurs below ground and Graff (1971) has estimated that up to 25% of the total A_h horizon in a temperate pasture could be turned over each year. The effects of this massive disturbance to soil structure through burrowing, litter consumption and gut passage are manifested in indirect effects on the microbial community.

INDIRECT EFFECTS OF INVERTEBRATES ON MICROBIAL POPULATIONS AND PROCESSES

Invertebrates have indirect effects on the structure and activities of bacteria and fungal communities through inoculum dispersal, grazing, litter comminution, gut passage and aggregate formation. Interactions at higher trophic levels also have indirect effects on microbial communities. The functional importance of these interactions for decomposition processes and plant nutrient uptake has not generally been demonstrated under field conditions and most will only be considered briefly.

Dispersal of fungal propagules

Fungal spores on invertebrate integuments are often those of common saprophytic fungi (*Mortierella*, *Cladosporium*, *Penicillium*, *Aspergillus*, etc.) found in the soil and litter horizons from which the animals were extracted. Visser (1985) reported a total of 120 taxa of fungi isolated from guts and whole-body squashes of one species of woodland collembolan (*Onychiurus subtenuis*) with higher mean numbers of propagules from animals extracted from the litter layers (2.5–5.5) than from the humus layers (2.5–3.1). There is, however, no conclusive evidence that spore carriage by microarthropods affects the composition or functioning of microbial communities colonizing resources, as judged from the exclusion experiments of Gourbière (1986), though the cortical invasion of rotten wood by

Table 2. *Colonization of* Pinus radiata *roots by the ectotrophic mycorrhizal fungus* Rhizopogon luteolus *in the presence of different numbers of soil amoebae (from Chakraborty, Theodorou & Bowen, 1985)*

Treatment	No. of *Saccamoeba* g^{-1} soil	Root length (mm)	Length of root colonized by *Rhizopogon* (mm)
R. luteolus	–	89	23.6
R. luteolus + Saccamoeba	846	100	5.8
R. luteolus + Saccamoeba	423	98	8.0
R. luteolus + Saccamoeba	85	102	11.3
R. luteolus + Saccamoeba	42	103	12.9

basidiomycetes may be facilitated by invertebrates (Swift & Boddy, 1984).

Grazing and Predation

A large number of studies have demonstrated important grazing effects in litter, soil and in the rhizosphere under laboratory conditions but as yet most of these phenomena have not been quantified in the field.

Laboratory and field studies

Mycophagous amoebae in pot experiments may reduce the colonization of *Pinus radiata* roots by the ectotrophic fungus *Rhizopogon luteus* (Table 2), while mycophagous nematodes can prevent the establishment of *Suillus granulatus* on *Pinus resinosa* (Sutherland & Fortin, 1968); no effects on seedling growth were demonstrated in these studies. Wiggins & Curl (1979) suggested that collembolans grazing in the rhizosphere may reduce the invasion of roots by pathogens and symbionts but Warnock, Fitter & Usher (1982) found no effects on the growth of leeks when the mycorrhizal fungus *Glomus fasciculatum* was grazed by *Folsomia candida*. Selective feeding on different species of fungi, particularly basidiomycetes, by collembolans can affect the saprophytic colonizing ability of fungi invading leaf litter (Parkinson, Visser & Whittaker, 1979) and hence the rate of litter decomposition (Newell, 1984a). However, Andrén & Schnürer (1985) found no effects of grazing on microbial respiration, biomass or mass loss from decomposing straw but suggested that the species of collembolan used in their experiments was not predominantly mycophagous.

Field studies using litter bags have shown both increased nutrient immobilization and release in the presence of microarthropods (Seastedt, 1984). Parker *et al.* (1984) showed that nitrogen immobilization around root litter treated with pesticides in a desert soil was about twice that of untreated soil over a 3 month period and was associated with an increase in fungal biomass. It has also been suggested that tydaeid mites may indirectly regulate the decomposition of buried leaf litter by consuming bacteriophagous nematodes (Whitford *et al.*, 1982). When these mites were eliminated by pesticide treatment, the nematode populations increased dramatically, thus reducing the bacteria and yeast populations and significantly reducing the initial rates of litter decomposition. A different approach was used by Verhoef & de Goede (1985) who established cages with and without populations of litter-living, entomobryid collembolans in a pine-forest soil. After 3 months the nitrogen concentration in the defaunated litter was more than twice that of normal litter. Selective grazing by the most abundant mycophagous collembolan species, *Onychiurus latus*, in a Sitka spruce forest also appeared to determine the distribution and abundance of two litter-decomposing basidiomycetes (Newell, 1984*b*). *Marasmius androsaceus*, the preferred food and faster litter-decomposing species, was restricted to the surface litter layers where collembolans were scarce. In contrast, *Mycena galopus*, which was less palatable, was associated with the underlying fermentation layers where collembolans were abundant.

Variables affecting invertebrate–microbial interactions

Estimates of the percentage of fungal biomass consumed per annum by microarthropods range from 2% for cryptostigmatid mites (Mitchell & Parkinson, 1976) to 86% for the entire mycophagous invertebrate community (McBrayer, Reichle & Witkamp, 1974). Soil habitat structure and the quality of the resources exploited by the fungus appear to be important variables underpinning these trophic dynamics, though both are poorly quantified under field conditions.

Haarløv (1960) estimated that the habitat area occupied by microarthropods in a pine-forest soil was $1/170$ to $1/5000$ of the total macropore surfaces, so despite the fact that hyphal lengths in pine forest-floor materials range from 10^3 to 10^5 m g^{-1} (Hunt & Fogel, 1983), only a small proportion of the active mycelium is likely to be exposed to grazing by animals. The sensitivity of the fungus to grazing, and the population responses of the invertebrates, appear to depend on the extent to which this feeding depletes the nutrient capital of the fungus.

Laboratory studies carried out under simplified and defined artificial culture conditions are surprisingly variable. Invertebrates may feed selectively or non-selectively on species of fungi and bacteria with stimulatory, neutral or inhibitory effects on microbial growth, metabolism and species interactions (Ingham et al., 1985; Visser, 1985). The responses of bacteria and fungal species to culture conditions, particularly nutrient concentrations in the media, appear to be important determinants of the variable and often contradictory results.

Park (1976) found that only 8 out of 43 fungal isolates exhibited the expected pattern in which the growth and activity of cellulose-decomposing fungi is inversely proportional to the nitrogen concentration of the media. The remaining isolates showed a range of negative, neutral and non-linear responses in culture. These optimum or sub-optimum conditions affect the physical and biochemical characteristics of the fungus and hence its palatability and nutritional value to soil invertebrates. Leonard (1984) has shown that inter- and intra-specific selection of fungi by collembolans can be changed by the growth medium (soil-, leaf- or malt-extract agar), solid or liquid culture conditions, the age of the culture and the nitrogen concentration of the media. Nutrient concentrations also determine whether grazing by collembolans stimulates or inhibits fungal respiration (Hanlon, 1981) and the growth and fecundity of animals feeding on the fungus (Booth & Anderson, 1979). The interactions of these variables often result in non-linear responses in which fungal growth and activity are stimulated at low grazing intensity but are inhibited above a threshold level (Hanlon & Anderson, 1979). A similar situation has been demonstrated in mixed cultures of protozoans and/or nematodes with bacteria or fungi (Ingham et al., 1985).

The implication of these responses is that while it may be possible to model these grazing effects in the laboratory, their prediction in the field requires quantification of microsite nutrient regimes and the availability of microbial tissues to invertebrates. Such data are currently unavailable.

Effects of gut transit

Several studies of earthworm gut bacteria suggest that the groups of organisms present are the same as those in the soil where the animals are living (Satchell, 1983). On the other hand, 73% of 473 bacterial strains isolated from the gut of the oligochaete *Eisenia lucens* by Marialigeti (1979) belonged to the genus *Vibrio* (including pathogenic strains of *V. cholerae, V. parahaemolyticus* and *V. alginolyticus*).

Descriminant analysis of isolates from the food, gut contents and faeces of a millipede (*Glomeris marginata*) and a woodlouse (*Oniscus asellus*) revealed that, although there was some overlap, the bacterial floras of the food litter and the guts were distinct (Ineson & Anderson, 1985). The most abundant isolated from the gut of both animals was *Klebsiella pneumoniae*, supporting the results of a number of similar studies on soil invertebrates, which might account for the occurrence of this coliform in sawdust, logs and other woody habitats (Ineson & Anderson, 1985). Griffiths & Woods (1985) found *Enterobacter agglomerans* was the most frequent organism isolated from the guts of *Oniscus asellus*. In other studies, *K. pneumoniae* and *E. agglomerans* were among the most abundant free-living nitrogen-fixing bacteria found in an oak woodland soil (Jones & Bangs, 1985). Irrespective of the specific identity of the faecal flora, these results suggest that, even in acid soils, the microbial populations colonizing litter and humus materials are dominated by bacteria, albeit transiently, as a consequence of invertebrate feeding activities.

The effects of gut passage are most pronounced in base-rich soils which are intensively worked by earthworms. The casts contain fewer fungal propagules and denitrifying bacteria, higher counts of total bacteria and more cellulolytic, hemicellulolytic, amylolytic and nitrifying bacteria than unworked soil (Loquet, Bhatnagar & Bouché, 1977). Bhatnagar (1975) estimated that 40% of all aerobic, free-living nitrogen-fixers, 13% of anaerobic nitrogen-fixers and 16% of denitrifying bacteria in the total soil volume were located in a narrow zone a few millimetres around earthworm burrows. Kretzschmar (1978) estimated that this zone constituted only 0.9% of the soil volume in a French pasture but contained 15% of the total soil microflora. Given the magnitude of their casting rates, earthworms must be modifying the dynamics of microbial processes in these soils but this has never been considered explicitly in calculations of microbial generation and turnover times.

Indirect effects of invertebrates on mineral nutrient fluxes

Any animal processes which modify the microbial environment, and hence alter the generation and turnover times of bacteria and fungi, will change the rates of microbial processes. These synergistic interactions have been demonstrated for microfauna (Ingham *et al.*, 1985) and mesofauna (Seastedt, 1984) but are most apparent for the effects of soil- and litter-feeding macrofauna on nitrogen mineralization. For example, the addition of millipedes to leaf litter does not produce a step-wise increment in nitrogen mineralization which might be attributable to excretion by the animals or the lysis of microbial biomass at the onset of feeding (Anderson & Ineson, 1984). Instead, the ammonium and nitrate losses from the material increase gradually over several weeks to levels 2–10 times higher than those found in the absence of animals, and if the animals are removed the enhanced rates of nitrogen mineralization slowly return to control levels. These responses are apparently unrelated to the species of saprotrophic fungi and invertebrates used in the laboratory systems and hence general models can be formulated which quantify the indirect effects of animals on nitrogen mineralization from leaf litter or soil organic matter (Anderson *et al.*, 1985*b*). Hence, cumulative ammonium-nitrogen mineralization from oak woodland leaf litter in the absence of animals (N_c) is given by the equation

$$N_c = 72.3\,(\pm 41.7) + 12.4\,(\pm 3.9)T$$

where T is the incubation temperature between 5 and 15 °C.

With litter-feeding animals present in replicate samples the cumulative ammonium-nitrogen mineralization function (N_A) becomes:

$$N_A = 27.8\,(\pm 49.7) + 17.1\,(\pm 4.6)T + 543\,(\pm 133)B + 14.7\,(\pm 12.3)BT$$

where B is the 'effective biomass', i.e. the animal biomass (g fresh wt) divided by the dry weight (g) of the resource (B = 0.1 to 0.3 in these experiments). The regression coefficients in N_c are not significantly different from the same partial regression coefficients in N_A. Hence the first two functions in N_A are an expression of microbial nitrogen mineralization and the B and BT functions represent the additional effects of animals. The animal-mediated flux is not corrected for excretory nitrogen and is therefore only a partial expression of indirect animal effects. Expressions of N_A for total nitrogen mineralization from litter and humus materials from three oak woodlands are shown in Fig. 3.

These multiple regression functions are not a general expression of nitrogen mineralization by animals and have to be derived empirically for each resource type, but they do show that invertebrates can have significant effects on net nitrogen mineralization, particularly during the early stages of litter decomposition when it is often assumed that net microbial immobilization of nitrogen occurs.

To link the laboratory and field experiments, a series of replicated soil and litter lysimeters was set up in an oak woodland (Anderson et al., 1985a). Treatments were: with or without macrofauna (millipedes, woodlice and earthworms added at the equivalent of 13 g fresh wt m^{-2}) and with or without living tree roots (introduced through ports in the sides of the lysimeters). Mineral element concentrations in leachates were monitored for 2 years. Cumulative mineral nitrogen losses from treatments with and without roots, and with and without macrofauna for the first 64 weeks of the experiment are shown in Table 3.

The results show that over the first 32 weeks, from spring to leaf fall, the fauna increased ammonium nitrogen losses by 5.4 kg ha^{-1} and nitrate nitrogen losses by 6.2 kg ha^{-1} in the lysimeters without roots; a total of 11.6 kg ha^{-1} or about a third of the annual nitrogen input in leaf litter. The macrofauna also increased the labile nitrogen pool in the litter and soil organic matter leading to greater mineral nitrogen losses, up to 3.8 kg N ha^{-1} wk^{-1}, than in the controls, following wetting and drying events during a hot, dry summer (Anderson et al., 1985a). However, after leaf fall in week 33 ammonium nitrogen losses were lower from the lysimeters with animals than those without them until soil temperatures increased the following spring. Roots took up approximately half of the mineral nitrogen flux in all treatments in contrast to undisturbed forest soils where mineral nitrogen was scavenged with high efficiency. None the less, these experiments do show that a small biomass of macrofauna, less than that found in many deciduous forest soils, can have significant effects on seasonal patterns of nitrogen immobilization and mineralization.

Anderson et al. (1985b) have predicted that the indirect effects of macrofauna will contribute significantly to nitrogen mineralization where the effective biomass for litter is at least 0.1. Higher values than this are associated with base-rich soils, high-quality litters which decompose rapidly and often a large biomass of earthworms. Under these conditions additional indirect effects on microbial processes are introduced by the burrowing activities of the worms.

Earthworm-worked soils have a higher pore volume, increased

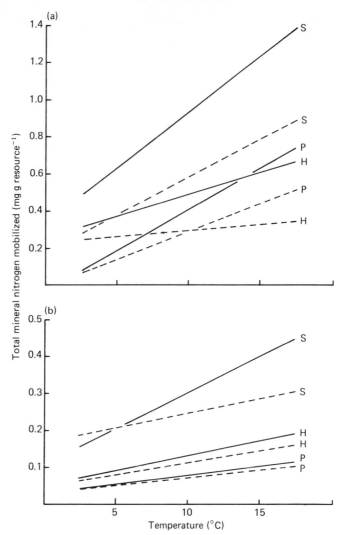

Fig. 3. Expressions of nitrogen mineralization from litter (a) and humus (b) from three oak woodlands (Perridge, P; Hillersdon, H; and Stoke, S) using mutiple regression functions based on animal biomass (including zero biomass) and temperature. Nitrogen mineralization with animal effects, at an effective biomass (B) of 0.3, is shown by the solid line, while microbial mineralization (when $B = 0$) is shown by the broken line. The same value of B was used for the litter and humus materials to emphasize the different magnitude of animal effects on the two resource types (note the difference in scale of nitrogen mineralization in a and b). (After Anderson et al., 1985b.)

field water-holding capacity, more water-stable aggregates and higher infiltration rates than soils without earthworms or with surface active species. The introduction of deep-burrowing lumbricid earthworms into New Zealand pastures has dramatically altered soil

Table 3. *Cumulative mineral nitrogen losses from a series of small (0.25 m²)
lysimeters in an oak woodland over two periods from April to October
(weeks 1–32) and November to June (weeks 33–64). Treatments are with
and without macrofauna and with and without oak tree root systems. Results
are calculated as kg N ha⁻¹ (±SE, n = 3). For details see text.*

| Period (weeks) | No roots | | Roots | |
	Without animals	With animals	Without animals	With animals
Ammonium nitrogen				
1–32	23.26 ± 1.09	28.68 ± 0.92	11.27 ± 3.63	9.89 ± 1.00
33–64	16.25 ± 1.13	10.03 ± 1.47	1.23 ± 0.37	0.63 ± 0.02
Nitrate nitrogen				
1–32	8.86 ± 1.91	15.06 ± 1.14	2.18 ± 0.33	2.27 ± 0.34
33–64	12.20 ± 0.87	13.46 ± 1.67	0.79 ± 0.03	0.46 ± 0.05

physical conditions and plant rooting depth and has raised pasture
productivity by up to 72% (Stockdill, 1982).

RESOLUTION OF MICROSITE PROCESSES

The variable responses of animal–microbial interactions in field and
laboratory studies are consistent with the hypothesis that soil organism communities comprise few species with very specific roles in
decomposition processes and nutrient cycling. Most functional
groups of invertebrates, fungi and bacteria contain a large number
of species with broadly overlapping, and often very plastic, trophic
capacities.

Swift (1976, 1984) has suggested that fungal communities have
a hierarchical structure built up of species groups, or 'unit communities' inhabiting individual resource units such as twigs or faecal pellets. The microorganisms in the unit community are assortatively
selected from the species pool so that they have the capacity to
decompose that resource; interactions between different or adjacent
units are minimized. Thus, no matter how much variation there may
be in floristic composition between unit communities, it is theoretically only necessary to understand the functioning of a single unit
community to establish the principles for decomposition as a whole.
The invertebrates are integral components of these unit communities
and have very variable associations with different microbial species

assemblages. Anderson (1978) found that the gut contents (higher plant material, bacteria, fungal hyphae and spores) of mites were related to the microhabitats they occupied. Furthermore, the composition of the ingested material showed less variation between different mite species occupying similar microhabitats, defined at a scale of about 0.1 mm, than between the same mite species in different soil and litter horizons.

Different sized organisms will use resources on different scales. Thus microhabitats for assemblages of bacteria, nematodes and protozoans may be defined in a soil aggregate or on a section of fine root, while at the other extreme a single individual of a cord-forming basidiomycete may invade a number of tree boles over an area of more than 100 m^2 (Thompson & Boddy, 1983).

Evidence from field and laboratory studies shows that the soil is made up of a mosaic of these different-sized patches in different phases of microbial growth; some patches, for example, immobilize nitrogen and others show a new release of nitrogen. Net nitrogen mineralization in bulk soil will only be detectable where microsites releasing nitrogen exceed those immobilizing nitrogen.

The relative importance of animal microbial interactions to these patch dynamics may be predicted from the size–resource relations of organisms in the community. Hence direct contributions to nutrient fluxes may be important where the organisms and resources are of similar size (e.g. amoebae/bacteria; earthworms/leaf litter). On the other hand indirect effects of invertebrates on microbial processes may be more functionally significant where the animal structurally modifies microbial microhabitats and the associated unit community (litter-feeding invertebrates and earthworms ingesting soil).

At an intermediate scale, such as microarthropods grazing on fungal hyphae, the dynamics of the interactions will depend on the animal population densities, the extent of the mycelium, microhabitat structure and the quality of the resource exploited by the fungus.

The micro-, meso- and macrofauna therefore represent different types and scales of perturbation which contribute to these microsite dynamics and hence to mineral element fluxes. While we can identify specific organism interactions in the laboratory, or the synchrony imposed on microbial activities in microsites by passage through the earthworm gut, we are currently unable to link these events to gross soil processes. Plant rooting systems resolve the soil mosaic at a level necessary to optimize nutrient uptake and thus provide an integrated measurement of the net products of these microsite processes.

As nitrogen availability declines, a greater proportion of new production is allocated below ground (McClaugherty, Aber & Melillo, 1982) and roots become more finely branched to exploit a larger soil volume (Barley, 1970; Fitter, 1985). Under conditions of very low nutrient availability in organic soils, roots become longer lived and, since nutrient uptake is mainly associated with the unsuberized regions of the proliferating tips, their absorptive capacity is reduced. Instead, nutrient uptake is facilitated by the plants reallocating carbon from short-lived root hairs to mycorrhizal associations which have comparatively long-lived and extensive hyphal ramifications through the soil and litter matrix. Microsites with low rates of nitrogen mineralization can thus be exploited efficiently by trees with rooting volumes as low as $0.13-5.3\,\mathrm{cm\,cm^{-3}}$ compared with $50\,\mathrm{cm\,cm^{-3}}$ for grasses and $5-25\,\mathrm{cm\,cm^{-3}}$ for intensely cultivated cereals (Bowen, 1984). In low fertility soils both tree roots and the hyphae of vesicular-arbuscular mycorrhizas show increased branching when they encounter local patches of organic matter within the soil matrix (St John, Coleman & Reid, 1983).

On the basis of these rooting patterns it is hypothesized that under conditions of moderate to high nutrient availability the effects of earthworms in promoting a favourable environment for roots are likely to be more important for plant growth than the activities of rhizosphere invertebrates in mediating processes of nitrogen mineralization.

CONCLUSIONS

The understanding of ecosystem and global fluxes of carbon and nitrogen is based on a knowledge of microbial physiological processes operating in microsite environments. Under steady state conditions the products of the microsites are always the net expression of animal–microbial interactions since there are no natural terrestrial environments from which soil invertebrates are excluded. If the system is disturbed then the whole community will undergo adjustment to some new equilibrium state and the reduced or increased activities of the animals will be part of the definition of the new dynamic state of the community. Thus it can be argued that establishing the conditions under which invertebrates make a major contribution to soil processes may provide greater insight into the responses of soils to major perturbations. At the other extreme, studies on the ecology of specific organisms mediating key nutrient transformations (nitrification, denitrification) or plant processes (root pathogens, nodulating bacteria and mycorrhizas) operate in microsites where animal effects may modify microbial processes expressed in axenic cultures. There is no doubt that animal–microbial interactions are real phenomena in all systems but the quantitative links between species and ecosystem processes cannot be resolved without more synergistic approaches. More integrated studies between soil microbiologists, zoologists, physicists and biochemists are required to assess whether these interactions have manipulative potential or can be regarded merely as a source of variation or 'noise' within extremely heterogeneous soil systems.

REFERENCES

ANDERSON, J. M. (1975). Succession, diversity and trophic relationships of some soil animals in decomposing leaf litter. *Journal of Animal Ecology*, **44**, 475–95.

ANDERSON, J. M. (1978). Competition between two unrelated species of soil cryptostigmata (Acari) in experimental microcosms. *Journal of Animal Ecology*, **47**, 787–803.

ANDERSON, J. M. HUISH, S. A., INESON, P., LEONARD, M. A. & SPLATT, P. R. (1985a). Interactions of invertebrates, microorganisms and tree roots in nitrogen and mineral element fluxes in deciduous woodland soils. In *Ecological Interactions in Soil*, ed. A. H. Fitter, D. Atkinson, D. J. Read & M. B. Usher, pp. 377–92. Oxford, Blackwell Scientific Publications.

ANDERSON, J. M. & INESON, P. (1984). Interactions between microorganisms and soil invertebrates in nutrient flux pathways of forest ecosystems. In *Invertebrate–Microbial Interactions*, ed. J. M. Anderson, A. D. M. Rayner & D. W. H. Walton, pp. 59–88. Cambridge, Cambridge University Press.

ANDERSON, J. M., INESON, P. & HUISH, S. A. (1983). Nitrogen and cation mobilization by soil fauna feeding on leaf litter and soil organic matter from deciduous woodlands. *Soil Biology and Biochemistry*, **15**, 463–7.

ANDERSON, J. M., LEONARD, M. A., INESON, P. & HUISH, S. (1985b). Faunal biomass: a key component of a general model of nitrogen mineralization. *Soil Biology and Biochemistry*, **17**, 735–7.

ANDERSON, J. M., RAYNER, A. D. M. & WALTON, D. W. H. (1984). *Invertebrate–Microbial Interactions*. Cambridge, Cambridge University Press.

ANDERSON, R. V., COLEMAN, D. C. & COLE, C. V. (1981). Effects of saprotrophic grazing on net mineralization. In *Terrestrial Nitrogen Cycles*. Ecological Bulletins No. 33, ed. F. E. Clark & T. Rosswall, pp. 201–16. Stockholm, NFR.

ANDRÉN, O. & SCHNÜRER, J. (1985) Barley straw decomposition with varied levels of microbial grazing by *Folsomia fimetaria* (L.) (Collembola, Isotomidae). *Oecologia*, **68**, 57–62.

BARLEY, K. P. (1970). The configuration of the root system in relation to nutrient uptake. *Advances in Agronomy*, **22**, 159–201.

BHATNAGAR, T. (1975). Lombriciens et humification un aspect nouveau de l'incorporation microbienne d'azote induite par les vers de terre. In *Biodégradation et Humification*, ed. G. Kilbertus, O. Reisinger, A. Mourey & J. P. Cancela da Fonseca, pp. 169–82. Sarreguemines, Pierron.

BOOTH, R. G. & ANDERSON, J. M. (1979) The influence of fungal food quality on the growth and fecundity of *Folsomia candida* (Collembola: Isotomidae). *Oecologia*, **38**, 317–23.

BOWEN, G. D. (1984). Tree roots and use of soil nutrients. In *Nutrition of Plantation Forests*, ed. G. D. Bowen and E. K. S. Nambiar, pp. 147–79. London & New York, Academic Press.

CHAKRABORTY, S., THEODOROU, C. & BOWEN, G. D. (1985). The reduction of root colonization by mycorrhizal fungi by mycophagous amoebae. *Canadian Journal of Microbiology*, **31**, 295–7.

CLARHOLM, M. (1985a). Interactions of bacteria, protozoa and plants leading to mineralization of soil nitrogen. *Soil Biology and Biochemistry*, **17**, 181–8.

CLARHOLM, M. (1985b). Possible roles for roots, bacteria, protozoa and fungi in supplying nitrogen to plants. In *Ecological Interactions in Soil*, ed. A. H. Fitter, D. Atkinson, D. J. Read & M. B. Usher, pp. 355–65. Oxford, Blackwell Scientific Publications.

ELLIOTT, E. T. & COLEMAN, D. C. (1977). Soil protozoan dynamics in a shortgrass prairie. *Soil Biology and Biochemistry*, **9**, 113–18.

ELLIOTT, E. T., COLEMAN, D. C. & COLE, C. V. (1979). The influence of amoebae on the uptake of nitrogen by plants in gnotobiotic soil. In *The Root–Soil Interface*, ed. J. L. Harley & R. Scott-Russell, pp. 221–9. New York and London, Academic Press.

FITTER, A. H. (1985). Functional significance of root morphology and root system architecture. In *Ecological Interactions in Soil*, ed. A. H. Fitter, D. Atkinson, D. J. Read & M. B. Usher, pp. 87–106. Oxford, Blackwell Scientific Publications.

FITTER, A. H., ATKINSON, D., READ, D. J. & USHER, M. B. (ed.) (1985). *Ecological Interactions in Soil*. Oxford, Blackwell Scientific Publications.

FRISSEL, M. J. & VAN VEEN, J. A. (1982). A review of models for investigating the behaviour of nitrogen in soil. *Philosophical Transactions of the Royal Society of London, Series B*, **296**, 341–9.

GOURBIÈRE, F. (1986). Méthode d'étude simulanée de la décomposition et des mycoflores des aiguilles de conifères (*Abies alba*). *Soil Biology and Biochemistry*, **18**, 155–60.

GRAFF, O. (1971). Stickstoff, Phosphor und Kalium in der Regenwurmlosung auf der Wiesenversuchsfläche des Sollingprojektes. In *IV Colloquim Pedobiologiae*, ed. J. d'Aguilar *et al.* pp. 503–11. Paris, INRA.

GRIFFITHS, B. S. & WOODS, S. (1985). Microorganisms associated with the hindgut of *Oniscus asellus* (Crustacea, Isopoda). *Pedobiologia*, **28**, 377–82.

HAARLØV, N. (1960). Microarthropods from Danish soils. *Oikos*, suppl. 3, 1–176.

HANLON, R. D. G. (1981). Influence of grazing by collembola on the activity of senescent fungal colonies grown on media of different nutrient concentration. *Oikos*, **36**, 362–7.

HANLON, R. D. G. & ANDERSON, J. M. (1979). The effects of collembola grazing on microbial activity in decomposing leaf litter. *Oecologia*, **38**, 93–9.

HUNT, G. A. & FOGEL, R. (1983). Fungal hyphal dynamics in a Western Oregon Douglas-fir stand. *Soil Biology and Biochemistry*, **15**, 641–9.

INESON, P. & ANDERSON, J. M. (1985). Aerobically isolated bacteria associated with the gut of the feeding animal macroarthropods. *Oniscus asellus* and *Glomeris marginata*. *Soil Biology and Biochemistry*, **17**, 843–9.

INGHAM, R. E., TROFYMOW, J. A., INGHAM, E. R. & COLEMAN, D. C. (1985).

Interactions of bacteria, fungi, and their nematode grazers: effects on nutrient cycling and plant growth. *Ecological Monographs*, **55**, 119–40.

JONES, K. & BANGS, D. (1985). Nitrogen fixation by free-living heterotrophic bacteria in an oak forest: the effect of liming. *Soil Biology and Biochemistry*, **17**, 705–9.

KRETZSCHMAR, A. (1978). Quantification écologiques des galeries de lombriciens. Techniques et premières estimations. *Pedobiologia*, **18**, 31–8.

KUBIENA, W. L. (1953). *The Soils of Europe*. Madrid, Thomas Murby.

LEE, K. E. (1983). The influence of earthworms and termites on soil nitrogen cycling. In *New Trends in Soil Biology*, ed. P. Lebrun, H. M. André, A. de Medts, C. Gregoire-Wibo and G. Wauthy, pp. 35–88. Louvain-la-Neuve, Dieu Brichart.

LEONARD, M. A. (1984). Observations on the influence of culture conditions on the fungal feeding preferences of *Folsomia candida* (Collembola: Isotomidae). *Pedobiologia*, **26**, 361–7.

LOQUET, M., BHATNAGAR, T. & BOUCHÉ, M. B. (1977). Essai' d'estimation de l'influence écologique des lombriciens sur les microorganisms. *Pedobiologia*, **17**, 400–17.

McBRAYER, J. F., REICHLE, D. E. & WITKAMP, M. (1974). *Energy Flow and Nutrient Cycling in a Cryptozoan Food-web*. Report EDFB – IBP – 73–8. Oak Ridge (Tennessee), Oak Ridge National Laboratories.

McCLAUGHERTY, C. A., ABER, J. D. & MELILLO, J. M. (1982). The role of fine roots in the organic matter and nitrogen budgets of two forested ecosystems. *Ecology*, **63**, 1481–90.

MARIALIGETI, K. (1979). On the community structure of the gut microbiota of *Eisenia lucens* (Annelida, Oligochaeta). *Pedobiologia*, **19**, 213–20.

MITCHELL, M. J. & PARKINSON, D. (1976). Fungal feeding of oribatid mites (Acari: Cryptostigmata) in an aspen woodland soil. *Ecology*, **57**, 302–12.

NEWELL, K. (1984*a*). Interactions between two decomposer basidiomycetes and a collembolan under Sitka spruce: distribution, abundance and selective grazing. *Soil Biology and Biochemistry*, **16**, 227–34.

NEWELL, K. (1984*b*). Interactions between two decomposer basidiomycetes and a collembolan under Sitka spruce: grazing and its potential effects on fungal distribution and litter decomposition. *Soil Biology and Biochemistry*, **16**, 235–40.

NYE, P. H. & TINKER, P. B. (1977). *Solute Movement in the Soil Root System*. Oxford, Blackwell Scientific Publications.

PARK, D. (1976). Nitrogen level and cellulose decomposition by fungi. *International Biodeterioration Bulletin*, **12**, 95–9.

PARKER, L. W., SANTOS, P. F., PHILLIPS, J. & WHITFORD, W. G. (1984). Carbon and nitrogen dynamics during the decomposition of litter and roots of a Chihuahuan desert annual, *Lepidium lasiocarpum*. *Ecological Monographs*, **54**, 339–60.

PARKINSON, D., VISSER, S. & WHITTAKER, J. B. (1979). Effects of collembolan grazing on fungal colonization of leaf litter. *Soil Biology and Biochemistry*, **11**, 529–35.

PERSSON, T. (1983). Influence of soil animals on nitrogen mineralization. In *New Trends in Soil Biology*, ed. P. Lebrun, H. M. André, A. de Medts, C. Gregoire-Wibo & G. Wauthy, pp. 117–26. Louvain-la-Neuve, Dieu-Brichart.

PETERSEN, H. & LUXTON, M. (1982). A comparative analysis of soil fauna populations and their role in decomposition processes. *Oikos*, **39**, 287–388.

ROSSWALL, T. & PAUSTIAN, K. (1984). Cycling of nitrogen in modern agricultural systems. *Plant and Soil*, **76**, 3–21.

RUSEK, J. (1985). Soil microstructures – contributions on specific soil organisms. *Quaestiones Entomologicae*, **21**, 497–516.

SATCHELL, J. E. (1974). Introduction: litter–interface of animate/inanimate matter. In *Biology of Plant Litter Decomposition*, ed. C. H. Dickinson & G. J. F. Pugh, pp. xiii–xliv. London and New York, Academic Press.

SATCHELL, J. E. (1983). *Earthworm Ecology*. London, Chapman and Hall.

SEASTEDT, T. R. (1984). The role of microarthropods in decomposition and mineralization processes. *Annual Review of Entomology*, **29**, 25–46.

ST JOHN, T. V., COLEMAN, D. C. & REID, C. P. P. (1983). Growth and spatial distribution of nutrient-absorbing organs: selective exploitation of soil heterogeneity. *Plant and Soil*, **71**, 487–93.

STOCKDILL, S. M. J. (1982). Effects of introduced earthworms on the productivity of New Zealand pastures. *Pedobiologia*, **24**, 29–35.

SUTHERLAND, J. R. & FORTIN, J. A. (1968). Effect of the nematode *Aphelenchus avenae* on some ectotrophic, mycorrhizal fungi and on a red pine mycorrhizal relationship. *Phytopathology*, **58**, 519–23.

SWIFT, M. J. (1976). Species diversity and the structure of microbial communities. In *The Role of Terrestrial and Aquatic Organisms in Decomposition Processes*, ed. J. M. Anderson & A. Macfadyen, pp. 185–222. Oxford, Blackwell Scientific Publications.

SWIFT, M. J. (1984). Microbial diversity and decomposer niches. In *Current Perspectives in Microbial Ecology*, ed. M. J. Klug & C. A. Reddy, pp. 8–16. Washington, American Society for Microbiology.

SWIFT, M. J. & BODDY, L. (1984). Animal–microbial interactions in wood decomposition. In *Invertebrate–Microbial Interactions*, ed. J. M. Anderson, A. D. M. Rayner & D. W. H. Walton, pp. 89–132. Cambridge, Cambridge University Press.

SWIFT, M. J., HEAL, O. W. & ANDERSON, J. M. (1979). *Decomposition in Terrestrial Ecosystems*. Oxford, Blackwell Scientific Publications.

THOMPSON, W. & BODDY, L. (1983). Decomposition of suppressed oak trees in even-aged plantations. II. Colonization of tree roots by cord- and rhizomorph-producing basidiomycetes. *New Phytologist*, **93**, 277–91.

USHER, M. B., BOOTH, R. G. & SPARKS, K. E. (1982). A review of progress in understanding the organization of communities of soil arthropods. *Pedobiologia*, **23**, 126–44.

VERHOEF, J. A. & DE GOEDE, R. G. M. (1985). Effects of collembolan grazing on nitrogen dynamics in a coniferous forest. In *Ecological Interactions in Soil*, ed. A. H. Fitter, D. Atkinson, D. J. Read and M. B. Usher, pp. 367–76. Oxford, Blackwell Scientific Publications.

VISSER, S. (1985). Role of the soil invertebrates in determining the composition of soil microbial communities. In *Ecological Interactions in Soil*, ed. A. H. Fitter, D. Atkinson, D. J. Read & M. B. Usher, pp. 297–317. Oxford, Blackwell Scientific Publications.

WARNOCK, A. J., FITTER, A. H. & USHER, M. B. (1982). The influence of a springtail *Folsomia candida* (Insecta, Collembola) on the mycorrhizal association of the leek *Allium porrum* and the vesicular-arbuscular mycorrhizal endophyte *Glomus fasciculatum*. *New Phytologist*, **90**, 285–92.

WHITFORD, W. G., FRECKMAN, D. W., PARKER, L. W., SCHAEFER, D., SANTOS, P. F. & STEINBERGER, Y. (1982). The contributions of soil fauna to nutrient cycles in desert systems. In *New Trends in Soil Biology*, ed. P. H. Lebrun, H. M. André, A. de Medts, C, Gregoire-Wibo & G. Wauthy, pp. 49–60. Louvain-la-Neuve, Dieu Brichart.

WIGGINS, E. A. & CURL, E. A. (1979). Interactions of Collembola and microflora of cotton rhizosphere. *Phytopathology*, **69**, 244–9.

ANALYSIS OF MICROBIAL COMMUNITIES WITHIN SEDIMENTS USING BIOMARKERS

R. J. PARKES

Scottish Marine Biological Association, PO Box 3, Oban, Argyll, Scotland

INTRODUCTION

In the last five years there has been a rapid advance in our knowledge of microbial interactions within sediments. This is due largely to the widespread use of radio-tracer techniques and specific inhibitors either separately or in combination, as pioneered by the elegant work of Cappenberg on freshwater sediments (Cappenberg, 1974; Cappenberg & Prins, 1974). Sediments are ideal for this type of approach as anaerobic reactions often dominate a few millimetres below the sediment surface (Revsbech, Sørensen & Blackburn, 1980) and these reactions characteristically involve interacting microbial communities where the product of one reaction becomes the substrate for another reaction (Fig. 1). Such techniques can also be

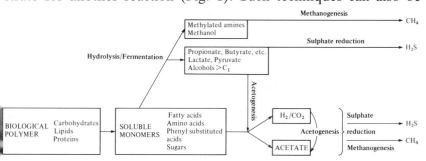

Fig. 1. A scheme of carbon flow through an anaerobic ecosystem showing some of the functional metabolic groups (bold type) and their interrelationships.

used to obtain quantitative data regarding the relative importance of various reactions within the sediment and indicate groups of microbes which are important within the sediment. For example Jørgensen (1982) demonstrated that anaerobic sulphate-reduction in inshore marine sediments was the dominant pathway of organic matter degradation, whilst in freshwater sediments, where sulphate concentrations are low, aerobic respiration, nitrate respiration and methanogenesis tend to be more important than sulphate-reduction (Jones & Simon, 1981; Lovley & Klug, 1983).

In addition to this advance in field microbial ecology, there has been a significant increase in the isolation of bacteria, especially anaerobic bacteria, with new metabolic reactions which have often provided support for reactions only previously demonstrated by field experiments or have indicated new reactions that might take place *in situ*. An important example of this is the work of Widdel and Pfennig (Widdel, 1980; Widdel & Pfennig, 1981, 1982; Widdel, Kohring & Mayer, 1983) who have isolated a large number of new types of sulphate-reducing bacteria, including *Desulfobacter* sp. which is able to catabolise acetate, an important metabolic intermediate within anaerobic systems (Fig. 1). There has been a similar increase in the isolation of fermentative and acetogenic bacteria which supply the low molecular weight substrates for the terminal oxidisers, e.g. sulphate-reducing and methanogenic bacteria (Fig. 1) and studies with these bacteria have indicated how tightly coupled the reactions between these organisms may be *in situ*. The types of association between fermenters and terminal oxidisers range from an increase in energy conservation by the fermenter if reaction products are kept at low concentrations by the oxidiser (Thauer & Morris, 1984) in particular the concentration of H_2 (Tewes & Thauer, 1980), to obligately syntrophic relationships between proton-reducing bacteria and hydrogen-utilising anaerobes (e.g. McInerney *et al.*, 1981; Stieb & Schink, 1985). The isolation of these new organisms with potentially very close interactions with other bacteria poses the question of how can we realistically identify these interacting microorganisms *in situ*, in order to elucidate their environmental significance?

The *in situ* identification of microorganisms has lagged behind other advances in microbial ecology and to a certain extent the present interest in types and rates of microbially mediated processes, may reflect the problems of identifying bacteria *in situ*. There are a large number of techniques available to identify microorganisms in the environment and these have been extensively reviewed, both in general (Rheinheimer, 1977; Costerton & Colwell, 1979; Jones, 1979; Van Es & Meyer-Reil, 1982; Parkes, 1982; Fry, 1982) and in respect to sediments (Collins, 1977; Litchfield & Seyfried, 1979; Parkes & Taylor, 1985). Although there are techniques available for the direct *in situ* identification of microorganisms such as the fluorescent antibody technique and autoradiography (which is substrate specific rather than organism specific), they have not had widespread use due to limited sensitivity and other problems (Parkes & Taylor, 1985). Viable count techniques provide an indirect measure of microorganisms *in situ*, as they rely on the growth of organisms in selective media, either solid to provide a colony count or in liquid to indicate positive growth in a Most Probable Number estimate (MPN). Viable count media by necessity have to be relatively rich so as to allow sufficient biomass production for a visual indication of growth, e.g. colony formation or turbidity. This will select for opportunistic bacteria and thus may result in a distorted representation of the *in situ* population. Although the viable count technique underestimates severely the total bacterial propulation (they represent only 0.0001 to 10% of the direct total bacterial count) (Jannasch & Jones, 1959; Hoppe, 1976; Meyer-Reil, 1977), it remains for a large number of organisms the only means of assessing their *in situ* population. Sediments are particularly difficult environments to assess microbial population *in situ*, as the population is heterogeneous, closely interacting and associated with films on sediment particles, which may be difficult to remove quantitatively. It has been estimated that even with homogenisation only about 60% of bacteria are removed from sediment particles (Moriarty, 1980; Newell & Fallon, 1982) and this may account for some of the underestimate when the viable count procedure is used with sediments. Similarly biomass estimates based on the commonly used total direct count procedures, e.g. acridine orange stain plus epifluorescence microscopy, will underestimate the total bacterial biomass because of the bacteria remaining on the sediment. There is therefore an obvious need for the development of new techniques for the identification and quantification of microbial populations *in situ* within sediments which overcome the limitations of the more traditional procedures.

BIOMARKERS

The use of microbial biomarkers which can be analysed directly from intact sediments seems a promising procedure for the estimation of *in situ* microbial populations, as it overcomes many of the limitations of the enrichment procedures for population analysis and due to modern analytical equipment can be an extremely sensitive procedure. Microbial biomarkers are chemical components of microorganisms which can be analysed directly from the environment and be interpreted both quantitatively and qualitatively in terms of *in situ* microbial biomass. Membrane lipids and their associated

fatty acids are particularly useful biomarkers as they are essential
components of every living cell and have great structural diversity
coupled with high biological specificity. Using these compounds to
identify microorganisms *in situ* is particularly appealing as the same
compounds are used extensively in bacterial taxonomy (Shaw, 1974;
Lechevalier, 1977; Keddie & Bousfield, 1980; Lechevalier, 1982;
Bousfield *et al.*, 1983) and hence there is a constantly growing data
base which can be used to interpret biomarkers. In addition the
biomarker approach can be applied with equal ease to both shallow-
water and deep-sea sediments as there are no pressure effects, in
contrast to techniques based on enrichment and growth.

The use of lipid markers is not just limited to microorganisms
but can be applied much more extensively. Organic geochemists
for example use lipid markers to characterise sedimentary environments in terms of type of input to the sediment, e.g. marine or
terrestrial, the contribution of the bacterial biomass to the total lipids
and prevailing diagenetic conditions, e.g. aerobic or anaerobic,
(Brassel & Eglinton, 1984). Also specific marker lipids if they are
incorporated into or influence the biochemical composition of the
predator can be used to follow predator–prey relationships within
a food chain.

The very chemical basis that makes the biomarker approach so
attractive also makes it difficult for non-chemists (who just want
to use it as another tool to study microbial ecology) to get into
as there are new terms, e.g. lipid classes, chemical nomenclature,
new techniques, e.g. extraction, derivatisation, separation and possibly new analytical equipment, e.g. capillary gas chromatography,
GC including GC mass spectrometry (GCMS) to come to terms
with. Such problems should be alleviated by collaboration between
chemists and microbiologists. This will be essential if this exciting
area is to fulfil its true promise. It is beyond the scope of this chapter
to cover the biochemistry of lipids and their distribution within microbes, which is the basis for the biomarker approach, or to provide
exact experimental details but such details can be found in the following references: Kates, 1972; Lechevalier, 1982; White, 1983; Harwood & Russel, 1984.

Nomenclature

A short-hand nomenclature for lipid fatty acids is used which is
in the form of numbers separated by a colon. The number before

the colon indicates the carbon chain length and the figure after corresponds to the number of double bonds. The position of the double bond is defined by the number of carbon units from the carboxyl end of the chain by the symbol $\triangle$ followed by a number, or 'ω' followed by the number of carbons from the methyl end. The geometry of the double bond is indicated by 'c' for cis and 't' for trans. The prefix 'i', 'a', and 'br' refer to iso, anteiso and methyl-branching of unconfirmed position respectively. Other methyl-branching is indicated in terms of its position from the carboxylic end of the chain, e.g. 10 Me 16 : 0. Methoxy fatty acids are represented by MeO. Cyclopropyl fatty acids are indicated by the symbol ∇.

Experimental procedures

Although there are many different procedures for the analysis of lipids from environmental samples or laboratory cultures (harvested by centrifugation), the first stage in the analysis is the efficient extraction of the lipid from other components. This is normally achieved with a single phase chloroform : methanol : water extraction (Bligh & Dyer, 1959) coupled with either homogenisation or sonication to improve extraction efficiency. At this stage an internal standard can be added to enable lipid fatty acids to be accurately quantified irrespective of possible small losses of fatty acids on subsequent handling. It is often prudent to split the sample in half, with the internal standard only being added to half the sample in case the compound used as internal standard is already in the sample. A useful internal standard is C19 : 0 as usually it is either not present or present only in small concentrations. The extraction is allowed to proceed for several hours and then the sediment removed by centrifugation. Water is added to produce a two-phase solution which is allowed to separate overnight, the lipids are in the organic phase relatively free of water-soluble contaminants. Either the fatty acids of the total lipid can then be released, methylated, and purified by preparative TLC (thin layer chromatography), prior to analysis by capillary GC (Taylor & Parkes, 1983; Taylor & Parkes, 1985); or the total lipid separated into a different lipid classes by silicic acid column chromatography (neutral lipid, glycolipid, and phospholipid) and each lipid class analysed separately (King, White & Taylor, 1977). However for bacteria most of the fatty acids will be in the phospholipid fraction (Harwood & Russell, 1984). An alternative procedure for the release and derivatization of the fatty acids to that outlined above is mild alkaline methanolysis (White *et al.*, 1979*a*) which, due to the mild conditions, helps to preserve cyclopropyl fatty acids from possible acid-sensitive ring-opening reactions (Moss, Lambert & Merwin, 1974). This procedure also leaves plasmalogens, amides and ether linkages intact and therefore these compounds have to be analysed separately. D. C. White and his colleagues have developed a complete suite of analytical methods based on the lipid extraction procedure to determine community biomass and structure, metabolic activity and storage products (White, 1983; Fig. 2).

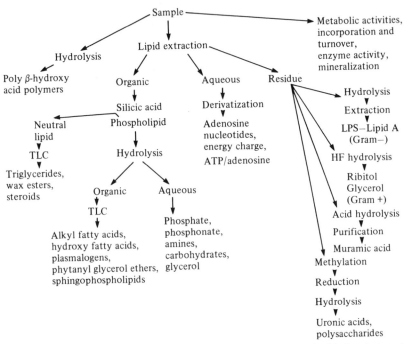

Fig. 2. The analytical scheme used by White and co-workers for the biochemical analysis of microbial consortia (White, 1983).

The more detailed the structural information obtained for the individual fatty acids (positional and geometrical isomers) the more useful the data will be and techniques are available to identify both the position and geometry of the double bond in unsaturated fatty acids (Anderson & Holman, 1974; Nichols, Shaw & Johns, 1985). In this context it may be important to consider that high performance liquid chromatography (HPLC) analysis of fatty acids may provide additional information to that obtained by GC due to improved separation (Bussel *et al.*, 1979). Another technique which has been explored recently by Nichols *et al.* (1985) is Fourier transform-infrared spectroscopy (FT-IR) which has the potential for non-destructive, *in situ* analysis of microbial communities, especially biofilms.

As it is often difficult or inconvenient to conduct lipid extraction in the field, samples for lipid analysis are normally stored frozen prior to analysis. There is some evidence, however, that this can result in significant losses in lipid phosphate (50%) and in many fatty acids including polyenoic fatty acids (Federle & White, 1982). These authors recommend either short term storage by refrigeration or long-term storage of sieved samples in formalin, although neither procedure is totally effective.

BIOMARKERS THAT CAN BE USED AS INDICATORS OF MICROBIAL POPULATIONS

Biomarkers fall into two broad categories, those which reflect either general biomass or the biomass of groups of organisms, and those that indicate the presence of specific organisms. These two categories are discussed separately.

General biomarkers

The major general biomarkers are outlined in Table 1. This does not include all biomarkers proposed for this category but only those which have general acceptance and supporting experimental evidence. Phospholipids are ubiquitous components of cellular membranes and as such provide a measure of total biomass of both eukaryotes and prokaryotes. They are not used as endogenous storage products and have a relatively rapid turnover in both living and killed cells added to the environment (King *et al.*, 1977; White *et al.*, 1979a), and therefore are a good measure of active biomass. The rate of turnover of phospholipids within sediments, as reflected by their half-life, was found to be 2 days under aerobic conditions and between 12 and 16 days under anaerobic conditions (White *et al.*, 1979a). The analysis of phospholipid is relatively straightforward and is based on the colorimetric determination of phosphate released from phospholipid (White *et al.*, 1979a). The sensitivity of the assay can be improved by concentrating the phospholipid in the extraction solvent prior to the release of inorganic PO_4. Using

Table 1. *General biomarkers of microbial biomass*

Biomarker	Measures	Reference
Phospholipid 16:0 Glycerol released from phospholipid br fatty acids Cyclopropyl fatty acids	Total biomass, eukaryotes and prokaryotes The biomass of some bacteria The biomass of some bacteria	White *et al.* (1979*a*) Harwood & Russel (1984) Gheron & White (1983) Leo & Parker (1966) Harwood & Russel (1984)
Lipopolysaccharide (LPS)	Gram negative bacteria	Rogers (1983)
Teichoic acid	Most Gram positive bacteria	Rogers (1983)
Muramic acid	Bacteria except Archaebacteria N.B. different amounts present in Gram positive and Gram negative bacteria	Rogers (1983)
Plasmalogens	Some anaerobic bacteria (including rumen bacteria)	Goldfine & Hagen (1972)
Hopanoids	Some bacteria	Rohmer, Bouvier-Nave & Ourisson (1984)
Phytanyl ether lipids	Archaebacteria	Harwood & Russel (1984)
Polyenoic fatty acids	Eukaryotes and gliding bacteria	Harwood & Russel (1984)
Phytopigments	Photoautotrophs	Gillan & Johns (1983)

this modification we have found that the recovery of various concentrations of bacteria added to a marine sediment was approximately 98% (unpublished data). Sensitivity can be improved further if the glycerol released from phospholipid is analysed by GC (Gehron & White, 1982). Palmitic acid (C16:0) is present in virtually all lipids and therefore can also be considered as a measure of total community biomass; however its accurate determination requires GC analysis.

The cell wall constituents peptidoglycan, lipopolysaccharide (LPS) and teichoic acid are unique to bacteria and therefore the analysis of these compounds *in situ* can provide a measure of their biomass. The muramic acid component of peptidoglycan has been used as a measure of bacterial biomass. Several different analytical procedures have been used: colorimetry (King & White, 1977), enzyme analysis (Moriarty, 1980), GC (Findlay, Moriarty & White, 1983) and HPLC (Moriarty, 1983; Mimura & Romano, 1985). A limitation of the technique for bacterial biomass estimation is that the content of muramic acid in Gram-positive bacteria is much higher and more variable than in gram-negative bacteria. Therefore, unless the proportion of the two types of bacteria in the sample is known, muramic acid analysis will provide an inaccurate estimate of bacterial biomass.

The proportion and biomass of Gram-positive and Gram-negative bacteria can be estimated chemically; LPS is specific for Gram-negative bacteria (including cyanobacteria) and teichoic acid is present in most Gram-positive bacteria. LPS can be measured using an extract of *Limulus* amoebocytes (Watson *et al.*, 1977; Jorgensen *et al.*, 1979; Melvaer & Fystro, 1982) or GC determination of hydroxy fatty acids released from the lipid A or LPS (Parker *et al.*, 1982). The latter technique is particularly useful as types of Gram-negative bacteria can often be distinguished by the patterns of their hydroxy fatty acids. Biomass estimates based on LPS are in good agreement with other estimates of bacterial biomass, such as ATP and direct bacterial counts (Watson & Hobbie, 1979), in environments where Gram-negative bacteria dominate, e.g. open ocean, mountain streams, and pristine rivers and lakes. However, in polluted inshore environments and sediments (Moriarty, 1980), significant concentrations of Gram-positive bacteria occur and hence the LPS assay will underestimate the total bacterial biomass. Gehron *et al.* (1984) have developed a sensitive assay for teichoic acid which has confirmed the work of Moriarty (1980) which showed that the proportion of Gram-positive bacteria increases with sediment depth.

Both branched and cyclopropyl fatty acids are found predominantly in bacteria (Harwood & Russel, 1984) and the presence of these fatty acids in the environment has been associated with bacterial activity (Leo & Parker, 1966; Cranwell, 1976; White *et al.*, 1980; Parkes & Taylor, 1983). However, there are a significant number of bacteria which do not contain these particular fatty acids (Lechevalier, 1982) and hence these compounds do not reflect total bacterial biomass. Some anaerobic bacteria and in particular Gram-positive fermentative bacteria found in the rumen contain relatively large amounts of plasmalogen lipids which have both ester and ether linkages (Harwood & Russel, 1984). They can account for between 2 and 10% of the phospholipids of soils and sediments and with soil depth there is a general increase in the proportion of plasmalogen phospholipid (White, 1985). Plasmalogens seem to be absent from the anaerobic sulphate-reducing bacteria (Taylor & Parkes, 1983; Edlund *et al.*, 1985; Dowling, Widdel & White, 1986), although preliminary results from our laboratory suggest that they may be present in some acetogenic bacteria (unpublished data).

The presence of hopanoids within sediments has been used to detect microorganisms (Brassell & Eglinton, 1984). A recent study, including some 100 strains of prokaryotes, confirmed the presence of this compound in about half the organisms studied, representing diverse taxonomic groups (Rohmer, Bouvier-Nave & Ourisson, 1984), however these compounds are not unique to bacteria.

Archaebacteria (methanogens, extreme halophiles, some thermoacidophiles and thermoalkalophiles), differ from all the bacteria previously discussed in that they contain exclusively phytanyl ether lipids rather than ester lipids (Harwood & Russel, 1984). As other members of the archaebacteria are unlikely to be present in significant amounts within sediments, these compounds may be ideal markers for methanogens. These lipids can be extracted from bacteria within sediments and assayed quantitatively by HPLC (Martz, Sebacher & White, 1983; Mancuso, Nichols & White, 1986), and their distribution within sediments seems to correspond with both methane fluxes from the sediment and methane concentrations in the pore water (Martz, Sebacher & White, 1983). It may be possible to differentiate between different groups of methanogenic bacteria by examination of their ether lipids.

A significant difference between bacteria and most other organisms is that bacteria do not usually have polyenoic fatty acids (more than one double bond in the fatty acid) and hence absence of these compounds within a sediment indicates a biomass dominated by bacteria. A notable exception are the gliding bacteria which all have very unusual fatty acids (Harwood & Russel, 1984), including polyenoic fatty acids, e.g. Johns & Perry (1977). These unusual fatty acids may be an adaptation to provide membrane flexibility necessary for their gliding locomotion. Phytopigments can be used to indicate the source of any phytobiomass reaching the sediment as well as to characterise the microflora within the sediment (Gillan & Johns, 1983), including photosynthetic bacteria.

Specific biomarkers

Despite the ubiquitous distribution of fatty acids within living organisms there are some fatty acids considered to be relatively specific to individual genera or species. Some of these are presented in Table 2. This list is rather conservative but even so the individual biomarkers have different degrees of specificity. The number of specific biomarkers is growing steadily as new isolates and their individual

Table 2. *Specific bacterial markers*

Bacteria	Characteristic fatty acid	Reference
Sulphate-reducing		
Desulfovibrio sp. except *D. gigas*	i 17 : 1*w*7	Taylor & Parkes (1983) Edlund *et al.* (1985)
Desulfobacter sp.	10 Me 16 : 0	Taylor & Parkes (1983) Dowling, Widell & White (1986)
Desulfobulbus sp.	17 : 1*w*6	Taylor & Parkes (1983) Taylor & Parkes (1985) Parkes & Calder (1985)
Bacillus sp.	br fatty acids, usually unsaturated	Kaneda (1977)
Methane-oxidising:		
Type 1	C16 monoenoic fatty acids	Nichols *et al.* (1985*a*)
Type 2	C18 monoenoic fatty acids	
Type 1 *Methylomonas* sp.	16 : 1*w*8c, 16 : 1*w*8t, 16 : 1*w*7t 16 : 1*w*5c and 16 : 1*w*5t	
Type 2 *Methylosinus trichosporium*	18 : 1*w*8c, 18 : 1*w*8t, 18 : 1*w*7c and 18 : 1*w*6c	
Thiobacillus sp.	10- 11-Me18 : 1*w*6, ▽19 : 0 (8,9) 10- 11-MeO18 : 0, 12- 13-MeO20 : 0, 20-OH▽16 : 0, 2-OH▽18 : 0, 11-OH & 13-OH19 : 0	Kerger *et al.* (1986)
Chemotypes	Nine different types of unspecified bacteria defined by specific fatty acids	Gillan & Hogg (1984)
Clostridia	▽15 : 1	Chan, Himes & Akagi (1971)

fatty acids are studied in more detail. Ideally a specific biomarker should be both unique to an individual genus or species and also represent a high proportion of the total fatty acids so that low concentrations of the bacteria can be detected by the analysis of its biomarker. The anaerobic sulphate-reducing bacteria (Table 2) contain some relatively unique fatty acids which are present in high (*Desulfovibrio*, *Desulfobulbus*) or significant concentrations (*Desulfobacter*), (Taylor & Parkes, 1983). Recent work has indicated that the biomarkers proposed originally by Taylor & Parkes, (1983) for individual species of sulphate-reducing bacteria can also be applied to the different genera of sulphate-reducing bacteria (Table 2,

Edlund *et al.*, 1985; Parkes & Calder, 1985; Dowling, Widdel & White, 1986). All *Bacillus* spp. contain major amounts of branched-chain plus small amounts of saturated fatty acids (Kaneda, 1977), although branched-chain fatty acids are by no means confined to this group of bacteria (Lechevalier, 1982). Unsaturated cyclopropane fatty acids e.g. cyclopropane 15:1 (Chan, Himes & Akagi, 1971) in representative clostridia, including mesophiles, thermophiles and psychrophiles may be potential biomarkers for these bacteria *in situ*. Recent work has shown the presence of some unusual fatty acids in both methane-oxidising bacteria and some *Thiobacillus* spp. which may be useful biomarkers for these organisms in environmental samples (Nichols *et al.*, 1985a; Kerger *et al.*, 1986). Finally Gillan & Hogg (1984) proposed nine chemotypes which should include all bacteria. However the individual fatty acids in several of these chemotypes occur within the same bacteria which complicates interpretation (Lechevalier, 1982; Parkes & Taylor, 1983).

EXAMPLES OF THE USE OF BIOMARKERS TO CHARACTERISE THE SEDIMENT POPULATION *IN SITU*

Direct sediment analysis

Several workers have analysed sedimentary lipid fatty acids and by comparison to either the fatty acids of bacterial cultures obtained from the same sediment or to fatty acids thought to be unique to bacteria (Table 1) have estimated the bacterial input to sediments (Johns, Perry & Jackson, 1977; Perry *et al.*, 1979; Van Vleet & Quinn, 1979; Gillan *et al.*, 1981; Parkes & Taylor, 1983; Federle *et al.*, 1983a; Gillan & Hogg, 1984). All of these studies show that bacteria make a significant contribution to the total biomass within the sediment and often represent the largest proportion of the biomass. The lipid fatty acids of the pure and mixed bacterial isolates obtained directly from the sediment in these studies were similar to the fatty acids generally considered to be indicators for bacteria. These include iso- and anteiso-branched chain fatty acids, 10 Me16:0, cyclopropyl 17:0 and 19:0 acids, 18:1 $w7$, the 15:1, 17:1 $w6$ and $w8$ isomers (especially when these occur in pairs), certain monoenoic branched chain fatty acids and possibly some trans fatty acids. Although the individual fatty acids are characteristic of bacteria they are not unique to these organisms (Perry, Volkman & Johns, 1979) and hence it is the combination of these fatty acids which can be considered as a general marker for bacteria within sediments. The relative proportions of these markers can also be used to analyse community structure (Parkes & Taylor, 1983; Gillan & Hogg, 1984) as they have different distributions within different bacterial types. An important aspect of this approach is the requirement to obtain bacteria representative of those within the sediment, cultured under appropriate conditions, so that the resultant fatty acids are truly representative of the bacteria or group of bacteria. These problems will be discussed in more detail later.

The study of deep-sea sediments by the biomarker approach seems to be particularly appropriate as more conventional methods of bacterial analysis require growth or metabolism of radiolabelled compounds, which is difficult to conduct under the appropriate conditions of temperature and pressure that exist within the deep sea (Jannasch & Taylor, 1984). The biomass and community structure of deep-sea sediments have been studied using the biomarker approach (Baird & White, 1985). The results of this study showed that the total sediment biomasses (as measured by the concentration of C16:0) at three Venezuelan Basin sites (water depth 3500–8400 m) were all low, and dominated by bacteria. Fatty acids believed to be markers for anaerobic organisms were also present in these sediments. In contrast, a high-energy site in the North Atlantic (HEBBLE site, 4800 m) and an estuarine sediment had much higher biomasses, 5 and 35 times respectively those of the Venezuela Basin, although these sediments seemed to be less dominated by bacteria. The estimate of bacterial concentrations for the basin sites agreed within an order of magnitude with direct bacterial counts (acridine orange stain plus epifluorescence microscopy) obtained in an earlier study. Other studies have also shown general agreement between estimates of biomass from biomarkers and measurements, such as ATP (White et al., 1979b) and oxygen uptake (White et al., 1979c), although there is a need for more comparisons.

Manipulation of environmental samples

Direct manipulation of natural microbial communities to enhance a specific group of microorganisms, followed by analysis of the resulting lipid fatty acids seems a realistic way of trying to overcome some of the problems of interpreting *in situ* distributions of biomarkers by reference to the fatty acids of laboratory grown cultures of microorganisms. Bobbie & White (1980) used this technique to stimulate the development of either prokaryotes or fungi in natural populations of marine organisms attached to Teflon squares. Fungal development was stimulated by maintaining the pH at 5.5, adding daily supplements of sucrose, nutrient broth, streptomycin and penicillin. Bacterial growth was stimulated by maintaining the pH at 7.8, and supplementing daily with NaH_2PO_4, glutamate and cyclohexylimide. Scanning electron micrographs confirmed that the desired populations had been obtained and lipid fatty acid analysis demonstrated the expected differences between the two communities. The bacterial community was enriched in branch-chain, cyclopropyl 17:0 and 18:1 $w7$ fatty acids whilst the fungus community was enriched in 18:2, and the 18 and 20 polyenoic fatty acids.

Taylor & Parkes (1985) used anaerobic sediment slurries supplemented with growth substrates for specific types of sulphate-reducing bacteria, to determine whether biomarkers previously found in pure cultures of these organisms (Taylor & Parkes, 1983) could be applied to natural sediments. They found that the biomarker for *Desulfobulbus* (Table 2), a propionate and H_2 utiliser was stimulated under conditions of propionate and H_2 utilisation via sulphate-reduction. Surprisingly the biomarker for the well-studied *Desulfovibrio* (Table 2) was not stimulated. This in part was due to the added lactate being partly fermented to substrates that *Desulfovibrio* could not use, but these results do tend to cast doubt on the environmental importance of this bacteria. Similar experiments with estuarine sediments have shown some increase in the biomarker for *Desulfovibrio* following lactate addition but again the *Desulfobulbus* biomarker increased the most (unpublished observations). Taylor & Parkes (1985) also concluded that *Desulfobacter* seemed to be the main acetate-utilising, sulphate-reducing bacteria, although this was based on indirect evidence as 10Me16:0 (Table 2) did not increase even though there was active sulphate-reduction at the expense of acetate. This work demonstrates the potential of lipid biomarkers for studying competition between sediment microbes under realistic incubation conditions.

Manipulation of mixed microbial communities under controlled laboratory conditions will enable any changes in the lipid fatty acids of the community to be related to the change in growth conditions, and this information may assist in the interpretation of lipid fatty acids within the environment. This approach was used by Parkes & Taylor (1983). In an attempt to obtain a bacterial community characteristic of the sediment environment, these authors used a multiple chemostat system to enrich bacterial communities at discrete stages along a redox gradient (Parkes & Senior, 1986). The bacterial community in the first vessel of this system was then subjected to different growth conditions within a single stage chemostat so as to select for three distinct types of communities: aerobes, facultative aerobes and facultative anaerobes. The lipid fatty acids of these communities were compared with the mixed culture of anaerobic sulphate-reducing bacteria which developed in the end vessel of the multiple chemostat system. All four communities could be distinguished in terms of their fatty acids and one of the most marked differences was the presence of significant amounts of cyclopropyl fatty acids in the aerobic communities but their absence in the anaerobic communities. These results suggest that cyclopropyl fatty acids are characteristic of aerobic sediment bacteria and this was supported by the presence of the highest concentrations of cyclopropyl fatty acids in the aerobic surface layers of the sediment, and their decrease with depth as anaerobic conditions prevailed.

A similar experiment by Guckert et al., (1985), but using batch growth to select bacterial communities of different respiratory types, found very different results to those of Parkes & Taylor (1983). These authors found anaerobic communities were enriched in cyclopropyl fatty acids, although in a repeat experiment under slightly different conditions aerobic communities were found to have slightly higher concentrations of cyclopropyl fatty acids than anaerobic communities. The different culture techniques used in these two studies may account for some of the differences in the results, as batch growth used by Guckert et al. (1985) combined with the longer incubation times used in the first experiment for anaerobes, may have favoured the development of cyclopropyl fatty acids as a result of an extended stationary phase, which is known to result in the accumulation of cyclopropyl fatty acids (Law, Zalkin & Kaneshiro, 1963). These effects would have been absent in the chemostat grown cultures of Parkes & Taylor (1983) as growth rate would have been constant. What ever the reason for the differences, these results demonstate how variable microbial responses can be even in relatively simple experiments. They also provide an indication of the potential dangers of extrapolating from laboratory experiments to the environment where conditions are very heterogeneous and differ greatly from laboratory cultures (see also Nedwell & Gray, this volume). Despite these problems, manipulation of environmental samples does seem a promising approach not only to 'calibrate' lipid markers but to gain further insight into the subtle microbial interactions that occur in the environment. There is an obvious requirement for more subtle manipulation if the results are to represent anything close to conditions *in situ* but this should be possible with a combination of labelled isotopes (both ^{14}C and ^{13}C), and the improvements that are taking place in lipid analysis.

Radiotracer incorporation into biomarkers

An attractive extension of the biomarker approach is to measure the rate of incorporation of labelled substrates into the various lipid components which could provide a measurement of microbial activity. By judicious selection of labelled substrate and lipid fraction to be analysed, a spectrum of microbial activities can be measured comparable to the spectrum of microbial biomasses that can be estimated using biomarkers (Tables 1 & 2). Using ^{32}P to measure total biomass activity and ^{14}C-acetate to measure bacterial activity, White *et al.* (1977) found good agreement between these methods and more conventional methods when they were used to assess the colonisation of oak leaves within an estuary.

A method for measuring ^{32}PO$_4$ incorporation into phospholipids has recently been developed to estimate bacterial productivity within sediments (Moriarty, White & Wassenberg, 1985). This technique gave bacterial productivity estimates that were equivalent to those obtained from rates of ^{3}H-thymidine incorporation into DNA. The advantages of this technique for sediments compared to the ^{3}H-thymidine procedure are (1) it can measure the productivity of anaerobic sulphate-reducing bacteria, which is not measured by ^{3}H-thymidine, as these organisms do not incorporate thymidine into their DNA (Moriarty, White & Wassenberg, 1985), and (2) the same technique can be used to measure the productivity of other groups of organisms e.g. ^{35}SO$_4$ incorporation into sulpholipids to measure eukaryotic activity (Moriarty, White & Wassenberg, 1985).

Sediment disturbance, macro-benthic feeding and nutritional status

Phospholipid synthesis seems to be enhanced rapidly by sediment disturbance whilst ^{3}H-thymidine incorporation into DNA is not immediately effected (Moriarty, White & Wassenberg, 1985). This provides a basis for estimating disturbance effects the addition of tracers to sediments may have on microbial rates of activity *in situ*. A similar estimate of sediment disturbance can be obtained from the relative rates of incorporation of ^{14}C-acetate into phospholipid fatty acids (PLFA) and endogenous storage lipid, poly-beta-hydroxy-alkanoate (PHA) (Findlay *et al.*, 1985). PLFA synthesis measures cellular growth and PHA synthesis measures carbon accumulation (unbalanced growth). Findlay *et al.* (1985) found that sediment disturbance prior to the introduction of the ^{14}C-acetate resulted in an increase in PLFA synthesis while PHA synthesis remained the same or was depressed, and therefore the ratio of PLFA/PHA could be used as measure of sediment disturbance. These authors found that disturbing the sediment with a garden rake 30 min before analysis, resulted in an increased PLFA/PHA ratio but only when the ^{14}C-acetate was injected into the sediment. When the label was introduced by the slurry technique the effect of raking the sediment surface could not be detected, as slurrying the sediment caused a bigger disturbance to the microbial community. Bioturbation caused by sand dollar feeding in an estuarine sediment could also be detected in an increased PLFA/PHA ratio (Findlay *et al.*, 1985).

Lipid fatty acid analysis has been used to demonstrate the significant effect epibenthic predators have on the microbial community structure within sediments (Federle *et al.*, 1983*b*). In this work, different population densities of predators were obtained by using cages either to include or exclude predators and then the microbial community of these areas, as determined by lipid fatty acid analysis, were compared with uncaged areas. Lipid fatty acid analysis has also shown that amphipod grazing causes an increase in microbial biomass and activity, and demonstrates selective feeding by amphipods with different mouthparts (White, 1983).

The use of the PLFA/PHA ratio of rate of synthesis to indicate sediment disturbance has already been discussed (Findlay *et al.*, 1985). The relative concentration of PHA can also be used directly to indicate the recent nutritional status of the prokaryotic community. In a detrital microbiota with a proper balance of carbon, nitrogen and phosphorus, a rapid turnover of PHA was found which coincided with increased microbial biomass; however during unbalanced growth PHA was formed rapidly (Nickels, King & White, 1979). In addition recent work has suggested that the trans/cis ratio of monoenoic fatty acids may be used as a 'starvation' or 'stress' index for determining the nutritional status of bacteria and as a

consequence address the question of bacterial dormancy in natural
aquatic environments (Guckert, Hood & White, 1986).

Microbial interactions in the laboratory

It has been emphasised that most microbial activity in the environment results from the action of a mixture of interacting organisms
(Bull, 1980) and hence laboratory experiments should reflect this
microbial heterogeneity if they are to relate to the natural environment. The study of such mixed microbial populations in the laboratory is difficult as microorganisms are difficult to identify either
directly by microscopy or indirectly by culture. To overcome this
problem, mixed cultures limited to a few bacteria that can be readily
identified are used (e.g. Laanbroek et al., 1984). The use of lipid
biomarkers to differentiate and quantify the microbial community
may allow much less limited and more realistic microbial communities to be used in laboratory studies. In our laboratory, we have
used this technique to follow the bacterial changes that take place
during the selective enrichment of sulphate-reducing bacteria from
marine sediments (unpublished observations). It has been noted that
the enrichment of sulphate-reducing bacteria on a medium containing lactate results in the isolation of bacteria belonging to the genera
Desulfovibrio (Jørgensen, 1977; Laanbroek & Pfennig, 1981), identification being based on cell morphology and substrate utilisation.
This observation was investigated using the biomarkers for sulphate-
reducing bacteria (Table 2). The ratio of br17.1/17.1 was used as
a measure of the proportions of *Desulfovibrio* and *Desulfobulbus*
respectively in the enrichment culture, as both of these sulphate-
reducing bacteria can use lactate as a growth substrate, and with
this substrate the biomarker represents a similar percentage of the
total fatty acids for each organism (Taylor & Parkes, 1983). After
7 days growth, there was significant sulphide production and the
ratio of br17.1/17.1 was 0.4, this ratio decreasing to 0.16 after 22
days. Therefore, during the enrichment the concentration of *Desulfovibrio* was significantly lower than that of *Desulfobulbus*. This
experiment was repeated and changes in substrate, bacterial numbers
and sulphide concentrations measured (Fig. 3). The lactate in the
media was rapidly fermented to propionate and acetate, substrates
that *Desulfovibrio* could not use. The propionate was subsequently
oxidised to acetate with sulphide being produced. It seems likely

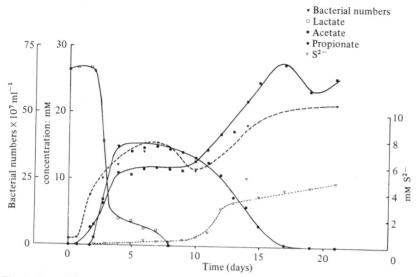

Fig. 3. Bacterial and chemical changes during the enrichment of sulphate-reducing bacteria on Postgate's B medium using Loch Eil sediment as an inoculum.

that this activity was due to *Desulfobulbus*, which can utilise propionate. The ratio of br17.1/17.1 after 14 days was 0.2. All these results are consistent with *Desulfobulbus* being the dominant sulphate-reducing bacteria during enrichment on lactate and especially during the latter stages of enrichment when viable counts are usually conducted. Results of similar experiments with a number of different sediments indicate that the relative proportions of *Desulfovibrio* and *Desulfobulbus* depends on whether the lactate is initially oxidised or fermented and therefore although lactate is a good substrate for enriching sulphate-reducing bacteria it does not result in the specific enrichment of *Desulfovibrio* spp.

Recent developments

Fourier transform-infrared (FT-IR) spectroscopy has been used for rapid and nondestructive analysis of bacteria, bacterial polymer mixtures, digester samples and microbial films (Nichols *et al.*, 1985*b*). A flow-through attachment allowed the development of a microbial film to be studied. In an attempt to reduce problems of loss of viability and possible metabolic changes due to successive subculture involved in the purification and growth of bacterial biomass for lipid fatty acid analysis, a procedure for fatty acid analysis of single bacterial colonies from the primary enrichment has been developed (Brondz

& Olsen, 1986). This type of approach may help to reduce the problems involved in obtaining realistic cultures for the identification of potential biomarkers in the environment.

PROBLEMS ASSOCIATED WITH THE USE OF BIOMARKERS TO CHARACTERISE SEDIMENT MICROBIAL COMMUNITIES

Biomarkers characteristic in situ of microbial populations

The successful use of lipid biomarkers to quantify and characterise microbial communities *in situ* relies to a large extent on the isolation of bacteria which are representative of those *in situ* and their subsequent growth under appropriate conditions to enable their characteristic lipids to be analysed. The percentage of the total direct bacterial count that is represented by viable bacterial counts is between 0.0001 and 10% (Jannasch & Jones, 1959; Hoppe, 1976; Meyer-Reil, 1977). This suggests that biomarkers may be used for a relatively small portion of the total bacterial population. However, another view of the relevance of laboratory cultures to the environment might be based on a comparison of the different physiological types of bacteria in the environment and in culture, as it may be that similar types of bacteria are present both in culture and in the environment but only a small proportion of each type is able to grow in the rich isolation media. On this basis, if we compare the types of processes thought to be mediated microbially in the environment with the different physiological types of bacteria capable of catalysing these environmental processes already in culture, then it seems that we do have a representative collection of bacterial isolates.

Irrespective of the above considerations the isolation of microorganisms more representative of those *in situ* will extend greatly the usefulness of the biomarker approach. The use of different enrichment techniques such as chemostats (Parkes, 1982), and gel-stabilised systems (Wimpenny, Lovitt & Coombs, 1983), together with the use of mixed microbial communities should assist in this aim. It should be noted, however, that the isolation of microorganisms is not a prerequisite for the use of biomarkers. The correlation of a particular microbial activity with the presence of a specific lipid within environmental samples and the manipulation of environmental samples to enhance a particular microbial activity coupled with biomarker analysis (previously discussed) are both ways of 'calibrating' biomarkers without the requirement for the isolation of microorganisms.

Although there are some bacteria which can be characterised by the presence of specific lipid fatty acids (Table 2), the fatty acid composition of other bacteria is not a stable characteristic and is rather susceptible to growth medium, temperature, aeration, or other cultural conditions as well as growth phase (Lechevalier, 1982). Therefore, even if representative microorganisms are isolated their growth under inappropriate conditions will cause problems in the use of their lipids as biomarkers. To reduce these problems, the following recommendations for growth of bacterial cultures for lipid fatty acid analysis are suggested:

(1) They should be grown on well-defined media in the absence of complex constituents such as yeast extract and trypticase as these often contain fatty acids that can be incorporated directly into bacterial fatty acids. If this is not possible then the constituent should be extracted with organic solvent to remove any fatty acids;

(2) As the growth substrate may have a profound influence on the cellular fatty acids of a bacteria (Cerniglia & Perry, 1975), bacteria should be grown on more than one substrate;

(3) Temperature often affects fatty acids in order to maintain cell fluidity (Kaneda, 1972; Lechevalier, 1982). To prevent temperature-related changes in fatty acids biasing bacterial fatty acid profiles and to make the comparison of bacteria with different temperature optimums more meaningful, bacteria should be grown at their temperature optimum;

(4) The growth status of cells analysed can be an important factor in the fatty acid profile of certain bacteria (Lechevalier, 1982). The greatest changes, if they occur, are usually during the exponential growth phase and hence early stationary phase cultures should be analysed. There can also be changes during the stationary phase, one of the most common being the conversion of monoenoic fatty acids into cyclopropyl fatty acids (Law, Zalkin & Kaneshiro, 1963). The growth of bacteria in chemostats at low dilution rates could overcome these problems. It may also be more appropriate than batch culture, facilitating comparison with fatty acids of bacteria in the environment, which are also considered to be slow growing.

Not all the bacterial fatty acid profiles quoted in the literature have been obtained under the appropriate growth conditions and therefore should be used with caution. Also when biomarkers are used to characterise microbial communities *in situ*, differences between the culture conditions under which bacterial markers were obtained and the growth conditions *in situ* should be considered. This is because factors which influence bacterial fatty acid composition in laboratory grown cultures will probably also affect fatty acid composition *in situ*, and if these factors are not considered incorrect conclusions may result. An example of the difficulties in interpreting lipid fatty acids in environmental samples is provided by the cyclopropyl fatty acids. Although these fatty acids are characteristic of some bacteria, the situation is complicated as they have also been related to stress or starvation in bacteria (Thomas & Batt, 1969; Guckert, Hood & White, 1986), limitation of growth by oxygen availability (Rogers, 1983), an indicator of aerobic sediment bacteria (Parkes & Taylor, 1983) and indicative of stationary phase cultures (Law, Zalkin & Kaneshiro, 1963). All of these may be correct for particular bacteria or a particular environment; the difficulty is knowing which bacteria and which environment when dealing with environmental samples!

Microbial marker studies are also essential to help explain the various inputs and diagenetic changes within sediments. For example the trans/cis ratio of fatty acids has been used by geochemists as an indication of diagenesis of the cis-monoenoic fatty acids and an increase in this ratio for the total lipid fatty acids with sediment depth has been attributed to preferential degradation of cis-isomers or clay-catalysed isomerisation to the trans-isomer (Van Vleet & Quinn, 1979). The demonstration of the presence of trans-fatty acids in bacteria (Gillan *et al.*, 1981) and that the proportion of trans-fatty acids increases in bacteria during starvation (Guckert, Hood & White, 1986) provides other possible explanations for these changes.

The rate of turnover of microbial markers within sediments

For biomarkers to measure successfully living biomass they must be degraded rapidly once the organism dies. If the biomarker is only degraded slowly after death then its *in situ* concentration will reflect cell debris as well as biomass and hence the *in situ* biomarker would not measure accurately living biomass. We have measured the rate of degradation of killed but intact ^{14}C-labelled bacterial fatty acids within marine sediments under both aerobic and anaerobic conditions (unpublished results). The results of these experiments show (Fig. 4), that the rate of degradation of bacterial fatty acids

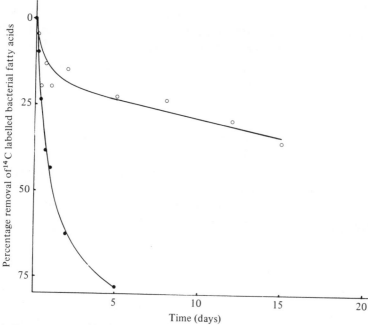

Fig. 4. Degradation of ^{14}C-labelled fatty acids in sediments from Loch Etive (E9) under aerobic and anaerobic conditions at 25 °C. ○ Anaerobic; ● Aerobic.

was rapid under aerobic conditions with over 75% of the added material being degraded within 5 days. The pattern of degradation under anaerobic conditions was, however, quite different with two distinct rates. An initial rapid rate was followed after about 2 days by a much slower rate (Fig. 4). The half-life for the bacterial fatty acids under aerobic conditions was 0.8 days but under anaerobic conditions was 24 days. These results were similar to those reported for the degradation of phospholipids by White et al., (1979a), and indicate that bacterial fatty acids in aerobic sediments should be a good measurement of living biomass but in anaerobic sediments lipid fatty acid measurements may include a certain amount of cell debris. Recently, Harvey, Fallon & Patton (1986) have shown that in addition to aerobic and anaerobic conditions, lipid degradation is influenced by the type of lipid and the organic content of the sediment. Ether lipids of a methanogen were degraded very slowly under anaerobic conditions, which questions their use as effective measures for methanogenic biomass under anaerobic conditions.

Another factor that can create difficulties in both the quantitative and qualitative interpretation of lipid fatty acid markers within sediments is the physiological state of the cells. The physiological state of bacteria *in situ* may be very different from the bacteria grown in the laboratory to 'calibrate' the microbial markers. Starved bacterial cells lose phospholipid rapidly (Oliver & Stringer, 1984; Guckert, Hood & White, 1986) and hence the conversion of phospholipid concentration to bacterial numbers may not be appropriate. These bacteria also lose fatty acids rapidly, but there is a relative increase in cyclopropyl fatty acids and it has been postulated that the cellular membrane in this form may be more stable to turnover and degradation (Gucket, Hood & White, 1986). The problem is that the biomarkers of these bacteria can now be measured but how do we interpret the data in terms of microbial biomass? A similar situation exists with muramic acid, as Moriarty (1980) estimated that about 40% of muramic acid in sediments at 20 cm depth might be present in empty cell wall sacs.

CONCLUSIONS

Microbial biomarkers are an important additional tool for the microbial ecologist, which in combination with microbial activity measurements will allow our knowledge of the microbe and its environment to be advanced significantly. The technique is particularly useful in sediments where the characteristic heterogeneity of sediments (such as the presence of mixed microbial communities, microbial films on sediment particles, and heterogeneous physical and chemical conditions e.g. aerobic and anaerobic conditions) does not present problems for the effective analysis of microbial biomass. There are problems in the interpretation of lipid fatty acids in terms of characterising the community *in situ* but progress is being made as more environmentally relevant bacteria are analysed, more detailed structural analysis of lipids is becoming possible and new approaches to the 'calibration' of biomarkers are being developed. The biomarker approach offers the potential of a uniquely detailed picture of microbial interaction which does not seem possible with other methods.

A particularly useful aspect of the biomarker approach is that labelled isotopes can be readily incorporated into the procedure which provides measurements of microbial activity with the same fine detail as possible for biomass determination. In addition, the incorporation of labelled substrates into different lipid components provides the potential for investigating the physiological state of sediment microbes and ultimately unravel the different strategies for life within the sediment.

The majority of this work was conducted in collaboration with Dr Jim Taylor who introduced me to an exciting area of research that would have been difficult to enter without his guidance and encouragement. Will Buckingham provided excellent technical assistance throughout and also helped in the production of this manuscript. Cameron Young assisted in the experiments to measure the degradation rates of bacterial fatty acids.

REFERENCES

ANDERSON, B. A. & HOLMAN, R. T. (1974). Pyrrolidides for mass spectrometric determination of the position of the double bond in monounsaturated fatty acids. *Lipids*, **9**. 185–90.

BAIRD, B. H. & WHITE, D. C. (1985). Biomass and community structure of the abyssal microbiota determined from the ester-linked phospholipids recovered from Venezuela basin and Puerto Rico trench sediments. *Marine Geology*, **68**, 217–31.

BLIGH, E. G. & DYER, W. J. (1959). A rapid method of total lipid extraction and purification. *Canadian Journal of Biochemistry and Physiology*, **37**, 911–17.

BOBBIE, R. J. & WHITE, D. C. (1980). Characterization of benthic microbial community structure by high-resolution gas chromatography of fatty acid methyl esters. *Applied and Environmental Microbiology*, **39**, 1212–22.

BOUSFIELD, I. J., SMITH, G. L., DANDO, T. R. & HOBBS, G. (1983). Numerical analysis of total fatty acid profiles in the identification of coryneform, nocardioform and some other bacteria. *Journal of General Microbiology*, **129**, 375–94.

BRASSELL, S. C. & EGLINTON, G. (1984). Lipid indicators of microbial activity in marine sediments. In *Heterotrophic Activity in the Sea*, ed. J. E. Hobbie & P. J. le B. Williams, pp. 481–503. New York: Plenum Press.

BRONDZ, I. & OLSEN, I. (1986). Chemotaxonomy at a crossroads? Gas chromatographic analyses of a single colony from the bacterium *Haemophillus aprophilus*. *Journal of Chromatography*, **374**, 119–24.

BULL, A. T. (1980). Biodegradation: some attitudes and strategies of microorganisms and microbiologists. In *Contemporary Microbial Ecology*, ed. D. C. Ellwood, J. N. Hedger, M. J. Latham, J. M. Lynch & J. H. Slater, pp. 105–36. London: Academic Press.

BUSSELL, N. E., MILLER, R. A., SETTERSTROM, J. A. & GROSS, A. (1979). High pressure liquid chromatography in the analysis of fatty acid composition of oral streptococci and its comparison to gas chromatography. In *Biological/Biomedical Applications of Liquid Chromatography*, ed. G. L. Hawk, pp. 57–89. New York: Marcel Dekker Inc.

CAPPENBERG, T. E. (1974). Interrelations between sulfate-reducing and methane-producing bacteria in bottom deposits of a fresh-water lake. II. Inhibition experiments. *Antonie Van Leeuwenhoek. Journal of Microbiology and Serology*, **40**, 297–306.

CAPPENBERG, T. E. & PRINS, R. A. (1974). Interrelations between sulfate-reducing and methane producing bacteria in bottom deposits of a fresh-water lake. III. Experiments with ¹⁴C-labelled substrates. *Antonie Van Leeuwenhoek. Journal of Microbiology and Serology*, **40**, 457–69.

CERNIGLIA, C. E. & PERRY, J. J. (1975). Metabolism of n. propylamine, isopropylamine, and 1,3-propane diamine by *Mycobacterium convolutum*. *Journal of Bacteriology*, **124**, 285–9.

CHAN, M., HIMES, R. H. & AKAGI, J. M. (1971). Fatty acid composition of thermophilic, mesophilic and psychrophilic clostridia. *Journal of Bacteriology*, **106**, 876–81.

172 R. J. PARKES

COLLINS, V. G. (1977). Methods in sediment microbiology. In *Advances in Aquatic Microbiology*, vol. 1, ed. M. R. Droop & H. W. Jannasch. London: Academic Press.

COSTERTON, J. W. & COLWELL, R. R. (1979). *Native Aquatic Bacteria: Enumeration, activity and ecology, ASTM STP 695*. Philadelphia: American Society for Testing and Materials.

CRANWELL, P. A. (1976). Decomposition of aquatic biota and sediment formation: organic compounds in detritus resulting from microbial attack on the alga *Ceratium hirundinella*. *Freshwater Biology*, **6**, 41–8.

DOWLING, N. J., WIDDEL, F. & WHITE, D. C. (1986). Comparison of the phospholipid ester-linked fatty acid biomarkers of acetate-oxidising sulphate-reducers and other sulphide-forming bacteria. *Journal of General Microbiology*, **132**, 1815–25.

EDLUND, A., NICHOLS, P. D. ROFFEY, R. & WHITE, D. C. (1985). Extractible and lipopolysaccharide fatty acid and hydroxy acid profiles from *Desulfovibrio* species. *Journal of Lipid Research*, **26**, 982–8.

FEDERLE, T. W. & WHITE, D. C. (1982). Preservation of estuarine sediments for lipid analysis of biomass and community structure of microbiota. *Applied and Environmental Microbiology*, **44**, 1166–9.

FEDERLE, T. W., HULLAR, M. A., LIVINGSTON, R. J., MEETER, D. A. & WHITE, D. C. (1983a). Spatial distribution of biochemical parameters indicating biomass and community composition of microbial assemblies in estuarine mud flat sediments. *Applied and Environmental Microbiology*, **45**, 58–63.

FEDERLE, T. W., LIVINGSTON, R. J., MEETER, D. A. & WHITE, D. C. (1983b). Modifications of estuarine sedimentary microbiota by exclusion of epibenthic predators. *Journal of Experimental Marine Biology and Ecology*, **73**, 81–94.

FINDLAY, R. H., MORIARTY, D. J. W. & WHITE, D. C. (1983). Improved method of determining muramic acid from environmental samples. *Geomicrobiology*, **3**, 135–50.

FINDLAY, R. H., POLLARD, P. C., MORIARTY, D. J. W. & WHITE, D. C. (1985). Quantitative determination of microbial activity and community nutritional status in estuarine sediments: evidence for a disturbance artifact. *Canadian Journal of Microbiology*, **31**, 493–8.

FRY, J. C. (1982). The analysis of microbial interactions and communities *in situ*. In *Microbial Interactions and Communities*, ed. A. T. Bull & J. H. Slater, pp. 103–52. London: Academic Press.

GEHRON, M. J. & WHITE, D. C. (1982). Quantitative determination of the nutritional status of detrital microbiota and the grazing fauna by triglyceride glycerol analysis. *Journal of Experimental Marine Biology and Ecology*, **64**, 145–58.

GEHRON, M. J. & WHITE, D. C. (1983). Sensitive assay of phospholipid glycerol in environmental samples. *Journal of Microbiological Methods*, **1**, 23–32.

GEHRON, M. J., DAVIS, J. D., SMITH, G. A. & WHITE, D. C. (1984). Determination of the gram-positive bacterial content of soils and sediments by analysis of teichoic acid components. *Journal of Microbiological Methods*, **2**, 165–76.

GILLAN, F. T. & JOHNS, R. B. (1983). Normal-phase HPLC analysis of microbial carotenoids and neutral lipids. *Journal of Chromatographic Science*, **21**, 34–8.

GILLAN, F. T. & HOGG, R. W. (1984). A method for the estimation of bacterial biomass and community structure in mangrove-associated sediments. *Journal of Microbiological Methods*, **2**, 275–93.

GILLAN, F. T., JOHNS, R. B., VERHEYEN, T. V., VOLKMAN, J. K. & BAVOR, H. J. (1981). Trans-monounsaturated acids in a marine bacterial isolate. *Applied and Environmental Microbiology*, **41**, 849–56.

GOLDFINE, H. & HAGEN, P. O. (1972). Bacterial plasmalogens. In *Ether Lipids: Chemistry and Biology*, ed. F. Snyder, pp. 329–50. New York: Academic Press.

ANALYSIS OF MICROBIAL COMMUNITIES   173

GUCKERT, J. B., ANTWORTH, C. P., NICHOLS, P. D. & WHITE, D. C. (1985). Phospholipid, ester-linked fatty acid profiles as reproducible assays for changes in prokaryotic community structure of estuarine sediments. *FEMS Microbiology Ecology*, **31**, 147–58.

GUCKERT, J. B., HOOD, M. A. & WHITE, D. C. (1986). Phospholipid, ester-linked fatty acid profile changes during nutrient deprivation of *Vibrio cholerae*: increase in the *trans/cis* ratio and proportions of cyclopropyl fatty acids. *Applied and Environmental Microbiology*, **151**, (in press).

HARWOOD, J. L. & RUSSELL, N. J. (1984). *Lipids in Plants and Microbes*. Boston, London & Sydney: George Allen & Unwin.

HARVEY, H. R., FALLON, R. D. & PATTON, J. S. (1986). The effect of organic matter and oxygen on the degradation of bacterial membrane lipids in marine sediments. *Geochimica et Cosmochimica Acta*, **50**, 795–804.

HOPPE, H. G. (1976). Determination and properties of actively metabolizing heterotrophic bacteria in the sea, investigated by means of micro-autoradiography. *Marine Biology*, **36**, 291–302.

JANNASCH, H. W. & JONES, G. E. (1959). Bacterial populations in seawater as determined by different methods of enumeration. *Limnology and Oceanography*, **4**, 128–39.

JANNASCH, H. W. & TAYLOR, G. D. (1984). Deep sea microbiology. *Annual Review of Microbiology*, **38**, 487–514.

JOHNS, R. B. & PERRY, G. J. (1977). Lipids of the marine bacterium *Flexibacter polymorphus*. *Archives of Microbiology*, **114**, 267–71.

JOHNS, R. B., PERRY, G. J. & JACKSON, K. S. (1977). Contribution of bacterial lipids to recent marine sediments. *Estuarine and Coastal Marine Science*, **5**, 521–9.

JONES, J. G. (1979). *A guide to methods for estimating microbial numbers and biomass in fresh water*. Kendal: Freshwater Biological Association, No. 39.

JONES, J. G. & SIMON, B. M. (1981). Differences in microbial decomposition processes in profundal and littoral lake sediments. *Journal of General Microbiology*, **123**, 297–312.

JØRGENSEN, B. B. (1977). The sulfur cycle of a coastal marine sediment (Limfjorden, Denmark). *Limnology and Oceanography*, **22**, 814–32.

JØRGENSEN, B. B. (1982). Mineralization of organic matter in the sea bed – the role of sulphate-reduction. *Nature*, **296**, 643–5.

JØRGENSEN, J. H., LEE, J. C., ALEXANDER, G. A. & WOLF, H. W. (1979). Comparison of *Limulus* assay, standard plate count, and total coliform count for microbiological assessment of renovated wastewater. *Applied and Environmental Microbiology*, **37**, 928–31.

KANEDA, T. (1972). Positional preference of fatty acids in phospholipids of *Bacillus cereus* and its relation to growth temperature. *Biochimica et Biophysica Acta*, **280**, 297–305.

KANEDA, T. (1977). Fatty acids of the genus *Bacillus*: an example of branched-chain preference. *Bacteriological reviews*, **41**, 391–418.

KATES, M. (1972). *Techniques of lipidology: isolation, analysis and identification of lipids*. Amsterdam & Oxford: North-Holland Publishing Company.

KEDDIE, R. M. & BOUSFIELD, I. J. (1980). Cell wall composition in the classification and identification of Coryneform bacteria. In *Microbiological Classification and Identification*, SAB Symposium Series No. 8, ed. M. Goodfellow & R. G. Board, pp. 167–88. London & New York: Academic Press.

KERGER, B. D., NICHOLS, P. D., ANTWORTH, C. P., SAND, W., BOCK, E., COX, J. C., LANGWORTHY, T. A. & WHITE, D. C. (1986). Signature fatty acids in the polar lipids of acid producing *Thiobacillus* sp.: Methoxy, cyclopropyl, alpha-hydroxy-cyclopropyl and branched and normal monoenoic fatty acids. *FEMS*

Microbiology Ecology, **32**, (in press).

KING, J. D. & WHITE, D. C. (1977). Muramic acid as a measure of microbial biomass in estuarine and marine samples. *Applied and Environmental Microbio logy*, **33**, 777–83.

KING, J. D., WHITE, D. C. & TAYLOR, C. W. (1977). Use of lipid composition and metabolism to examine structure and activity of estuarine detrital microflora *Applied and Environmental Microbiology*, **33**, 1177–83.

LAANBROEK, H. J. & PFENNIG, N. (1981). Oxidation of short-chain fatty acids by sulfate-reducing bacteria in freshwater and in marine sediments. *Archives of Mic robiology*, **128**, 330–5.

LAANBROEK, H. J., GEERLIGS, H. J., SIJTSMA, L. & VELDKAMP, H. (1984). Competi tion for sulfate and ethanol among *Desulfobacter, Desulfobulbus* and *Desulfovi brio* species isolated from intertidal sediments. *Applied and Environmental Microbiology*, **47**, 329–34.

LAW, J. H., ZALKIN, H. & KANESHIRO, T. (1963). Transmethylation reactions in bacterial lipids. *Biochimica et Biophysica Acta*, **70**, 143–51.

LECHEVALIER, M. P. (1977). Lipids in bacterial taxonomy – a taxonomists view *Critical Reviews in Microbiology*, **5**, 109–210.

LECHEVALIER, M. P. (1982). Lipids in bacterial taxonomy. In *Handbook of Micro biology*, 2nd edition, vol. 4, ed. A. I. Laskin & H. A. Lechevalier, pp. 435–541 Boca Raton, Florida: CRC Press.

LEO, R. G. & PARKER, P. L. (1966). Branched chain fatty acids in sediments *Science*, **152**, 649–50.

LITCHFIELD, C. D. & SEYFRIED, P. L. (1979). *Methodology for biomass determina tions and microbial activities in sediments, ASTM STP* 673. Philadelphia: Ameri can Society for Testing and Materials.

LOVLEY, D. R. & KLUG, M. J. (1983). Sulfate reducers can outcompete methanogens at freshwater sulfate concentrations. *Applied and Environmental Microbiology* **45**, 187–92.

MANCUSO, C. A., NICHOLS, P. D. & WHITE, D. C. (1986). A method for the sepa ration and characterization of archaebacterial signature ether lipids. *Journal of Lipid Research*, **27**, 49–56.

MARTZ, R. F., SEBACHER, D. I. & WHITE, D. C. (1983). Biomass measurement of methane forming bacteria in environmental samples. *Journal of Microbiologi cal Methods*, **1**, 53–61.

MCINERNEY, M. J., BRYANT, M. P., HESPELL, R. B. & COSTERTON, J. W. (1981). *Syntrophomonas wolfei* gen. nov., sp. nov., an anaerobic syntrophic fatty acid oxidizing bacterium. *Applied and Environmental Microbiology*, **41**, 1029–39.

MELVAER, K. L. & FYSTRO, D. (1982). Modified micromethod of the *Limulus* amoebocyte lysate assay for endotoxin. *Applied and Environmental Microbio logy*, **43**, 493–4.

MEYER-REIL, L.-A. (1977). Bacterial growth rates and biomass production. In *Mic robial Ecology of a Brackish Water Environment*, ed. G. Rheinheimer, pp. 223–36. Berlin, Heidelberg & New York: Springer-Verlag.

MIMURA, T. & ROMANO, J.-C. (1985). Muramic acid measurements for bacterial investigations in marine environments for high-pressure liquid chromatography. *Applied and Environmental Microbiology*, **50**, 229–37.

MORIARTY, D. J. W. (1980). Measurement of bacterial biomass in sandy sediments. In *Biogeochemistry of Ancient and Modern Environments*, ed. P. A. Trudinger, M. R. Walker & B. J. Ralph, pp. 131–8. Berlin: Australian Academy of Science, Canberra & Springer-Verlag.

MORIARTY, D. J. W. (1983). Measurements of muramic acid in marine sediments by high performance liquid chromatography. *Journal of Microbiological Methods*,

1, 111–17.

MORIARTY, D. J. W., WHITE, D. C. & WASSENBERG, T. J. (1985). A convenient method for measuring rates of phospholipid synthesis in seawater and sediments: its relevance to the determination of bacterial productivity and the disturbance artifacts introduced by measurements. *Journal of Microbiological Methods*, **3**, 321–30.

MOSS, C. W., LAMBERT, M. A. & MERWIN, W. H. (1974). Comparison of rapid methods for analysis of bacterial fatty acids. *Applied and Environmental Microbiology*, **28**, 80–5.

NEWELL, S. Y. & FALLON, R. D. (1982). Bacterial productivity in the water column and sediments of the Georgia (USA) coastal zone: estimates via direct counting and parallel measurement of thymidine incorporation. *Microbial Ecology*, **8**, 333–46.

NICHOLS, P. D., SHAW, P. M. & JOHNS, R. B. (1985). Determination of the double bond position and geometry in monoenoic fatty acids from complete microbial and environmental samples by capillary GC-MS of their Dies-Alder adducts. *Journal of Microbiological Methods*, **3**, 311–20.

NICHOLS, P. D., SMITH, G. A., ANTWORTH, C. P., HANSON, R. S. & WHITE, D. C. (1985a). Phospholipid and lipopolysaccharide normal and hydroxy fatty acids as potential signatures for methane-oxidising bacteria. *FEMS Microbiology Ecology*, **31**, 327–35.

NICHOLS, P. D., HENSON, J. M., GUCKERT, J. B., NIVENS, D. E. & WHITE, D. C. (1985b). Fourier transform-infrared spectroscopic methods for microbial ecology: analysis of bacteria-polymer mixtures and biofilms. *Journal of Microbiological Methods*, **4**, 79–94.

NICKELS, J. S., KING, J. D. & WHITE, D. C. (1979). Poly-beta-hydroxybutyrate metabolism as a measure of unbalanced growth of estuarine detrital microbiota. *Applied and Environmental Microbiology*, **37**, 459–65.

OLIVER, J. D. & STRINGER, W. F. (1984). Lipid composition of a psychrophylic marine *Vibrio* sp. during starvation-induced morphogenesis. *Applied and Environmental Microbiology*, **47**, 461–6.

PARKER, J. H., SMITH, G. A., FREDRICKSON, H. L., VESTAL, J. R. & WHITE, D. C. (1982). Sensitive assay, based on hydroxy fatty acids from lipopolysaccharide lipid A, for gram-negative bacteria in sediments. *Applied and Environmental Microbiology*, **44**, 1170–7.

PARKES, R. J. (1982). Methods for enriching, isolating, and analysing microbial communities in laboratory systems. In *Microbial Interactions and Communities*, vol. 1, ed. A. T. Bull & J. H. Slater, pp. 45–101. London: Academic Press.

PARKES, R. J. & TAYLOR, J. (1983). The relationship between fatty acid distributions and bacterial respiratory types in contemporary marine sediments. *Estuarine, Coastal and Shelf Science*, **16**, 173–89.

PARKES, R. J. & CALDER, A. G. (1985). The cellular fatty acids of three strains of *Desulfobulbus*, a propionate utilising sulphate-reducing bacterium. *FEMS Microbiology Ecology*, **31**, 361–3.

PARKES, R. J. & TAYLOR, J. (1985). Characterisation of microbial populations in polluted marine sediments. *Journal of Applied Bacteriology Symposium Supplement*, 155S–73S.

PARKES, R. J. & SENIOR, E. G. (1986). Multi-stage chemostats and other models for studying anoxic ecosystems. In *Handbook of Laboratory Model Systems for Microbial Ecosystem Research*, ed. J. W. T. Wimpenny, (in press).

PERRY, G. J., VOLKMAN, J. K., JOHNS, R. B. & BAVOR, H. J. JN (1979). Fatty acids of bacterial origin in contemporary marine sediments. *Geochimica et Cosmochimica Acta*, **43**, 1715–25.

REVSBECH, N. P., SØRENSEN, J. & BLACKBURN, T. H. (1980). Distribution of oxygen in marine sediments measured with microelectrodes. *Limnology and Oceanography*, **25**, 403–11.

RHEINHEIMER, G. (1977). Microbial ecology of a brackish water environment. Berlin, Heidelberg & New York: Springer-Verlag.

ROGERS, H. J. (1983). Bacterial cell structure. In *Aspects of Microbiology*, **6**, ed. J. A. Cole, C. J. Knowles & D. Schlessinger. Wokingham: Van Nostrand Reinhold (UK) Co. Ltd.

ROHMER, M., BOUVIER-NAVE, P. & OURISSON, G. (1984). Distribution of hopanoid triterpenes in prokaryotes. *Journal of General Microbiology*, **130**, 1137–50.

SHAW, N. (1974). Lipid composition as a guide to classification of bacteria. *Advances in Applied Microbiology*, **17**, 63–108.

STIEB, M. & SCHINK, B. (1985). Anaerobic oxidation of fatty acids by *Clostridium bryantii* sp. nov., a spore forming, obligately syntrophic bacterium. *Archives of Microbiology*, **140**, 387–90.

TAYLOR, J. & PARKES, R. J. (1983). The cellular fatty acids of the sulphate-reducing bacteria, *Desulfobacter* sp., *Desulfobulbus* sp. and *Desulfovibrio desulfuricans*. *Journal of General Microbiology*, **129**, 3303–9.

TAYLOR, J. & PARKES, R. J. (1985). Identifying different populations of sulphate-reducing bacteria within marine sediment systems, using fatty acid biomarkers. *Journal of General Microbiology*, **131**, 631–42.

TEWES, F. J. & THAUER, R. K. (1980). Regulation of ATP-synthesis in glucose fermenting bacteria involved in interspecies hydrogen transfer. In *Anaerobes and Anaerobic Infections*, ed. G. Gottschalk, N. Pfennig & Werner, pp. 97–104. Stuttgart: Fischer Verlag.

THAUER, R. K. & MORRIS, J. G. (1984). Metabolism of chemotrophic anaerobes: old views and new aspects. In *The Microbe 1984 II: Prokaryotes and Eukaryotes*, ed. D. P. Kelly & N. G. Carr, pp. 123–68. Cambridge University Press.

THOMAS, T. D. & BATT, R. D. (1969). Degradation of cell constituents by starved *Streptococcus lactis* in relation to survival. *Journal of General Microbiology*, **58**, 347–62.

VAN ES, F. B. & MEYER-REIL, L.-A. (1982). Biomass and metabolic activity of heterotrophic marine bacteria. *Advances in Microbial Ecology*, **6**, 111–70.

VAN VLEET, E. S. & QUINN, T. G. (1979). Early diagenesis of fatty acids and isoprenoid alcohols in estuarine and coastal sediments. *Geochimica et Cosmochimica Acta*, **43**, 289–303.

WATSON, S. W. & HOBBIE, J. E. (1979). Measurement of bacterial biomass as Lipopolysaccharide. In *Native Aquatic Bacteria: Enumeration, Activity and Ecology*, ASTM STP 695, ed. J. W. Costerton & R. R. Colwell, pp. 82–8. Philadelphia: American Society for Testing and Materials.

WATSON, S. W., NOVITSKY, T. J., QUINBY, H. L. & VALOIS, F. W. (1977). Determination of bacterial number and biomass in the marine environment. *Applied and Environmental Microbiology*, **33**, 940–6.

WHITE, D. C. (1983). Analysis of microorganisms in terms of quantity and activity in natural environments. In *Microbes in their Natural Environment*, ed. J. H. Slater, R. Whittenbury & J. W. T. Wimpenny, pp. 37–66. Cambridge University Press.

WHITE, D. C. (1985). Quantitative physical-chemical characterization of bacterial habitats. In *Bacteria in Nature II: Methods of Bacterial Ecology*, ed. J. Poindexter & E. Leadbetter. New York: Plenum Press.

WHITE, D. C., BOBBIE, R. J., MORRISON, S. J., OOSTERHOFF, D. K., TAYLOR, C. W. & MEETER, D. A. (1977). Determination of microbial activity of estuarine detritus by relative rates of lipid biosynthesis. *Limnology and Oceanography*,

22, 1089–99.

WHITE, D. C., DAVIS, W. M., NICKELS, J. S., KING, J. D. & BOBBIE, R. J. (1979a). Determination of the sedimentary microbial biomass by extractible lipid phosphate. *Oecologia*, **40**, 51–62.

WHITE, D. C., BOBBIE, R. J., KING, J. D., NICKELS, J. S. & AMOE, P. (1979b). Lipid analysis of sediments for microbial biomass and structure. In *Methodology for Biomass Determinations and Microbial Activities in Sediments*: ASTM STP 673, ed. C. D. Litchfield & P. L. Seyfried, pp. 87–103. Philadelphia: American Society for Testing and Materials.

WHITE, D. C., BOBBIE, R. J., HERRON, J. S., KING, J. D. & MORRISON, S. J. (1979c). Biochemical measurements of microbial mass and activity from environmental samples. In *Native Aquatic Bacteria: Enumeration, Activity and Ecology*, ASTM STP 695, ed. J. W. Costerton & R. R. Colwell, pp. 69–81. Philadelphia: American Society for Testing and Materials.

WHITE, D. C., BOBBIE, R. J., NICKELS, J. S., FAZIO, S. D. & DAVIS, W. M. (1980). Nonselective biochemical methods for the determination of fungal mass and community structure in estuarine detrital microflora. *Botanica Marina*, **23**, 239–50.

WIDDEL, F. (1980). Anaerober Abbau von Fettsauren und Benzoesaure durch neu isolierte arten sulfat-reduzierender Bakterien. Doctoral thesis, University of Gottingen, F.R.G.

WIDDEL, F. & PFENNIG, N. (1981). Studies on dissimilatory sulfate-reducing bacteria that decompose fatty acids. I. Isolation of new sulfate-reducing bacteria enriched with acetate from saline environments. Description of *Desulfobacter postgatei* gen. nov., sp. nov. *Archives of Microbiology*, **129**, 394–400.

WIDDEL, F. & PFENNIG, N. (1982). Studies on dissimilatory sulfate-reducing bacteria that decompose fatty acids. II. Incomplete oxidation of propionate by *Desulfobulbus propionicus* gen. nov., sp. nov. *Archives of Microbiology*, **131**, 360–5.

WIDDEL, F., KOHRING, G. W. & MAYER, F. (1983). Studies on dissimilatory sulfate-reducing bacteria that decompose fatty acids III. Characterization of the filamentos gliding *Desulfonema limicola* gen. nov., sp. nov., and *Desulfonema magnum* sp. nov. *Archives of Microbiology*, **134**, 286–94.

WIMPENNY, J. W. T., LOVITT, R. W. & COOMBS, J. P. (1983). Laboratory model systems for the investigation of spatially and temporally organized microbial ecosystems. In *Microbes in their Natural Environments*, ed. J. H. Slater, R. Whittenbury & J. W. T. Wimpenny, pp. 67–117. Cambridge University Press.

COMMUNITY STRUCTURE AND INTERACTIONS AMONG COMMUNITY MEMBERS IN HOT SPRING CYANOBACTERIAL MATS

DAVID M. WARD, TIM A. TAYNE, KAREN L. ANDERSON and MARY M. BATESON

Department of Microbiology, Montana State University, Bozeman, MT 59717, USA

INTRODUCTION

Laminated cyanobacterial mats occur in many different aquatic settings, but are often found in extreme environments such as hypersaline lagoons, salt ponds, salt lakes or thermal streams. The general nature of the cyanobacterial mats in such systems has been reviewed recently (see Cohen, Castenholz & Halvorson, 1984). Because extreme environmental conditions limit the range of organisms which may be present, cyanobacterial mat communities are often less complex than conventional aquatic communities. Only a limited number of plants, animals (Gerdes & Krumbein, 1984; Javor & Castenholz, 1984) and eukaryotic microorganisms (Stolz, 1984) are present, and photosynthesis is often carried out by a few dominant cyanobacteria (Krumbein, Cohen & Shilo, 1977; Javor & Castenholz, 1981; Bauld, 1984; Stolz, 1984).

The most extreme environments in which cyanobacterial mats are found are thermal springs. At temperatures above 49–50 °C animals are absent (Wickstrom and Castenholz, 1973, 1985), and above 56–62 °C eukaryotic microorganisms are absent (Brock, 1978). In mats of neutral to alkaline springs eukaryotic microorganisms are seldom seen above about 50 °C. Thus, above about 50 °C the mat communities are totally microbial, and often are composed entirely of prokaryotic microorganisms. The number of cyanobacterial species inhabiting these mats also decreases with increasing temperature (Brock, 1978), so that often only one or a few cyanobacteria are present. These features make hot spring cyanobacterial mats attractive model systems in which to study the basic questions of the composition and structure of microbial communities and the interactions among the various community members.

Fortunately, two other factors have increased the appeal of studying hot spring microbial communities, and as a result, more information has been obtained which can be applied to help answer the basic ecological questions raised above. Since hot spring cyanobacterial mats are natural high-temperature environments where organic matter decomposes, there has been an interest in discovering thermophilic processes, microorganisms and biochemicals which might have value in industrial processes. For example, an interest in the potential use of thermophilic microorganisms in the production of fuels and chemical feedstocks (Zeikus, 1979; Wiegel, 1980) led to the discovery of new thermophilic fermentative bacteria which may be important members of hot spring cyanobacterial mat communities.

Microbial mats are also modern analogs of stromatolites (Walter, Bauld & Brock, 1972; Doemel & Brock, 1974), the fossilized laminated mat-like communities which dominate the record of the Earth's biota during the Precambrian period (Schopf, Hayes & Walter, 1983; Awramik, 1984). Studies of modern microbial mats of various types, including cyanobacterial mats, have been carried out in an effort to learn more about the possible nature of these ancient communities. Though most ancient mats formed in coastal seas, hot spring mats are more representative of the totally microbial (and perhaps totally prokaryotic) biota thought to have predominated in Precambrian times.

Various types of mats formed by photosynthetic microorganisms are found in different hot spring environments (Castenholz, 1984). A variety of photosynthetic prokaryotes, including photosynthetic bacteria (Giovanonni *et al.*, in press) and thermophilic cyanobacteria (Castenholz, 1969), can form laminated mats in hot spring waters. The types of microorganisms present are, of course, determined by environmental conditions such as the temperature and chemistry of the hot spring water.

In this review we will consider in detail a single type of hot spring cyanobacterial mat which grows in the effluent channels or pools of neutral to alkaline hot springs of Yellowstone National Park. In particular, the mats of Octopus Spring, Mushroom Spring and other springs in the Lower Geyser Basin have been studied in detail with respect to community composition, chemistry and interactions among community members, and serve as a model for cyanobacterial mats in general. These mats are typically found between 70–73 °C (the upper temperature limit for photosynthetic microorganisms) and 40 °C (the upper temperature limit for animals in these springs

(Wiegert & Mitchell, 1973)). Other mats which occur in neutral springs of the Mammoth Terrace Group (Yellowstone National Park) and in Oregon hot springs will also be considered. The ecology of the latter exhibits some interesting differences due to the presence of high concentrations of sulfate in the spring water. Wherever possible, reference will be given to analogous findings from cyanobacterial mats of non-thermal systems.

COMMUNITY COMPOSITION

Microscopic and enrichment culture approaches have revealed that microorganisms of various physiological types inhabit hot spring communities. Previous reviewers have compiled lists of microorganisms observed in and isolated from such communities (Brock, 1978; Castenholz, 1979). Here we focus attention on the microorganisms known to be present in a single type of hot spring community – the cyanobacterial mats of alkaline siliceous springs such as Octopus Spring (see Table 1).

Mat-forming microorganisms

The mat in Octopus Spring shows a typical vertical sequence of phototrophic microorganisms with cyanobacteria overlying photosynthetic bacteria (Castenholz, 1984). Although numerous thermophilic cyanobacteria dominate various different hot spring cyanobacterial mats (Castenholz, 1969), a single cyanobacterium, *Synechococcus lividus*, recognized by a combination of its unique shape and the red autofluorescence of its chlorophyll *a*, is present in the Octopus Spring mat. *S. lividus* is restricted to the upper 1 mm of the mat, presumably due to self-shading, as it exists at population densities estimated to be as high as 10^{10} cells ml^{-1} (Bauld & Brock, 1974). It is the only organism in this mat which is capable of oxygenic photosynthesis. Its oxygen production reduces the importance of anoxygenic photosynthesis in the mat (see below) so that *S. lividus* is the main primary producer in the mat community.

The filamentous photosynthetic bacterium *Chloroflexus aurantiacus* (Pierson & Castenholz, 1974) has also been isolated from the Octopus Spring mat (Bauld & Brock, 1973; Tayne & Ward, 1983). It can be recognized as an abundant member of hot spring cyanobacterial mats by the autofluorescence of its bacteriochlorophylls in the

Table 1. *Microorganisms present in the Octopus Spring cyanobacterial mat*

Organism	Physiological type	Abundance	Reference
Synechococcus lividus	Cyanobacterium	High[a] ($ca\ 10^{10}$ml^{-1})	Bauld & Brock, 1974
Chloroflexus aurantiacus	Photosynthetic eubacterium	High[a]	Bauld & Brock, 1973 Tayne & Ward, 1983
Thermus aquaticus	Aerobic heterotrophic eubacterium	Unknown[b]	Brock & Freeze, 1969
Isocystis pallida	Aerobic heterotrophic eubacterium	High[a]	Giovanonni & Schabtach, 1983 Doemel & Brock, 1977
Thermobacteroides acetoethylicus	Anaerobic fermentative eubacterium	High[c] ($ca\ 10^{7}$ ml^{-1})	Ben-Bassat & Zeikus, 1981 Zeikus, Ben-Bassat & Hegge, 1980
Thermoanaerobium brockii	Anaerobic fermentative eubacterium	Low[d]	Zeikus, Hegge & Anderson, 1979 Zeikus, Ben-Bassat & Hegge, 1980
Thermoanaerobacter ethanolicus	Anaerobic fermentative eubacterium	Unknown[b]	Wiegel & Ljungdahl, 1981
Clostridium thermo-hydrosulfuricum	Anaerobic fermentative eubacterium	Low[d] ($<10^{3}$ ml^{-1})	Zeikus, Ben-Bassat and Hegge, 1980 Wiegel, Ljungdahl & Rawson, 1979
Clostridium thermo-sulfurogenes	Anaerobic fermentative eubacterium	Low[d] ($<10^{3}$ ml^{-1})	Schink & Zeikus, 1983
Thermodesulfobacterium commune	Sulfate-reducing eubacterium	Unknown[b]	Zeikus et al., 1983
Methanobacterium thermoautotrophicum	Methane-producing archaebacterium	High[c] ($ca\ 10^{7}$ ml^{-1})	Zeikus, Ben-Bassat and Hegge, 1980 Sandbeck & Ward, 1982

[a] Direct microscopic observation
[b] Enriched from undiluted sample
[c] Enriched from highly diluted sample
[d] Enriched from low dilution samples

infrared region (Pierson & Howard, 1972), and by staining with an immunofluorescent probe specific for *Chloroflexus* strains (Tayne & Ward, 1983). Not all filamentous microorganisms in the Octopus Spring mat react with this probe, however, suggesting the presence of other numerically significant filamentous bacteria in the community. Though other filamentous bacteria, such as the photosynthetic *Heliothrix oregonensis* (Pierson, Giovanonni & Castenholz, 1984), the sulfur-oxidizer *Beggiatoa* (Nelson & Castenholz, 1981) and the chemoorganotrophic *Herpetosiphon geysericolus* (Lewin, 1970) have been isolated from other hot spring cyanobacterial mats, the presence of such organisms in the Octopus Spring mat has not as yet been reported.

C. *aurantiacus* is capable of anoxygenic photoautotrophic growth, but grows better photoheterotrophically (Madigan & Brock, 1975). It is also able to grow by aerobic respiration in darkness (Pierson & Castenholz, 1974), and is thus well adapted to the fluctuating light and oxygen levels which are typical of cyanobacterial mats (see below).

Several unsuccessful attempts have been made to isolate other photosynthetic bacteria from the Octopus Spring cyanobacterial mat (Mike Madigan, personal communication). Purple sulfur bacteria commonly form layers beneath cyanobacteria in mats of marine and hypersaline environments (Javor & Castenholz, 1981; Bauld, 1984; Stolz, 1984). Their lack of prevalence in the Octopus Spring mat may be due to the relatively low quantities of sulfide. A thermophilic *Chromatium* has been isolated (Madigan, 1984, 1986) from the source of high-sulfide hot springs, and appears to be a component of mats which form in high-sulfate thermal waters (see Castenholz, 1977). In mats of high-sulfate springs, as in marine and hypersaline environments, sulfide is supplied through intensive sulfate reduction (Ward & Olson, 1980).

Mat-decomposing microorganisms

Phototrophic microorganisms support the growth of microorganisms involved in mat decomposition either directly through excreted products or indirectly as a result of death upon burial beneath the photic zone. The bacteria involved in aerobic decomposition processes have not been studied thoroughly. *Thermus aquaticus* has been isolated from the Octopus Spring cyanobacterial mat, but its abundance in the mat is not known because it has only been enriched from undiluted samples. *Thermus* spp. can be found at densities of up to 10^2 cells ml^{-1} in water flowing above Icelandic hot spring mats (Kristjansson & Alfredsson, 1983), and have been reported at densities of 10^2 to 2×10^4 cells ml^{-1} in 'overgrowths' (presumably mats) occurring at about 50–55 °C in the effluent channels of Russian hot springs (Egorova & Loginova, 1975). Another aerobic chemoorganotrophic bacterium, *Isocystis pallida*, which has been isolated from Oregon and Yellowstone hot spring cyanobacterial mats (Giovanonni & Schabtach, 1983), is judged to be a significant community member on the basis of microscopic evidence of cells with its distinctive morphology in the Octopus Spring mat (Doemel & Brock, 1977). The facultatively anaerobic bacterium *Bacillus stearothermophilus* has also been isolated from the Octopus Spring mat, but its importance in the community is suspected to be slight, as Gram-positive cells are rarely seen in stained mat preparations (Brock, 1978). A number of other aerobic bacteria have been isolated from hot spring waters and muds and these might be members of the cyanobacterial mat community. Included are *Thermoleophilum album*, a hydrocarbon-oxidizer (Zarilla & Perry, 1984), *Thermomicrobium* (Jackson, Ramaley & Meinschein, 1973*a*), and the hydrogen-oxidizer *Hydrogenobacter thermophilus* (Kawasumi *et al.*, 1984; Kristjansson, Ingason & Alfredsson, 1985).

More is known of the microorganisms which participate in anaerobic decomposition of the Octopus Spring cyanobacterial mat (Ward *et al.*, 1984). Several obligately anaerobic fermentative bacteria have been isolated from this mat. One of these, *Thermobacteroides acetoethylicus*, was isolated from high-dilution enrichments, suggesting that it was numerically important in the mat community. *Thermoanaerobium brockii*, *Clostridium thermohydrosulfuricum* and *C. thermosulfurogenes* were isolated from low-dilution enrichments and may be only minor community members. The relative abundance of *Thermoanaerobacter ethanolicus* is at present unknown. *Thermonaerobacter* strains which ferment xylan have been isolated from Yellowstone cyanobacterial mats, though it is uncertain whether these were isolated from the Octopus Spring mat (Weimer *et al.*, 1984; Weimer, 1985). Other thermophilic anaerobic fermentative bacteria have been isolated from New Zealand (Patel, Morgan & Daniel, 1985) and Japanese (Saiki *et al.*, 1985) hot springs, but these were not described as being members of cyanobacterial mat communities.

Under dark anaerobic conditions the products of fermenting bacteria may be consumed by acetogenic bacteria, sulfate-reducing bacteria, or methanogenic bacteria (McInerney & Bryant, 1981; Ward

& Winfrey, 1985). Acetogenic bacteria which degrade butyric acid have been observed in enrichment culture from the Octopus Spring mat (Tayne, 1983). *Clostridium thermoautotrophicum*, which converts hydrogen and carbon dioxide to acetate, has been isolated from a water sample from an alkaline siliceous spring in Yellowstone, but it is not known to be a member of the Octopus Spring mat community (Wiegel, Braun & Gottschalk, 1981).

A sulfate-reducing bacterium, *Thermodesulfobacterium commune*, has been isolated from the Octopus Spring mat (Zeikus *et al.*, 1983) though, as described below, it is likely to be of limited importance in the mat due to the low sulfate concentration. Its abundance is not known as it has only been enriched from undiluted samples. Sulfate reduction is a more important process in high-sulfate mat systems including both hot spring (Ward & Olson, 1980; Ward *et al.*, 1984) and marine and hypersaline cyanobacterial mats (Jørgensen & Cohen, 1977; Skyring, Chambers & Bauld, 1983; Skyring, 1984). In fact, in marine and hypersaline cyanobacterial mats there is often a black layer of iron sulfides deep within the mat (Bauld *et al.*, 1980).

Methanogenic bacteria can be tentatively identified directly in Octopus Spring mat material by the green autofluorescence of their unique coenzymes (Mink & Dugan, 1977; Ward, unpublished results). *Methanobacterium thermoautotrophicum* has been isolated from high-dilution enrichments (most probable number estimates of $ca\,10^7$ cells ml^{-1}) from the Octopus Spring mat (Ward, 1978; Zeikus, Ben-Bassat and Hegge, 1980; Sandbeck & Ward, 1982), suggesting its numerical importance in the mat community.

BIOCHEMICAL MARKERS AND COMMUNITY STRUCTURE

While microscopic and enrichment culture approaches have revealed many of the inhabitants of the Octopus Spring cyanobacterial mat, it is unlikely that these techniques can reveal the complete collection of community members. Only a few microorganisms can be recognized directly by microscopy as a result of their distinctive morphologic features. Enrichment and isolation of community members depends on preconception of an organism's niche and on the ability of the investigator to devise culture conditions which duplicate the niche successfully. Thus, the organisms we know from a community are those which we can grow. The extent to which a collection of pure cultures obtained from a community represents the true community composition and structure is unknown.

Another approach to understanding community composition and structure is to observe biochemical components characteristic of specific microorganisms or specific microbial groups. Such an approach eliminates the need to isolate the community member in order to recognize its presence, thus avoiding the biases of the enrichment and pure culture approach. We have begun to investigate the composition of the Octopus Spring cyanobacterial mat using two types of 'biochemical marker' compounds – lipids and 16S ribosomal RNA. Lipids appear to be group-specific biochemical markers and are more useful for observing major types of microorganisms (e.g. cyanobacteria, methanogenic bacteria) rather than specific microorganisms. Extensive work on the 16S rRNA molecules of many organisms has shown that the primary nucleotide sequence of this molecule is unique for each microbial species (Balch *et al.*, 1979; Woese, 1981, 1982), indicating the potential of 16S rRNA as a species-specific biochemical marker.

Lipid biochemical markers

Fig. 1 summarizes the free lipids present in the 0–3 mm depth interval of the Octopus Spring cyanobacterial mat (Dobson, Ward, Robinson & Eglinton, unpublished). This depth interval would include all phototrophic community members, but may exclude some of the mat-decomposing microorganisms which may be active to a depth of about 5 mm in the mat (see below). The dominant free hydrocarbons (Fig. 1a) include heptadecane, 7-methylheptadecane, and hop-22(29)-ene which are thought to be contributed by cyanobacteria (Han, McCarthy & Calvin, 1968; Gelpi *et al.*, 1970; Blumer, Guillard & Chase, 1971; Ourisson, Albrecht & Rohmer, 1979; Rohmer, Bouvier-Nave & Ourisson, 1984). In addition, the abundance of phytene (Fig. 1a) and phytol (Fig. 1c) presumably reflects the importance of chlorophyll *a*, present only in *Synechococcus lividus* in this mat. The dominant fatty acids of the mat, hexadecanoic acid and octadecanoic acid (Fig. 1d), are similar to those of *S. lividus* in culture (Miller, 1976; Fork, Murata & Sato, 1979).

The lipids of *Chloroflexus* also appear to be among the dominant free lipids of the 0–3 mm interval of the mat. As with *S. lividus*, the dominant fatty acids produced by *Chloroflexus* in culture are hexadecanoic and octadecanoic acids (Kenyon & Gray, 1974; Knudsen *et al.*, 1982). The wax esters, among the more abundant free

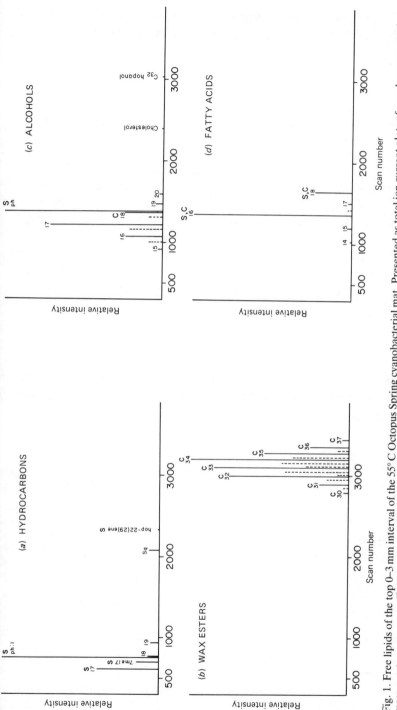

Fig. 1. Free lipids of the top 0–3 mm interval of the 55° C Octopus Spring cyanobacterial mat. Presented as total ion current plots of gas chromatography–mass spectrometry runs (Dobson, Ward, Robinson & Eglinton, unpublished). Numbers correspond to the carbon number of straight-chain compounds. Dashed lines indicate branched compounds of carbon number equivalent to the straight-chain compound which immediately follows. me = methyl; ph:1 = phytene; Sq = squalene; ph = phytol; S = suspected component of the cyanobacterium *Synechococcus lividus*. C = suspected component of *Chloroflexus aurantiacus*.

lipids (Fig. 1b), were of a similar molecular dimension to the wax esters C. *aurantiacus* produces in culture (Edmunds, 1982; Knudsen *et al.*, 1982). Degradation of the unique major bacteriochlorophyll of *Chloroflexus* (Gloe & Risch, 1978) may explain the abundance of octadecanol among the mat's free alcohols (Fig. 1c).

Lipids characteristic of decomposing microorganisms were less apparent among the mat's free lipids. For instance, the branched-chain fatty acids produced by *Thermus aquaticus* and *Bacillus stearothermophilus* in culture (Shen *et al.*, 1970; Ray, White & Brock, 1971; Jackson, Ramaley & Meinschein, 1973b; Oshima & Miyagawa, 1974) were not major free lipids of the top 0–3 mm of the Octopus Spring mat. The novel ether-linked lipids of the sulfate-reducing bacterium *Thermodesulfobacterium commune* (Langworthy *et al.*, 1983) and of methanogenic bacteria (Tornabene & Langworthy, 1978; Tornabene *et al.*, 1978) have been investigated in separate studies. The non-isoprenoid glyceryl ethers of *T. commune* are present in lower amounts than the isoprenoid glyceryl ethers of methanogens in the Octopus Spring mat (Ward, Brassell & Eglinton, 1985; Zeng & Eglinton, unpublished; see below), consistent with the domination of methanogenesis over sulfate reduction in this mat (Ward *et al.*, 1984).

Although it will be necessary to study the lipids of each mat isolate in order to confirm the correlations suggested above, several generalizations seem justified based on the studies which have been completed.

1. The simplicity of the free lipids of the Octopus Spring microbial mat is consistent with the presumption that high temperature restricts community diversity. This is particularly evidenced by the low abundance and variety of steroid compounds in the mat, compared to the extensive sterols and structurally related compounds produced by sterol diagenesis found in cyanobacterial mats in Solar Lake (Boon *et al.*, 1983; Edmunds & Eglington, 1984) and marine environments (e.g. Boudou *et al.*, in press). Mats in such locations contain a more diverse community than hot spring mats and this is reflected in the greater variety of photosynthetic microorganisms present (including cyanobacteria, eukaryotic algae, and photosynthetic bacteria) and by the presence of some animal grazers.
2. The lipids of phototrophic microorganisms appear to be dominant among the lipids of the mat. This may reflect the trophic structure

of the community with primary producing organisms dominant over the decomposers.
3. The lipid distribution reflects the known distribution of micro-organisms in microbial mats. The best evidence for this is the correlation between archaebacterial ethers and the known or suspected distribution of archaebacteria in different hot spring algal or cyanobacterial mats (Ward, Brassell and Eglinton, 1985).

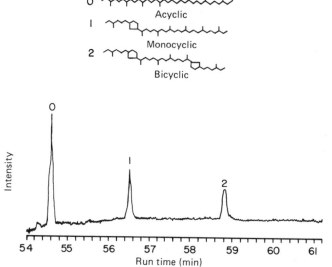

Fig. 2. Glass capillary gas chromatogram of biphytanes released by ether cleavage of archaebacterial glyceryl ethers in the 55 °C Octopus Spring cyanobacterial mat. (Modified from Ward, Brassell & Eglinton, 1985.)

4. The mat contains lipids of unknown origin, for example hopanols, branched-chain wax esters and cyclic biphytanyl ethers. These either reflect *unknown lipids* of known mat organisms, or may reflect the lipids of *unknown mat organisms*. For example, the occurrence of cyclic biphytanyl ethers in the Octopus Spring mat (Fig. 2) suggests the possible presence of a non-methanogenic archaebacterium which has not yet been isolated from this mat, since methanogens are characterized by acyclic biphytanyl ethers (Langworthy, Tornabene & Holzer, 1982).

16S Ribosomal RNA as a Biochemical Marker

We are currently developing methods for using 16S rRNA as a biochemical marker for the recognition of individual members of the

Octopus Spring cyanobacterial mat community. The method, which has been described by Pace *et al.* (1985), involves the use of recombinant DNA techniques to obtain the various genes encoding the primary sequences of 16S rRNA from the DNA of an entire community. Nucleotide sequences of individual 16S rRNA genes, each representative of a single community member, can then be compared to the sequences of microorganisms isolated from the mat to reveal the presence in the community of both known and unknown community members. The sequences of unknown community members can further be compared to larger 16S rRNA sequence collections (those used to examine phylogenetic relationships among widely divergent types of microorganisms: e.g. Fox *et al.*, 1980; Woese, 1981, 1982; Stackebrandt & Woese, 1984) to reveal the possible type of each unknown organism. Similar techniques employing 5S rRNA as a biochemical marker have been used to reveal the nature of bacteria which inhabit the boiling source pool of Octopus Spring (Stahl *et al.*, 1985) and the trophosome of deep-sea-vent tube worms (Stahl *et al.*, 1984). In each case, enrichment cultures failed to reveal any of the microbial community members, while rRNA sequences allowed an opportunity to learn the nature of the predominating microorganisms.

This technique should permit an analysis of how well the collection of community members in culture (Table 1) reflects the true community composition. It should also reveal the nature of community members which have so far eluded enrichment and isolation.

VERTICAL STRATIFICATION OF MICROORGANISMS AND PROCESSES

Low-sulfate springs

Cells with the distinctive morphology and autofluorescence of *Synechococcus lividus* can only be observed in the upper 1 mm of the mats of alkaline siliceous springs such as Octopus Spring (Doemel & Brock, 1977). This is consistent with the vertical distribution of chlorophyll *a* (Fig. 3a). Through the use of microelectrodes it has been possible to measure the vertical distribution of oxygenic photosynthesis by *S. lividus* in the Octopus Spring mat (Revsbech & Ward, 1984*a*) (Fig. 3*b*). Oxygenic photosynthesis is limited to the uppermost 1 mm of the mat apparently due to self-shading within the densely packed top mat layer. Photosynthetic rates are extremely

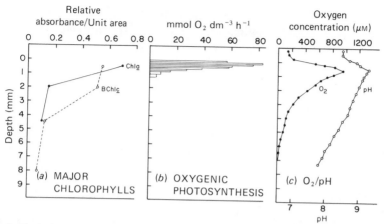

Fig. 3. Vertical profiles of photopigments, oxygenic photosynthesis, oxygen and pH in cyano-bacterial mats of alkaline siliceous hot springs. (a) Twin Butte Vista Spring (from Bauld & Brock, 1973). (b) and (c) Octopus Spring (from Revsbech & Ward, 1984a).

high (up to 150μmol $O_2 dm^{-3} h^{-1}$, see Revsbech & Ward, 1984a), causing accumulation of oxygen to very high levels, and a pH shift of about 1 pH unit compared to overflowing spring water (presumably due to intensive carbon dioxide consumption), as shown in Fig. 3(c).

The vertical position of *Chloroflexus aurantiacus* is best demonstrated by the distribution of its major bacteriochlorophyll, Bchl *c*, shown in Fig. 3(a). Although *Chloroflexus* coexists with *S. lividus* at the mat surface, it is adapted to use complementary regions of the light spectrum (Castenholz, 1984) and to grow at low light intensities (Madigan & Brock, 1977). Thus, it is also present in the 1–2 mm depth interval beneath the *S. lividus* layer. *C. aurantiacus* cannot grow below the photic zone (estimated to be no greater than 1–2 mm, see Doemel & Brock, 1977) because it can only grow aerobically in the dark. The presence of a photosynthetic bacterial layer underlying a cyanobacterial layer is characteristic of most hot spring cyanobacterial mats (Castenholz, 1984) and, as described above, of cyanobacterial mats in non-thermal environments which typically exhibit a purple undermat layer due to dense populations of purple sulfur bacteria.

Doemel & Brock (1977) studied the interaction between mat-forming and mat-decomposing microorganisms by evaluating changes in thickness and protein content above and between layers of carborundum which were added to the mat as inert dated markers. Their results suggest that these types of mats accrete at rates of between 3 and 110 μm per day. The upward accretion of the mat is balanced by decomposition in deeper layers so that there is no net accumulation and the mat community is in a steady state. Decomposition of the mat occurs most rapidly near the mat surface. Aerobic chemoorganotrophic microorganisms, such as the morphologically distinctive *Isocystis pallida*, are limited to the top mat region (Doemel & Brock, 1977), consistent with the abundance of oxygen near

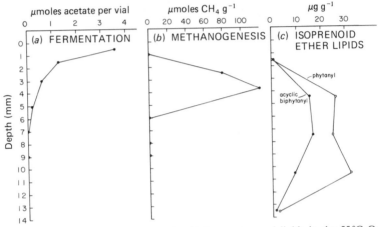

Fig. 4. Vertical profiles of anaerobic microbial processes and lipids in the 55 °C Octopus Spring cyanobacterial mat. (a) Acetate accumulation in core samples after 54 h dark anaerobic incubation (modified from Anderson & Ward, 1983). (b) Methane accumulation in core samples after 174 h dark anaerobic incubation (from Ward, 1978). (c) Isoprenoid ether cleavage products (from Ward, Brassell & Eglinton, 1985).

the mat surface. Fig. 4 illustrates the vertical distributions of anaerobic microbial processes. Fermentation (as measured during dark anaerobic conditions) occurs mainly in the top few millimeters. Methanogenesis, the dominant terminal anaerobic process in the Octopus Spring mat, occurs mainly in the 2–5 mm interval (as measured under continuous dark anaerobic incubation). The vertical distribution of lipids characteristic of methanogenic bacteria also suggests the important of methanogens below the top few millimeters. Lipids characteristic of methanogens appear to persist below the zone of intensive methanogenesis which extends only to about 5 mm below the mat surface.

The vertical profile of oxygen (and pH) varies depending upon light intensity (Fig. 5) so that the vertical distribution of aerobic

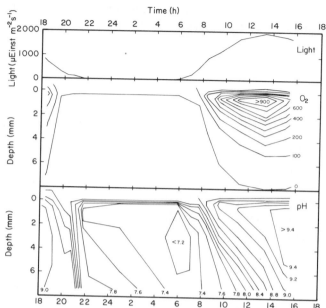

Fig. 5. Diurnal variation of light intensity, oxygen and pH in the 55 °C Octopus Spring cyano-bacterial mat. Presented as isopleths of oxygen (μmol^{-1}) and pH. (From Revsbech & Ward, 1984a.)

and anaerobic microorganisms and the decomposition processes they catalyze can also be expected to vary. Motile microorganisms may reposition themselves as environmental conditions change. For example, at night oxygen is only present in the top few hundred micrometers of the mat due to the combined effects of diffusion from overlying water and consumption in the mat. *Chloroflexus* can only grow aerobically in darkness and is thought to glide upwards at night, a process which may be essential to upward accretion of the mat (Doemel & Brock, 1977). Non-motile microorganisms may vary their activities in response to changing environmental para-meters. For example, anaerobic processes are likely to cycle diurnally in the mat, since oxygen usually penetrates to a depth of a least 5 mm during the midday period, presumably causing a survival stress on anaerobic bacteria. Methanogenic bacteria which were isolated from the Octopus Spring mat were able to withstand exposure for up to several hours of aeration and oxygenation (Revsbech & Ward, 1984a). Results obtained from other cyanobacterial mats (Skyring, 1984; Cohen, personal communication) suggest the simultaneous

occurrence of oxygenic photosynthesis and sulfate reduction, implying that some anaerobes may not only survive oxygen exposure,
but also metabolize in the presence of oxygen.

High-sulfate springs

In cyanobacterial mats of high-sulfate hot springs, such as Bath Lake
and Painted Pool (located at Mammoth Hot Springs in Yellowstone
National Park), vertical profiles of oxygenic photosynthesis and the
resultant oxygen and pH profiles are similar. However, decomposition processes are distinctly different, as sulfate reduction dominates
methanogenesis as the terminal process involved in anaerobic
decomposition (Ward & Olson, 1980; Ward et al., 1984). The vertical
distribution of sulfate reduction is similar to the vertical distribution
of methanogenesis in mats of low sulfate springs, occurring mainly
in the top 5 mm. In such mats intensive sulfate reduction leads to
the production of high levels of sulfide which can have a significant
effect on community structure. For example, depending on temperature, purple sulfur bacteria and sulfur-oxidizing chemolithotrophic
bacteria may form dense accumulations. This is also typical of mat
communities in high-sulfate marine and hypersaline systems.

The vertical distribution of sulfide in mats of high-sulfate springs
is a function of the amount of oxygen in the mat (itself a consequence
of photosynthetic activity in the top mat layer), as shown for the
cyanobacterial mat at Hunters Hot Spring (Oregon) in Fig. 6. In
darkness (Fig. 6, bottom), oxygen only enters the mat by diffusion
from overlying water. Intensive sulfate reduction leads to accumulation of hydrogen sulfide throughout the mat. As light intensity
increases during the day (Fig. 6, top), the sulfide–oxygen boundary
is driven deeper into the mat. This is likely to be due to the combined
effects of metabolism of hydrogen sulfide during anoxygenic photosynthesis by photosynthetic bacteria (and cyanobacteria (see, for
example, Jørgensen, Cohen & Revsbech, 1986)), and by aerobic
sulfur-oxidizing bacteria. At midday light intensities (Fig. 6, middle)
the intersect of sulfide and oxygen gradients is driven below the
photic zone so that only oxygenic photosynthesis is possible. Oxygen
may also inhibit sulfate reduction, thus explaining the low sulfide
levels observed in the experiment shown in Fig. 6 (middle). Sulfate
reduction was found to be sensitive to light (presumably to photosynthetically derived oxygen) in the Bath Lake cyanobacterial mat
(Winfrey & Ward, unpublished).

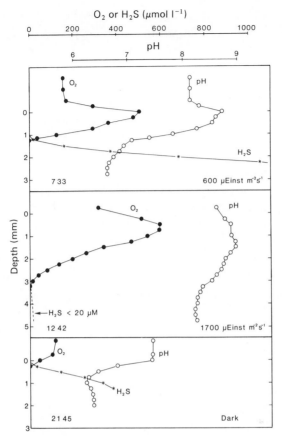

Fig. 6. Vertical profiles of oxygen, sulfide and pH in a 39–45 °C region of the Hunters Spring cyanobacterial mat, measured in early morning (0733 h), at midday (1242 h), and at night (2145 h). (From Revsbech & Ward, 1984*b*.)

Some microorganisms are able to follow the changing position of sulfide and/or oxygen. For example, the sulfur-oxidizing bacterium *Beggiatoa* (which occurs at about 40° C in some hot spring mats) exhibits vertical migrations toward the mat surface at night and into the mat in daylight (Nelson & Castenholz, 1981). Purple bacteria also carry out vertical migrations, coming to the mat surface at night to take advantage of sulfide and low light levels in early morning and returning to a subsurface position as light intensity increases and cyanobacterial photosynthesis generates sufficient oxygen to force sulfide deeper in the mat (Castenholz, personal communication). Some thermophilic cyanobacteria, such as *Oscillatoria terebriformis*, also exhibit vertical migrations which may relate to the need for sulfide (Richardson & Castenholz, 1983). Similar vertical migrations also occur in coastal sediments (Jørgensen, 1982).

In either low- or high-sulfate springs the active portion of hot spring cyanobacterial mat communities is compressed into a vertical distance of about 5 mm. Similar vertical profiles of oxygenic photosynthesis and resultant oxygen, pH and sulfide chemistry have been shown in the cyanobacterial mats of Solar Lake (Jørgensen, Blackburn & Cohen, 1981; Jørgensen, Revsbech & Cohen, 1983; Revsbech *et al.*, 1983) and Laguna Guerro Negro, Baja, California (Jørgensen, personal communication). Shallow sediments which receive light also exhibit similar profiles of oxygenic photosynthesis and chemistry (Revsbech *et al.*, 1979). The vertical distribution of sulfate reduction, the dominant terminal anaerobic process in cyanobacterial mats of Solar Lake (Jørgensen & Cohen, 1977) and coastal margins (Skyring, 1984) is also similar to that found in high-sulfate hot spring microbial mats. It seems fair to generalize that a compression of both mat-forming and mat-decomposing processes to a scale of a few millimeters to centimeters exists in all cyanobacterial mats.

INTERACTIONS AMONG COMMUNITY MEMBERS

An overview of the various processes carried out by the cyanobacterial mat communities of alkaline siliceous hot springs such as Octopus Spring, and of the interactions among community members in such mats, is shown in Fig. 7. Photosynthetic fixation of carbon dioxide in these mats is carried out almost exclusively by the cyanobacterium *Synechococcus lividus*. Although the photosynthetic bacterium *Chloroflexus aurantiacus* is capable of photoautotrophy using sulfide as an electron donor (Madigan & Brock, 1975), the oxygen produced during cyanobacterial photosynthesis excludes sulfide from the photic zone (Revsbech & Ward, 1984*b*) so that light and sulfide are seldom simultaneously present to support photoautotrophic growth of *Chloroflexus*. Although autoradiographic studies have shown that filaments resembling *Chloroflexus* incorporate carbon dioxide when DCMU (3-3,4-dichlorophenyl)-1,1-dimethylurea) is added to inhibit oxygenic photosynthesis (Doemel & Brock, 1977), such treatment would alter mat chemistry, permitting the accumulation of sulfide. These experiments probably show the potential for photoautotrophy by *Chloroflexus*, a process which probably only occurs to a limited extent during low light periods when oxygen does not penetrate

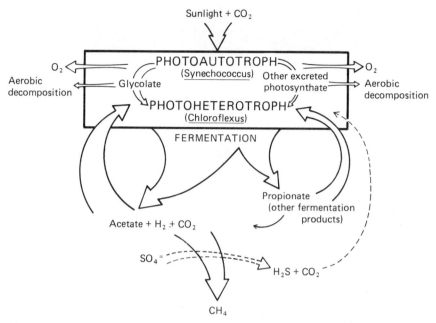

Fig. 7. Working model for microbial processes and interactions in low-sulfate alkaline siliceous hot spring cyanobacterial mats. Solid lines indicate major processes, whereas dashed lines indicate minor processes. Widths of arrows are in rough proportion to the flow of carbon along that path.

throughout the photic zone (see Fig. 5). During the day *C. aurantiacus* grows principally as a photoheterotroph in hot spring cyanobacterial mats.

Bauld & Brock (1974) showed that *S. lividus* photoexcretes up to 12% of its photosynthetically fixed carbon, and suggested that the excreted carbon may be cross-fed to heterotrophic community members. As mentioned above, the intensive photosynthetic activity of *S. lividus* leads to the development of extremely oxygen-rich and alkaline (presumably carbon dioxide-depleted) conditions in the photic zone (see Figs. 3 and 5). These conditions would favor production of glycolic acid as a major excretion product due to the oxygenase activity of ribulose diphosphate carboxylase/oxygenase (Chollet, 1977). The percentage of photosynthate which is excreted by natural *S. lividus* populations is increased by high oxygen or low carbon dioxide concentration, and under conditions similar to those of midday illumination, glycolic acid may account for up to 60% of the carbon which is excreted (Bateson & Ward, 1985) (see Fig. 8). Glycolic acid is, however, not the only photoexcretion product,

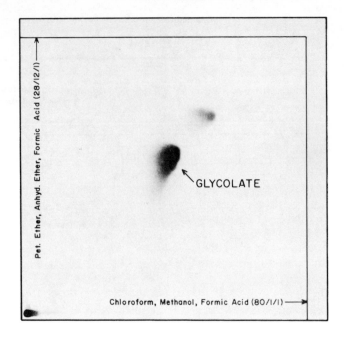

Fig. 8. Autoradiogram of a thin layer chromatogram showing the importance of glycolic acid among acidic compounds photoexcreted in the top layer of the Mushroom Spring cyanobacterial mat. Mat sample was labelled with $^{14}CO_2$.

as other neutral and acid products can also be detected. Several lines of evidence suggest that there is a tight link between the production of excreted photosynthate and its consumption by heterotrophic microorganisms. Filaments which resemble *Chloroflexus* are the principal organisms which incorporate glycolic acid (Bateson & Ward, 1985).

Aside from a direct flux of metabolic products between photosynthetic and heterotrophic microorganisms in the mat community, the polymeric organic components of the cells of photosynthetic organisms fuel decomposition within the mat. Obvious microscopic evidence of this is the total absence of *S. lividus* cells below 1 mm in the mat. Very little is known about the role of aerobic microorganisms in mat decomposition. They must, however, be restricted to the upper few millimeters of mat where oxygen is present, and could only be continuously active just near the mat surface.

In contrast, much is known of the various microorganisms involved in anaerobic decomposition. As in other environments (see Ward & Winfrey, 1985) fermentation leads mainly to the production of acetate, hydrogen and carbon dioxide (Anderson & Ward, unpublished). Smaller amounts of more reduced end products (e.g. ethanol, lactate, propionate, butyrates and valerates) are also produced. The predominance of acetate, hydrogen and carbon dioxide is probably due to interspecies hydrogen transfer (see Ward & Winfrey, 1985) since at higher temperatures (e.g. 65–70° C) where hydrogen accumulates, more reduced fermentation products such as ethanol are produced in greater amounts. In most anaerobic environments the reduced fermentation products are catabolized to acetate, hydrogen and carbon dioxide by acetogenic bacteria. In these mats, the potential exists for catabolism of some reduced fermentation products, such as butyrate, by acetogenic microorganisms (Tayne & Ward, 1983), but the net accumulation of butyrates, valerates, propionate and ethanol under dark anaerobic conditions suggests that acetogenesis is not an important process.

In most anaerobic environments, bacteria which terminate the anaerobic food chain consume the major fermentation products, acetate and hydrogen. In low-sulfate environments (e.g. <1mM sulfate), methanogenic bacteria dominate sulfate-reducing bacteria, whereas, in high-sulfate environments (e.g. >1mM sulfate), the opposite is true (Ward & Winfrey, 1985). Similar findings have been reported for hot spring microbial mats. In the cyanobacterial mat of the low-sulfate Octopus Spring, for example, both sulfate reduction (Doemel & Brock, 1976; Zeikus *et al.*, 1983; Ward *et al.*, 1984) and methanogenesis (Ward, 1978; Sandbeck & Ward, 1981, 1982) occur. Reported rates of methanogenesis (Ward, 1978) are about 25 times higher than the single reported rate of sulfate reduction (Zeikus *et al.*, 1983). However, the reported sulfate reduction rate is likely to be an overestimate of the true rate since the actual sulfate concentration at the site of sulfate reduction within the mat cannot be determined accurately (see Ward & Winfrey, 1985, for a discussion of the problem of determining sulfate reduction rates in low-sulfate environments where sulfate profiles occur over such small vertical intervals that they are immeasurably steep). Sulfate reduction must occur to a limited extent in the Octopus Spring mat, since addition of sulfate results in immediate inhibition of methanogenesis (Ward *et al.*, 1984). Sulfate reduction presumably occurs at a slightly higher position in the mat than does methanogenesis and the depletion of sulfate then permits methanogenesis in the 2–5 mm interval, as described above (see Fig. 4). In sediments, methane-producing and sulfate-reducing bacteria compete for acetate and hydrogen (see

Ward & Winfrey, 1985). However, in the Octopus Spring mat these microbial groups appear to compete only for hydrogen (Ward *et al.*, 1984). This is consistent with the finding that essentially all methane produced in the mat is derived from reduction of carbon dioxide (Sandbeck & Ward, 1981). Though acetate is a major methane precursor in most environments, it, like the more reduced fermentation products, is not catabolized but accumulates during dark anaerobic incubation of hot spring microbial mats.

In mats of high-sulfate hot springs, sulfate reduction dominates methanogenesis (Ward & Olson, 1980; Ward *et al.*, 1984). Although methanogenic bacteria are present in low numbers, sulfate reduction is the dominant terminal process as sulfate reduction rates are about one thousand times higher than rates of methanogenesis. As shown by the dashed arrows in Fig. 7, sulfate reduction results in formation of hydrogen sulfide, which at low light intensities may support anoxygenic photosynthesis by *Chloroflexus*. In mats of high-sulfate springs, high levels of hydrogen sulfide may support photoautotrophy by purple photosynthetic bacteria, and sulfide oxidation by chemolithotrophic bacteria. As mentioned above, most cyanobacterial mats of non-thermal environments exist in high-sulfate marine or evaporitic settings. Anaerobic decomposition of these mats is dominated by the sulfate reduction process, and the resultant sulfide can have a major impact on community structure and function.

Since acid fermentation products are not degraded under dark anaerobic conditions in hot spring cyanobacterial mats, they accumulate during darkness. This may explain why the mat pH decreases to become nearly one pH unit more acidic than the overflowing water during the night (see Fig. 5). As the mat becomes illuminated in early morning, these acids are available for assimilation by photoheterotrophic microorganisms. Indeed, the major fate of fermentation acids and alcohols in hot spring cyanobacterial mats is photoassimilation (Table 2). Autoradiography studies have shown that filamentous microorganisms which resemble *Chloroflexus* are the major microorganisms responsible for photoassimilation (Sandbeck & Ward, 1981; Tayne & Ward, 1983; Ward *et al.*, 1984) (Fig. 9). Tayne & Ward (1983) used an immunofluorescence probe which was specific for *C. aurantiacus*, in combination with autoradiography to show that *Chloroflexus* is the microorganism responsible for acetate photoassimilation (Plate 1). Fatty acid levels decrease during the morning hours in the Octopus Spring mat, presumably due to both photoassimilation and the inhibition of fermentation by oxygen

Table 2. *Fate of radiolabelled compounds in a hot spring cyanobacterial mat (Painted Pool, 42 °C)*

		Percent of labelled compound in fraction[b]		
Compound	Condition[a]	cells	CO_2	filtrate
2-[^{14}C]-acetate	Light	33.0	0.2	67.8
	Dark	3.0*	0.0	97.0
1-[^{14}C]-propionate	Light	47.4	0.1	52.5
	Dark	1.0*	0.03	99.0
1-[^{14}C]-butyrate	Light	21.0	0.3	78.7
	Dark	2.7*	0.13	97.2
1-[^{14}C]-lactate	Light	31.2	1.1	67.7
	Dark	0.8*	0.6	98.6
1-[^{14}C]-ethanol	Light	26.0	0.03	74.0
	Dark	0.9*	0.07	98.4
1-[^{14}C]-glycolic acid	Light	22.4	1.6	76.0
	Dark	1.1*	0.0	98.9

[a] Two hour incubation *in situ* in direct sunlight with (dark) or without (light) preincubation in darkness for 24 h. (Photoincorporation was also significantly reduced if samples were darkened only during incubation with radiolabel.)
[b] For determination of radioactivity within each fraction see Sandbeck & Ward (1981).
* $p \leqslant 0.05$.

produced during photosynthesis. It is not known whether fermentation products are catabolized or photoassimilated in non-thermal cyanobacterial mats, but the potential for photoassimilation would seem high due to the likely juxtaposition of fermenting and photoheterotrophic microorganisms. Similar interactions might also occur in shallow sediments where photosynthetic and fermentative bacteria should also be present within a similar vertical depth interval.

Though carbon and sulfur cycling have been investigated in hot spring cyanobacterial mats, there is relatively little information on nitrogen cycling. Nitrogen fixation is known to occur at temperatures below 55 °C where thermophilic nitrogen-fixing cyanobacteria, such as *Mastigocladus laminosus* and *Calothrix*, occur (Stewart, 1970; Wickstrom, 1980). Little is known of the importance of other nitrogen cycle processes.

SUMMARY

Hot spring cyanobacterial mats are among the best studied microbial communities. Significant progress has been made in determining the

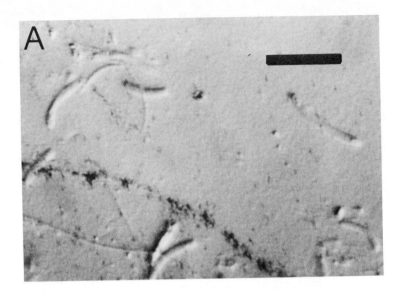

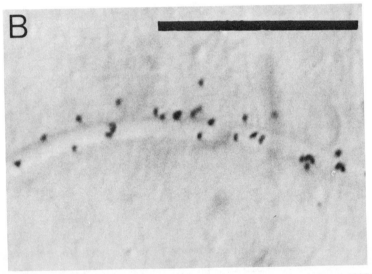

Fig. 9. Autoradiogram of cell material from the 55 °C Octopus Spring (A) and 40 °C Bath Lake (B) cyanobacterial mats which had been radiolabelled with 2-[^{14}C]-acetate. Silver grains surround cells which have incorporated [^{14}C]-acetate. Note the absence of silver grains around large curved rods of *Synechococcus lividus*. Bars indicate 15 μm.

microorganisms which are present, their physiology, and how they position themselves and function within the community. Yet the basic question of which microorganisms comprise the community remains largely unanswered, due to limitations of the enrichment

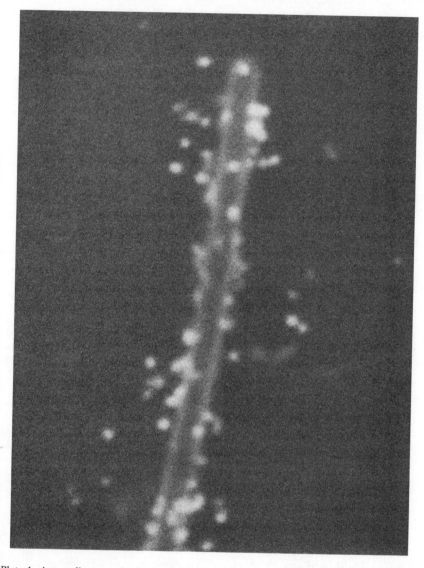

Plate 1. Autoradiogram of cell material from the 50 °C Octopus Spring cyanobacterial mat which was radiolabelled with 2-[^{14}C]-acetate and also stained with antiserum specific for *Chloroflexus aurantiacus*. Viewed by combined epifluorescence/darkfield microscopy. Silver grains are yellow while the filament is stained green by fluorescein isothiocyanate-conjugated anti-*Chloroflexus* antiserum. Magnification × 8500.

culture approach. New approaches which employ biochemical markers may help solve this problem. Some microbial groups, such as aerobic chemoorganotrophic microorganisms, have not yet been studied thoroughly. Much is known about how the activities of component microorganisms influence the chemical environment within mats. Though carbon, oxygen and sulfur cycling are relatively well understood, there is little known about cycling of other important elements, such as nitrogen. We are now learning that the chemical changes caused by microorganisms in the mats in turn cause secondary effects on the activities of the same or other microorganisms. Such effects are the bases for interactions among the members of mat communities. Continued studies of these relatively simple, totally microbial communities should help us develop the approaches needed to understand the composition and structure of natural microbial communities, and to understand the fundamentals of how microorganisms interact in natural environments.

ACKNOWLEDGEMENTS

We thank our colleagues who have collaborated in this work: Geoffrey Eglinton, Simon Brassell, Neil Robinson, Gary Dobson, Fernando Cassani (and others of the Organic Geochemistry Unit, University of Bristol), and Richard Castenholz (University of Oregon). This work was supported by grants from the Southern California Edison Company, the National Science Foundation, the Montana Department of Natural Resources and Conservation, and the Guest Scientist Program of the Royal Society of London. We also thank the US National Park Service for granting permission to work in Yellowstone National Park and for their co-operation and help.

REFERENCES

AWRAMIK, S. M. (1984). Ancient stromatolites and microbial mats. In *Microbial Mats: Stromatolites*, ed. Y. Cohen, R. W. Castenholz & H. O. Halvorson, pp. 1–22. New York, Alan R. Liss, Inc.

BALCH, W. E., FOX, G. E., MAGRUM, L. J., WOESE, C. R. & WOLFE, R. S. (1979). Methanogens: reevaluation of a unique biological group. *Microbiological Reviews*, **43**, 260–96.

BATESON, M. M. & WARD, D. M. (1985). The importance of glycolate as a photo-excretion product in a hot spring cyanobacterial mat. *Abstracts, Annual Meeting of the American Society for Microbiology*, p. 163.

BAULD, J. (1984). Microbial mats in marginal marine environments: Shark Bay, Western Australia, and Spencer Gulf, South Australia. In *Microbial Mats: Stromatolites*, ed. Y. Cohen, R. W. Castenholz & H. O. Halvorson, pp. 39–58. New York, Alan R. Liss, Inc.

BAULD, J. & BROCK, T. D. (1973). Ecological studies of *Chloroflexis*, a gliding photosynthetic bacterium. *Archiv für Mikrobiologie*, **92**, 267–84.

BAULD, J. & BROCK, T. D. (1974). Algal exretion and bacterial assimilation in hot spring algal mats. *Journal of Phycology*, **10**, 101–6.

BAULD, J., BURNE, R. V., CHAMBERS, L. A., FERGUSON, J. & SKYRING, G. W. (1980). Sedimentological and geobiological studies of intertidal cyanobacterial mats in north-eastern Spencer Gulf, South Australia. In *Biogeochemistry of Ancient and Modern Environments*, ed. P. A. Trudinger, M. R. Walter & B. J. Ralph, pp. 157–66. Canberra, Australian Academy of Science.

BEN-BASSAT, A. & ZEIKUS, J. G. (1981). *Thermobacteriodes acetoethylicus* gen. nov. and spec. nov., a new chemoorganotrophic, anaerobic, thermophilic bacterium. *Archives of Microbiology*, **128**, 365–70.

BLUMER, M., GUILLARD, R. R. L. & CHASE, T. (1971). Hydrocarbons of marine phytoplankton. *Marine Biology*, **8**, 183–9.

BOON, J. J., HINES, H., BURLINGAME, A. L., KNOK, J., RIJPSTRA, W. I. C., DELEEUW, J. W., EDMUNDS, K. & EGLINTON, G. (1983). Organic geochemical studies of Solar Lake laminated cyanobacterial mats. In *Advances in Organic Geochemistry – 1981*, ed. M. Bjorøy, pp. 207–27. Chichester, Wiley and Sons, Ltd.

BOUDOU, J. P., TRICHET, J., ROBINSON, N. & BRASSELL, S. C. (in press). Profile of aliphatic hydrocarbons in a recent Polynesian microbial mat. *International Journal of Environmental Analytical Chemistry*.

BROCK, T. D. (1978). *Thermophilic Microorganisms and Life at High Temperatures.* New York, Springer-Verlag.

BROCK, T. D. & FREEZE, H. (1969). *Thermus aquaticus* gen. n. and sp. n., a non-sporulating extreme thermophile. *Journal of Bacteriology*, **98**, 289–97.

CASTENHOLZ, R. W. (1969). Thermophilic blue-green algae and the thermal environment. *Bacteriological Reviews*, **33**, 476–504.

CASTENHOLZ, R. W. (1977). The effect of sulfide on the blue-green algae of hot springs. II. Yellowstone National Park. *Microbial Ecology*, **3**, 79–105.

CASTENHOLZ, R. W. (1979). Evolution and ecology of thermophilic microorganisms. In *Strategies of Microbial Life in Extreme Environments*, ed. M. Shilo, pp. 373–92. Berlin, Dahlem Konferenzen.

CASTENHOLZ, R. W. (1984). Composition of hot spring microbial mats: a summary. In *Microbial Mats: Stromatolites*, ed. Y. Cohen, R. W. Castenholz & H. O. Halvorson, pp. 101–19. New York, Alan R. Liss, Inc.

CHOLLET, R. (1977). The biochemistry of photorespiration. *Trends in Biochemical Sciences*, **2**, 155–9.

COHEN, Y., CASTENHOLZ, R. W. & HALVORSON, H. O. (eds) (1984). *Microbial Mats: Stromatolites.* New York, Alan R. Liss.

DOEMEL, W. N. & BROCK, T. D. (1974). Bacterial stromatolites: origin of laminations. *Science*, **184**, 1083–5.

DOEMEL, W. N. & BROCK, T. D. (1976). Vertical distribution of sulfur species in benthic algal mats. *Limnology and Oceanography*, **21**, 237–44.

DOEMEL, W. N. & BROCK, T. D. (1977). Structure, growth, and decomposition of laminated algal–bacterial mats in alkaline hot springs. *Applied and Environmental Microbiology*, **34**, 433–52.

EDMUNDS, K. L. H. (1982). Organic geochemistry of lipids and carotenoids in the Solar Lake microbial mat sequence. Ph.D. thesis, University of Bristol, UK.

EDMUNDS, K. L. H. & EGLINTON, G. (1984). Microbial lipids and carotenoids and their early diagenesis in the Solar Lake laminated microbial mat sequence. In *Microbial Mats: Stromatolites*, ed. Y. Cohen, R. W. Castenholz & H. O. Halvorson, pp. 343–89. New York, Alan R. Liss, Inc.

EGOROVA, L. A. & LOGINOVA, L. G. (1975). Distribution of extreme-thermophilic nonsporogenous bacteria in Tadzhikistan hot springs. *Microbiology*, **44**, 848–52.

FORK, D. C., MURATA, N. & SATO, N. (1979). Effect of growth temperature on the lipid and fatty acid composition, and the dependence on temperature of light-induced redox reactions of cytochrome *f* and of light energy redistribution in the thermophilic blue-green alga *Synechococcus lividus. Plant Physiology*, **63**, 524–30.

FOX, G. E., STACKEBRANDT, E., HESPELL, R. B., GIBSON, J., MANILOFF, J., DYER, T. A., WOLFE, R. S., BALCH, W. E., TANNER, R. S., MAGRUM, L., ZABLEN, L., BLAKEMORE, R., GUPTA, R., BONEN, L., LEWIS, B. J., STAHL, D. A., LUEHRSEN, K. R., CHEN, K. N. & WOESE, C. R. (1980). The phylogeny of prokaryotes. *Science*, **209**, 457–63.

GELPI, E., SCHNEIDER, H., MANN, J. & ORO, J. (1970). Hydrocarbons of geochemical significance in microscopic algae. *Phytochemistry*, **9**, 603–12.

GERDES, G. & KRUMBEIN, W. E. (1984). Animal communities in recent potential stromatolites of hypersaline origin. In *Microbial Mats: Stromatolites*, ed. Y. Cohen, R. W. Castenholz & H. O. Halvorson, pp. 59–83. New York, Alan R. Liss, Inc.

GIOVANONNI, S. J., REVSBECH, N. P., WARD, D. M. & CASTENHOLZ, R. W. (In Press) Obligately phototrophic *Chloroflexus*: primary production in hot spring microbial mats free of cyanobacteria. *Archives of Microbiology*.

GIOVANONNI, S. J. & SCHABTACH, E. (1983). Budding in the filamentous thermophile, *Isocystis pallida. Abstracts, Annual Meeting of the American Society for Microbiology*, p. 144.

GLOE, A. & RISCH, N. (1978). Bacteriochlorophyll c_s, a new bacteriochlorophyll from *Chloroflexus aurantiacus. Archives of Microbiology*, **118**, 153–6.

HAN, J., MCCARTHY, E. D. & CALVIN, M. (1968). Hydrocarbon constituents of the blue-green algae *Nostoc muscorum, Anacystis nidulans, Phormidium luridum*, and *Chlorogloea fritschii. Journal of the Chemical Society, Part C*, 2785–91.

JACKSON, T. J., RAMALEY, R. F. & MEINSCHEIN, W. G. (1973a). *Thermomicrobium*, a new genus of extremely thermophilic bacteria. *International Journal of Systematic Bacteriology*, **23**, 28–36.

JACKSON, T J., RAMALEY, R. F. & MEINSCHEIN, W. G. (1973b). Fatty acids of a non-pigmented, thermophilic bacterium similar to *Thermus aquaticus. Archiv für Mikrobiologie*, **88**, 127–33.

JAVOR, B. & CASTENHOLZ, R. W. (1981). Laminated microbial mats, Laguna Guerrero Negro, Mexico. *Geomicrobiol. J.*, **3**, 237–73.

JAVOR, B. & CASTENHOLZ, R. W. (1984). Invertebrate grazers of microbial mats, Laguna Guerrero Negro, Mexico. In *Microbial Mats: Stromatolites*, ed. Y. Cohen, R. W. Castenholz & H. O. Halvorson, pp. 85–94. New York, Alan R. Liss, Inc.

JØRGENSEN, B. B. (1982). Ecology of the bacteria of the sulphur cycle with special reference to anoxic–oxic interface environments. *Philosophical Transactions of the Royal Society of London, Series B*, **298**, 543–61.

JØRGENSEN, B. B., BLACKBURN, T. H. & COHEN, Y. (1981). Diurnal cycle of oxygen and sulfide microgradients and microbial photosynthesis in a cyanobacterial mat sediment. *Applied and Environmental Microbiology*, **38**, 46–58.

JØRGENSEN, B. B. & COHEN, Y. (1977). Solar Lake (Sinai). 5. The sulfur cycle of the benthic cyanobacterial mats. *Limnology and Oceanography*, **22**, 657–66.

JØRGENSEN, B. B., COHEN, Y. & REVSBECH, N. P. (1986). Transition from anoxygenic to oxygenic photosynthesis in a *Microcoleus chthonoplastes* cyanobacterial mat. *Applied and Environmental Microbiology*, **51**, 408–17.

JØRGENSEN, B. B., REVSBECH, N. P. & COHEN, Y. (1983). Photosynthesis and structure of benthic microbial mats: microelectrode and SEM studies of four cyanobacterial communities. *Limnology and Oceanography*, **28**, 1075–93.

KAWASUMI, T., IGARASHI, Y., KODAMA, T. & MINODA, Y. (1984). *Hydrogenobacter thermophilus* gen. nov., sp. nov., an extremely thermophilic, aerobic, hydrogen-oxidizing bacterium. *International Journal of Systematic Bacteriology*, **34**, 5–10.

KENYON, C. N. & GRAY, A. M. (1974). Preliminary analysis of lipids and fatty acids of green bacteria and *Chloroflexus aurantiacus*. *Journal of Bacteriology*, **120**, 131–8.

KNUDSEN, E., JANTZEN, E., BRYN, K., ORMEROD, J. G. & SIREVAG, R. (1982). Quantitative and structural characteristics of lipids in *Chlorobium* and *Chloroflexus*. *Archives of Microbiology*, **132**, 149–54.

KRISTJANSSON, J. K. & ALFREDSSON, G. A. (1983). Distribution of *Thermus* spp. in Icelandic hot springs and a thermal gradient. *Applied and Environmental Microbiology*, **45**, 1785–9.

KRISTJANSSON, J. K., INGASON, A. & ALFREDSSON, G. A. (1985). Isolation of thermophilic obligately autotrophic hydrogen-oxidizing bacteria, similar to *Hydrogenobacter thermophilus*, from Icelandic hot springs. *Archives of Microbiology*, **140**, 321–5.

KRUMBEIN, W. E., COHEN, Y. & SHILO, M. (1977). Solar Lake (Sinai). 4. Stromatolitic cyanobacterial mats. *Limnology and Oceanography*, **22**, 635–56.

LANGWORTHY, T. A., HOLZER, G., ZEIKUS, J. G. & TORNABENE, T. G. (1983). Iso- and anteiso-branched glycerol diethers of the thermophilic anaerobe *Thermodesulfotobacterium commune*. *Systematic Applied Microbiology*, **4**, 1–17.

LANGWORTHY, T. A., TORNABENE, T. G. & HOLZER, G. (1982). Lipids of archaebacteria. *Zentralblatt für Bakteriologie, Parasitenkunde, Infektionskrankheiten und Hygiene, Abteil I, C3*, 228–44.

LEWIN, R. A. (1970). New *Herpetosiphon* species (Flexibacterales). *Canadian Journal of Microbiology*, **16**, 517–20.

MCINERNEY, M. J. & BRYANT, M. P. (1981). Basic principles of bioconversions in anaerobic digestion and methanogenesis. In *Biomass Conversion Processes for Energy and Fuels*, ed. S. S. Sofer & O. R. Zaborsky, pp. 277–96. New York, Plenum.

MADIGAN, M. T. (1984). A novel photosynthetic purple bacterium isolated from a Yellowstone hot spring. *Science*, **225**, 313–15.

MADIGAN, M. T. (1986). *Chromatium tepidum* sp. nov., a thermophilic photosynthetic bacterium of the Family *Chromatiaceae*. *International Journal of Systematic Bacteriology*, **36**, 222–7.

MADIGAN, M. T. & BROCK, T. D. (1975). Photosynthetic sulfide oxidation by *Chloroflexus aurantiacus*, a filamentous, photosynthetic, gliding bacterium. *Journal of Bacteriology*, **122**, 782–4.

MADIGAN, M. T. & BROCK, T. D. (1977). Adaptation by hot spring phototrophs to reduced light intensities. *Archives of Microbiology*, **113**, 111–20.

MILLER, L. S. (1976). Effect of carbon dioxide on pigment and membrane content in the thermophilic blue-green alga, *Synechococcus lividus*. *Dissertation Abstracts International, B*, **37**, (4), 1561–2.

MINK, R. W. & DUGAN, P. R. (1977). Tentative identification of methogenic bacteria by fluorescence microscopy. *Applied and Environmental Microbiology*, **33**, 713–17.

NELSON, D. C. & CASTENHOLZ, R. W. (1981). Light responses of *Beggiatoa*. *Archives of Microbiology*, **131**, 146–55.

OSHIMA, M. & MIYAGAWA, A. (1974). Comparative studies on the fatty acid composition of moderately and extremely thermophilic bacteria. *Lipids*, **9**, 476–80.

OURISSON, G., ALBRECHT, P. & ROHMER, M. (1979). The hopanoids: palaeochemistry and biochemistry of a group of natural products. *Pure and Applied Chemistry*, **51**, 709–29.

PACE, N. R., STAHL, D. A., LANE, D. J. & OLSEN, G. J. (1985). Analyzing natural microbial populations by rRNA sequences. *ASM News*, **51**, 4–12.

PATEL, B. K. C., MORGAN, H. W. & DANIEL, R. M. (1985). *Fervidobacterium nodosum* gen. nov. and spec. nov., a new chemoorganotrophic, caldoactive, anaerobic bacterium. *Archives of Microbiology*, **141**, 63–9.

PIERSON, B. K. & CASTENHOLZ, R. W. (1974). A phototrophic gliding filamentous bacterium of hot springs, *Chloroflexus aurantiacus*. *Archives of Microbiology*, **100**, 5–24.

PIERSON, B. K., GIOVANONNI, S. J. & CASTENHOLZ, R. W. (1984). Physiological ecology of a gliding bacterium containing bacteriochlorophyll *a*. *Applied and Environmental Microbiology*, **47**, 576–84.

PIERSON, B. K. & HOWARD, H. M. (1972). Detection of bacteriochlorophyll-containing micro-organisms by infra red fluorescence photomicrography. *Journal of General Microbiology*, **73**, 359–63.

RAY, P. H., WHITE, D. C. & BROCK, T. D. (1971). Effect of temperature on the fatty acid composition of *Thermus aquaticus*. *Journal of Bacteriology*, **106**, 25–30.

REVSBECH, N. P., JØRGENSEN, B. B., BLACKBURN, T. H. & COHEN, Y. (1983). Microelectrode studies of the photosynthesis and O_2, H_2S and pH profiles of a microbial mat. *Limnology and Oceanography*, **28**, 1062–74.

REVSBECH, N. P., SØRENSEN, J., BLACKBURN, T. H. & LOMHOLT, J. P. (1979). Distribution of oxygen in marine sediments measured with a microelectrode. *Limnology and Oceanography*, **25**, 403–11.

REVSBECH, N. P. & WARD, D. M. (1984*a*). Microelectrode studies of interstitial water chemistry and photosynthetic activity in a hot spring microbial mat. *Applied and Environmental Microbiology*, **48**, 270–5.

REVSBECH, N. P. & WARD, D. M. (1984*b*). Microprofiles of dissolved substances and photosynthesis in microbial mats measured with microelectrodes. In *Microbial Mats: Stromatolites*, ed. Y. Cohen, R. W. Castenholz & H. O. Halvorson, pp. 171–88. New York, Alan R. Liss, Inc.

RICHARDSON, L. L. & CASTENHOLZ, R. W. (1983). Dark anaerobic metabolism and growth in the thermophilic cyanobacterium *Oscillatoria terebriformis. Abstracts, Annual Meeting of the American Society for Microbiology*, p. 282.

ROHMER, M., BOUVIER-NAVE, P. & OURISSON, G. (1984). Distribution of hopanoid triterpenes in prokaryotes. *Journal of General Microbiology*, **130**, 1137–50.

SAIKI, T., KOBAYASHI, Y., KAWAGOE, K. & BEPPU, T. (1985). *Dictyoglomus thermophilum* gen. nov., sp. nov., a chemoorganotrophic, anaerobic thermophilic bacterium. *International Journal of Systematic Bacteriology*, **35**, 253–9.

SANDBECK, K. A. & WARD, D. M. (1981). Fate of immediate methane precursors in low-sulfate, hot spring algal–bacterial mats. *Applied and Environmental Microbiology*, **41**, 775–82.

SANDBECK, K. A. & WARD, D. M. (1982). Temperature adaptations in the terminal processes of anaerobic decomposition of Yellowstone National Park and Icelandic hot spring microbial mats. *Applied and Environmental Microbiology*, **44**, 844–51.

SCHINK, B. & ZEIKUS, J. G. (1983). *Clostridium thermosulfurogenes* sp. nov., a new thermophile that produces elemental sulphur from thiosulphate. *Journal of General Microbiology*, **129**, 1149–58.

SCHOPF, J. W., HAYES, J. M. & WALTER, M. R. (1983). Evolution of earth's earliest ecosystems: recent progress and unsolved problems. In *Earth's Earliest Biosphere, its Origin and Evolution*, ed. J. W. Schopf, pp. 361–84. Princeton University Press.

SHEN, P. Y., COLES, E., FOOTE, J. L. & STENESH, J. (1970). Fatty acid distribution in mesophilic and thermophilic strains of the genus *Bacillus. Journal of Bacteriology*, **103**, 479–81.

SKYRING, G. W. (1984). Sulfate reduction in marine sediments associated with cyano-bacterial mats in Australia. In *Microbial Mats: Stromatolites*, ed. Y. Cohen, R. W. Castenholz & H. O. Halvorson, pp. 265–75. New York, Alan R. Liss, Inc.

SKYRING, G. W., CHAMBERS, L. A. & BAULD, J. (1983). Sulfate reduction in sedi-ments colonized by cyanobacteria, Spencer Gulf, South Australia. *Australian Journal of Marine and Freshwater Research*, **34**, 359–74.

STACKEBRANDT, E. & WOESE, C. R. (1984). The phylogeny of prokaryotes. *Micro-biological Science*, **1**, 117–22.

STAHL, D. A., LANE, D. J., OLSEN, G. J. & PACE, N. R. (1984). Analysis of hydrothermal vent-associated symbionts by ribosomal RNA sequences. *Science*, **224**, 409–11.

STAHL, D. A., LANE, D. J., OLSEN, G. J. & PACE, N. R. (1985). Characterization of a Yellowstone hot spring microbial community by 5S rRNA sequences. *Applied and Environmental Microbiology*, **49**, 1379–84.

STEWART, W. D. P. (1970). Nitrogen fixation by blue-green algae in Yellowstone thermal areas. *Phycologia*, **9**, 261–8.

STOLZ, J. F. (1984). Fine structure of the stratified microbial community at Laguna Figueroa, Baja California, Mexico. II. Transmission electron microscopy as a diagnostic tool in studying microbial communities *in situ*. In *Microbial Mats: Stromatolites*, ed. Y. Cohen, R. W. Castenholz & H. O. Halvorson, pp. 23–38. New York, Alan R. Liss, Inc.

TAYNE, T. A. (1983). The fate of fermentation products and glycollate in hot spring microbial mats with emphasis on the role played by *Chloroflexus aurantiacus*. M.S. thesis, Montana State University, USA.

TAYNE, T. A. & WARD, D. M. (1983). Fate of fermentation products in Yellowstone hot spring microbial mats. *Abstracts, Annual Meeting of the American Society for Microbiology*, p. 169.

TORNABENE, T. G. & LANGWORTHY, T. A. (1978). Diphytanyl and dibiphytanyl glycerol ether lipids of methanogenic archaebacteria. *Science*, **203**, 51–3.

TORNABENE, T. G., WOLFE, R. S., BALCH, W. E., HOLZER, G., FOX, G. E. & ORO, J. (1978). Phytanyl-glycerol ethers and squalenes in the archaebacterium *Methanobacterium thermoautotrophicum*. *Journal of Molecular Evolution*, **11**, 259–66.

WALTER, M. R., BAULD, J. & BROCK, T. D. (1972). Siliceous algal and bacterial stromatolites in hot spring and geyser effluents of Yellowstone National Park. *Science*, **178**, 402–5.

WARD, D. M. (1978). Thermophilic methanogenesis in a hot-spring algal–bacterial mat (71 to 30 °C). *Applied and Environmental Microbiology*, **35**, 1019–26.

WARD, D. M., BECK, E., REVSBECK, N. P., SANDBECK, K. A. & WINFREY, M. R. (1984). Decomposition of hot spring microbial mats. In *Microbial Mats: Stro-matolites*, ed. Y. Cohen, R. W. Castenholz & H. O. Halvorson, pp. 191–214. New York, Alan R. Liss, Inc.

WARD, D. M., BRASSELL, S. C. & EGLINTON, G. (1985). Archaebacterial lipids in hot-spring microbial mats. *Nature*, **318**, 656–9.

WARD, D. M. & OLSON, G. J. (1980). Terminal processes in the anaerobic degrada-tion of an algal–bacterial mat in a high sulfate hot spring. *Applied and Environ-mental Microbiology*, **40**, 67–74.

WARD, D. M. & WINFREY, M. R. (1985). Interactions between methanogenic and sulfate-reducing bacteria in sediments. *Advances in Aquatic Microbiology*, **3**, 141–79.

WEIMER, P. J. (1985). Thermophilic anaerobic fermentation of hemicellulose and hemicellulose-derived aldose sugars by *Thermoanaerobacter* strain B6A. *Archives of Microbiology*, **143**, 130–6.

WEIMER, P. J., WAGNER, L. W., KNOWLTON, S. & NG, T. K. (1984). Thermophilic anaerobic bacteria which ferment hemicellulose: characterization of organisms and identification of plasmids. *Archives of Microbiology*, **138**, 31–6.

WICKSTROM, C. E. (1980). Distribution and physiological determinants of blue-green algal nitrogen fixation along a thermogradient. *Journal of Phycology*, **16**, 436–43.

WICKSTROM, C. E. & CASTENHOLZ, R. W. (1973). Thermophilic ostracod: aquatic metazoan with the highest known temperature tolerance. *Science*, **181**, 1063–4.

WICKSTROM, C. E. & CASTENHOLZ, R. W. (1985). Dynamics of cyanobacterial and ostracod interactions in an Oregon hot spring. *Ecology*, **66**, 1024–41.

WIEGEL, J. (1980). Formation of ethanol by bacteria: a pledge for the use of extreme thermophilic anaerobic bacteria in industrial ethanol fermentation processes. *Experientia*, **36**, 1434–46.

WIEGEL, J., BRAUN, M. & GOTTSCHALK, G. (1981). *Clostridium thermoautotrophicum* species novum, a thermophile producing acetate from molecular hydrogen and carbon dioxide. *Current Microbiology*, **5**, 255–60.

WIEGEL, J. & LJUNGDAHL, L. G. (1981). *Thermoanaerobacter ethanolicus* gen. nov., spec. nov., a new, extreme thermophilic, anaerobic bacterium. *Archives of Microbiology*, **128**, 343–8.

WIEGEL, J., LJUNGDAHL, L. G. & RAWSON, J. R. (1979). Isolation from soil and properties of the extreme thermophile *Clostridium thermohydrosulfuricum*. *Journal of Bacteriology*, **139**, 800–10.

WIEGERT, R. G. & MITCHELL, R. (1973). Ecology of Yellowstone thermal effluent systems: intersects of blue-green algae, grazing flies (*Paracoenia*, *Ephydridae*) and water mites (*Partununiella*, *Hydrachnellae*). *Hydrobiologia*, **41**, 251–71.

WOESE, C. R. (1981). Archaebacteria. *Scientific American*, , **244**, 98–122.

WOESE, C. R. (1982). Archaebacteria and cellular origins: an overview. *Zentralblatt für Bakteriologie, Parasitenkunde, Infektionskrankheiten und Hygiene, Abteil I*, *C3*, 1–17.

ZARILLA, K. A. & PERRY, J. J. (1984). *Thermoleophilum album* gen. nov. and sp. nov., a bacterium obligate for thermophily and *n*-alkane substrates. *Archives of Microbiology*, **137**, 286–90.

ZEIKUS, J. G. (1979). Thermophilic bacteria: ecology, physiology and technology. *Enzyme and Microbial Technology*, **1**, 243–52.

ZEIKUS, J. G., BEN-BASSAT, A. & HEGGE, P. W. (1980). Microbiology of methanogenesis in thermal, volcanic environments. *Journal of Bacteriology*, **143**, 432–40.

ZEIKUS, J. G., DAWSON, M. A., THOMPSON, T. E., INGVORSEN, K. & HATCHIKIAN, E. C. (1983). Microbial ecology of volcanic sulphidogenesis: isolation and characterization of *Thermodesulfobacterium commune* gen. nov. and sp. nov. *Journal of General Microbiology*, **129**, 1159–69.

ZEIKUS, J. G., HEGGE, P. W. & ANDERSON, M. A. (1979). *Thermoanaerobium brockii* gen. nov. and sp. nov., a new chemoorganotrophic, caldoactive, anaerobic bacterium. *Archives of Microbiology*, **122**, 41–8.

THE COMMUNITIES – AQUEOUS SYSTEMS

PHYSICAL AND CHEMICAL PROPERTIES OF AQUATIC ENVIRONMENTS

GEORGE A. JACKSON

Institute of Marine Resources, A-018,
Scripps Institution of Oceanography,
La Jolla, CA 92093, USA

INTRODUCTION

Physical and chemical processes define the nature of planktonic systems. The transmission, absorption, and scattering of light control the availability of light energy for photosynthesis. Fluid motion controls organism location, mass transport rates, and efficiency of grazer feeding. The standard approach for analyzing processes in planktonic systems treats physical and chemical phenomena as if the included organisms were continuously distributed properties of the water rather than discrete entities separated by relatively large distances. Processes such as light absorbance, carbon production, nutrient uptake, zooplankton grazing, and nutrient regeneration are normally analyzed as continuous water properties rather than as the sum of the interactions of individuals. Process rates measured in a volume of water using uniformly added tracers have been considered to be the same as environmental rates which involve exchanges between individuals. This approach to the study of planktonic systems has generally worked well and need not be incompatible with insights derived from study of individual interactions. However, recent work has suggested that this emphasis on continuous distributions ignores important consequences of the discrete nature of organisms and of their interactions. This chapter contains an analysis of interactions between discrete organisms and a discussion of the ways in which physical and chemical processes constrain them.

The effect of particle size on fall velocity is an example of how the behavior of individual organisms can be used to provide information for the continuous distribution approach. Because organisms tend to have densities greater than the surrounding waters, they tend to sink under the pull of gravity. Their fall velocities are affected by organism size, density, shape, and motility in ways that have been discussed by Hutchinson (1957) and Smayda (1970), among others. Results from studies of bioparticle fall velocities can be used to estimate average biomass loss rates from water parcels, regardless of whether organisms are assumed to be continuously distributed.

Munk & Riley (1952), in a study later extended by Gavis (1976), analyzed organisms, particularly microalgae, as discrete entities and provided useful insights into the process of nutrient absorption which have not been incorporated easily into the continuous distribution approach. Munk & Riley noted that the uptake of a substance can be limited by the rate of transfer by molecular diffusion from solution to the organism surface. As a result, there is a region near an organism with low substance concentration and a small concentration gradient. Water motion reduces the size of this depleted region, increases the concentration gradient, and thereby increases transport to the cell. Transport enhancement by water motion depends on water velocity, organism size, organism shape, and other factors. Vertical settling can provide the fluid motion for planktonic organisms. One conclusion of these studies is that smaller organisms are less limited by diffusion kinetics than are larger organisms. This has been an important qualitative insight into cell morphology and nutrient uptake (e.g. Malone, 1980; Sournia, 1982), but this has not been generally translated into quantitative terms which can be incorporated into other process studies. Because the analysis considered discrete particles and constant nutrient concentrations, it addressed only half of the spatial and temporal nature of organism interactions.

Undetected interactions between individual planktonic organisms was suggested by McCarthy & Goldman (1979) to explain what they perceived as a puzzling problem in nutrient exchange between animal excretion and plant uptake. They noted that nutrient regeneration is not a process that occurs uniformly though a volume of water but rather occurs at discrete points for short time periods, with locally high concentrations where animals happen to be when they excrete. They suggested that short-term, high-speed nutrient uptake by phytoplankton within a few micrometers of zooplankton excretion would not be observed in the liter-sized water samples usually collected but could, nevertheless, supply the nutrient needs for the growth they expected in oligotrophic ocean areas. Such algal nutrient uptake would be at concentrations much higher than the extremely low surrounding nutrient concentrations. Goldman (1984) also noted the difficulty of obtaining accurate measurements of nitrogen uptake rates by microalgae for short exposures but noted that short-term uptake rates are much faster than steady state conditions would demand.

Jackson (1980) and Williams & Muir (1981) noted that the rapidity of diffusion on the small scale makes the exposure time of an alga to an animal excretion pulse extremely short, allowing only a small total amount of nutrient to be taken up during a pulse event, even if the uptake rate during the pulse time was fast. Jackson (1980) and Currie (1984a) calculated that the frequency of algal exposure to such a pulse was very small. Currie (1984b) also argued that the non-linear nature of algal nutrients uptake kinetics implies that nutrient uptake at pulse concentrations is inherently less efficient than uptake at lower average concentrations. Lastly, Jackson (1980) and Currie (1984b) disputed the requirement for a pulse mechanism to explain phytoplankton growth.

Lehman & Scavia (1982a, b) were able to show that rapid uptake of nutrients released by zooplankton does occur. They also made more elaborate diffusion-based calculations and argued that nutrient uptake within a very short distance of release was a significant part of the recycling process. Other authors continue to explore the issue (e.g. Scavia et al., 1984).

The importance of pulse uptake in ecological situations remains unclear. While Lehman & Scavia (1982a, b) did demonstrate that enhanced uptake of nutrients from pulses can exist in laboratory conditions, they did not establish its significance in ecological situations where different species of zooplankton and phytoplankton are present in different sizes and different abundances. Diffusion calculations made to date omit important aspects of the system. Calculations by Jackson (1980) and by Williams & Muir (1981) show the importance of spatial scales in determining short exposure times but do not adequately address the nature of plume generation or nutrient uptake. Calculations by Lehman & Scavia (1982b) are more inclusive, but assume nutrient uptake to be uniformly distributed, as if the phytoplankton were not discrete entities widely separated. Individual algal cells may be 1–50 μm in diameter, separated by as much as 1000 μm. Because Lehman and Scavia's calculation considered decreases of pulse concentrations within 100 μm of release, such an assumption of continuous nutrient uptake is inadequate. Furthermore, the relation between nutrient concentration and algal uptake rate is confused by the possibilities that short-term rates may be higher than we now estimate or, conversely, that diffusion limitation to individual cells makes actual rates lower than those calculated from laboratory results. There remain many unanswered questions regarding the role of short-time uptake rates. The difficulties of conducting experiments directly on the small spatial scales involved or of making theoretical calculations that address enough of the system complexity are largely to blame.

Azam & Ammerman (1984) have suggested a second type of microorganism–microorganism interaction which depends on the discrete nature of the organisms. They noted that marine bacteria live in a low organic concentration but appear to grow rapidly. Microalgae, however, are surrounded by zones of higher organic concentrations because they leak organic matter. Azam and Ammerman proposed that the chemotactic abilities of bacteria allow them to find and stay in the high organic concentration regions around the algal cells. Such behaviour would allow the bacteria to live in regions with higher nutrient concentrations than those measured in bulk water samples.

The two types of interactions proposed by McCarthy & Goldman (1979) and Azam & Ammerman (1984) emphasize the potential importance of organism–organism interactions even at the micrometer scale. Both are difficult to study experimentally because of the extremely small scale on which they would occur. Furthermore, it is unclear how to extrapolate laboratory results to environmental conditions. Both types of interaction occur at scales where our physical intuitions are not well developed, the physical process of diffusion is extremely important, the discrete nature of the organism must be considered, and concentrations are so small that substances are better considered as a few molecules rather than as fractions of moles.

My goal in this chapter is to explore how diffusion and the discrete nature of organisms constrain the systems. First, aspects of diffusion to and from bioparticles are considered, where spherical particles represent idealized microalgal or bacterial particles. This is then extended to describe bacterial chemotaxis around leaky algal cells. The results show which ecological conditions are most suitable for the different interactions to be strongest and suggest ways to extrapolate laboratory results to environmental conditions.

DIFFUSION, DISCRETE CELLS, AND THE FATE OF PULSES

At its simplest, mass transport to a cell from solution is determined only by molecular diffusion. Water motion enhances mass transport, but with simple molecular diffusion, a pulse lasts longer, is more confined, and should have its greatest effect.

There are two standard approaches to solving molecular diffusion problems. The first is to use differential equations describing concentration changes. The same differential equations are also used to describe temperature distributions in solids, voltage fields in electrostatics, potential flow in fluids, and ground water flow in aquifers. Solutions to various simple geometries and starting conditions have been assembled (e.g. Crank, 1956; Carslaw & Jaeger, 1959). The second approach is to consider the motion of any given molecule to be a random walk. The statistics of this motion, such as the average and mean square molecule position, are calculated by analytical techniques or by computer simulations. The fraction of a large number of molecules originally at one location having a particular fate is equal to the probability that one molecule has that same fate. Studying the probabilities of the motion of one molecule can provide information about the fate of a discrete release, such as a pulse. The two different approaches, one involving a differential formulation and the other the random walk, complement each other by the ease with which they answer different aspects of the same problem.

Diffusion-determined concentrations around a sphere at steady state

The standard differential equation for diffusion to a spherical particle in the absence of fluid motion or diffuse sources or sinks is

$$\partial C/\partial t = D\nabla^2 C \tag{1}$$

where D is the diffusivity, C is the concentration, and ∇^2 equals $(\partial^2/\partial x^2 + \partial^2/\partial y^2 + \partial^2/\partial z^2)$.

For the simple steady state case of a sphere of radius a with concentration C_0 far away (at infinity) and concentration C_1 at the sphere's surface, the concentration at a distance r from the sphere center is:

$$C = (C_1 - C_0)ar^{-1} + C_0 \tag{2}$$

The flow of molecules towards or away from the cell per unit time, F, is given by

$$F = -4\pi Da(C_1 - C_0) \tag{3}$$

Maximum flow inward occurs when the concentration at a is the lowest possible value, 0. Munk & Riley (1952) assumed that C_1 was 0 in their study of mass transport and flow rate. Gavis (1976) modified their results by assuming that F was related to C_1 by a Monod relation. Either F or C_1 is sufficient to define the cellular concentration (eqn 3).

Eqn 2 can be rewritten in terms of F:

$$C = C_0 - F(4\pi D)^{-1}r^{-1} \tag{4}$$

The sphere has proven to be a useful approximation to a cellular microorganism because the spherical case is more easily solved and provides results similar to those for other more complicated but more realistic shapes. For studies of the flux into a cell absorbing some constituent, C_0 is usually assumed to be 0 because the result is the maximum inward flux.

Fate of a molecule released near a sphere

A molecule present at a distance r from a sphere of radius a has a finite probability of ever hitting the sphere as part of the random walk that is molecular diffusion. This probability (p_0) is (Berg, 1983)

$$p_0 = ar^{-1} \tag{5}$$

The probability varies with distance from the sphere, ranging from 1 at the sphere's surface to 0 when the molecule is infinitely far away. It falls to 0.5 when the molecule is a spherical radius away from the surface.

If molecules are released uniformly in space, then the probability, p_1, that molecules released within a distance R of the sphere's center will strike the sphere is determined by integration:

$$p_1(R) = (4\pi/3)^{-1}(R^3 - a^3)^{-1} \int_a^R (a/r)4\pi r^2 dr \tag{6}$$

$$= 1.5a(R + a)(R^2 + aR + a^2)^{-1} \tag{7}$$

$$\approx 1.5a/R \text{ for } R^2 \gg a^2 \tag{8}$$

The fraction of molecules released within a given distance from a sphere that ultimately diffuse to it varies inversely with the distance.

The fate of a molecule released in a suspension of perfectly absorbing spherical particles is more complicated to calculate. If the spheres are randomly distributed, what is the probability that the molecule will be absorbed within a distance R? The probability of the molecule being absorbed by a particular sphere at r is less than that in eqn 5, because the molecule could be absorbed by a different particle first. To solve the problem in which particles compete for the molecule is more difficult than the problem where they do not. If the particles do not compete and if ρ is the particle abundance (number/volume), then the probability, dp_p, of a particle being within a distance ranging from r to $(r + dr)$ is

$$dp_p = 4\rho\pi r^2 dr \qquad (9)$$

The probability of the molecule hitting a particle in the range of r to $(r + dr)$ distant is the probability of a particle being there, multiplied by the probability of the molecule hitting that particle:

$$dp_2 = dp_p \cdot p_0 \qquad (10)$$

$$= 4\pi\rho a r dr \qquad (11)$$

The total probability p_2 of a particle being taken up within a distance R of its initial position is then the sum (integral) of all the dp_2:

$$p_2(R) = \int_a^R 4\pi\rho a r dr \qquad (12)$$

$$= 2\pi\rho a R^2 \qquad (13)$$

This estimate is not the true probability but an overestimate because of the assumption that the particles do not compete for the molecule.

The distance $R_{1/2}$ at which p_2 is 0.5 is given by

$$R_{1/2} = (4\pi\rho a)^{-0.5} \qquad (14)$$

Diffusion, random walks, and planktonic particles

Bacteria and microalgae are neither perfectly absorbing nor are they spheres. However, using the result of calculations that assume they are so allows us to place an upper boundary on the nutrient uptake that is independent of uptake rate measurements. We can determine a minimum distance that a pulse expands before it is taken out of solution by algal uptake.

Environmental phytoplankton sizes and abundances vary widely. Selected examples from the Pacific Ocean show abundances ranging from 4×10^2 to 10^6 cells cm^{-3} and dominant algal radii ranging from 0.5 to $13.0\,\mu$m (Table 1). These situations demonstrate typical values for the fraction of nutrient input that could be consumed by the nearest microalgal cell ($p_1(S/2)$) and the distance that a spike spreads

Table 1. *Pulse fate for different environmental conditions. Values for algal abundance* (ρ) *and algal radius* (a) *are representative of reported values. Costa Rica Dome is an extremely eutrophic area. The populations from the area off Hawaii are of cyanobacteria* (5) *and* Chrysochromulina-*like cells* (6). *The biomass density of the latter was about 15 times greater than that of the former. The area off southern California had a* Prorocentrum micans *bloom. S is the separation distance,* $\rho^{-1/3}$. $R_{1/2}$ *is the distance a pulse travels before half is absorbed.* $p_1(S/2)$ *is the fraction of a pulse absorbed by the nearest alga, which is within* $S/2$ *of the pulse.*

Case	a (μm)	(cells cm^{-3})	S (μm)	$P_1(S/2)$	$R_{1/2}$ (cm)	Location	Citation
1	3.0	10^5	220	0.041	0.052	Laboratory	Lehman and Scavia, 1982a, b
2	2.0	4×10^2	1400	0.004	1.0	North Pacific Central Gyre (28°N 15°W)	Beers et al., 1975
3	0.5	10^6	100	0.015	0.040	Costa Rica Dome (9°25'N 93°30'W)	Li et al., 1983
4	0.5	10^4	460	0.003	0.40	Eastern tropical Pacific (9°45'N 83°45'W)	Li et al., 1983
5	0.5	3×10^3	690	0.002	0.73	Off Hawaii (~21°N 158°W)	Laws et al., 1984
6	2.0	10^3	1000	0.006	0.63	Off Hawaii (~21°N 158°W)	Laws et al. 1984
7	13.0	4×10^2	1400	0.028	0.39	Off southern California	Eppley et al., 1977

before half could be consumed ($R_{1/2}$) if the algal cells were perfect sinks. Potential uptake by the nearest alga ranged from 0.2 to 2.8%; distance for half uptake ranged from 0.04 to 1.0 cm. The shortest distance occurred for the very eutrophic situation with extremely small cells at the Costa Rica Dome (Table 1, case 3). A *Prorocentrum micans* bloom off southern California (case 7) and oligotrophic situations (cases 2, 5, 6) had widths at half uptake closer to 1 cm. The Lehman & Scavia (1982a, b) laboratory experiment (case 1) was more similar to the eutrophic Costa Dome situation than to those of the other oceanic areas.

This analysis has considered a phytoplankton population to have a single, well defined size or, as in cases 5 and 6, has considered two different populations separately. Phytoplankton in most communities occur with a range of sizes. A fuller treatment that includes a distribution of microalgal sizes can easily be made by modifying eqns 9–14 to include an integration over the algal size spectrum.

The results would be more complete but would not alter the importance of algal size and abundance. A more important effect that should be accounted for is algal competition for the molecules. This would result in much larger values of $R_{1/2}$.

These arguments imply that the uptake of a nutrient pulse is more localized for greater algal abundances and larger algal cells. Both cannot be increased indefinitely. What are the optimal conditions for uptake for a given biomass density? If the biomass of an individual algal cell is proportional to the a^2 (Mullin, Sloan & Eppley, 1966), then the abundance is proportional to a^{-2} and $R_{1/2}$ is inversely proportional to the square root of a. Small cells are more effective per unit biomass with respect to localized uptake of pulses. Small is better.

The relative importance of pulses to phytoplankton also depends on the actual concentrations of the pulses and of other nutrient sources. Without information about the actual concentration field around a zooplankter and data on the background concentrations in the water, it is difficult to make exact predictions. Maximum concentrations from a pulse decrease rapidly with distance from a zooplankter (Jackson, 1980). If the initial source is a plume $100\,\mu m$ thick, the maximum concentration $0.1\,cm$ from the plume is less than 10^{-3} of the initial concentration. The concentration at $0.04\,cm$, the $R_{1/2}$ of the Costa Rica Dome, is not much greater. At least half of nutrient uptake takes place when the concentration of a pulse is decreased by 1000 or more. If the size of the initial plume is smaller, as it would be if microzooplankton excretion were important, then the dilution at $R_{1/2}$ is even greater. The tendency of small algae to be eaten by small grazers suggests that this will be true for situations where the small algae dominate.

The conditions most conducive to pulse uptake are those where microalgae are most abundant. Such a condition is more characteristic of eutrophic than oligotrophic conditions.

BACTERIAL CHEMOTAXIS AROUND A LEAKING MICROALGAL CELL

Chemotactic behavior

Like most planktonic organisms, bacteria are not immobile particles whose locations are subject only to fluid motions but rather have limited ability to find and move to suitable conditions. One important need for a microorganism can be that for a better chemical environment. The ability of organisms to move, on average, towards a preferred chemical environment is known as chemotaxis. This behavior has been widely observed in aquatic bacteria and dinoflagellates (e.g. Fitt, 1985; Paerl & Gallucci, 1985; Spero, 1985) and intensively studied for those laboratory standards *Escherichia coli* (e.g. Berg & Brown, 1972; Berg & Purcell, 1977; Berg, Manson & Conley, 1982; Block, Segall & Berg, 1982, 1983) and *Salmonella typhimurium* (e.g., Koshland, 1979; Macnab & Han, 1983). As a result, much is known about the mechanism of swimming, the molecular nature of bacterial response, and bacterial response kinetics.

A bacterial cell is too small to be able to discern concentration gradients by sensing concentration differences over its length. Instead, a cell detects and responds to the temporal changes of concentrations surrounding it. A cell swimming up a concentration gradient sees an increase in concentration with time. The cell's motion translates a spatial gradient that it cannot sense to a temporal change to which it can respond.

Bacterial motion typically consists of two phases. The first is a run phase, in which a bacterium moves in a nearly straight trajectory. The forward motion of a run is interrupted by the transition to the second phase, the tumble, during which the bacterium rotates its orientation to a random new direction. After the tumble, it begins a new run. The length of a run is a random variable whose statistics are changed by increases or decreases in concentration of the sensed substance. While chemical concentrations improve, the run length, on average, lengthens; when chemical concentrations worsen, the run length, on average, shortens. The bacterial motion in the absence of chemical cues is a form of Brownian motion. In the presence of a chemical gradient, the changes in average run length give the cell a random walk with a bias which provides a net drift towards better conditions.

Brown & Berg (1974) developed a model to describe the influence of concentration history on run length. Mean bacterial run length depends on the rate of change of binding of the attractant to a protein receptor. Brown & Berg followed 'synthetic' bacteria as they swam through a 'synthetic' concentration field by using a Monte Carlo simulation to generate the random changes from run to tumble mode and to select new travel directions during tumbles. Their model results agreed with results from laboratory studies on bacteria.

Bacterial chemotaxis was invoked to explain bacterial–algal interactions as early as 1894, when Engelmann observed bacteria clustering around photosynthesizing microalgae under a microscope coverslip. In this case, bacteria in anoxic water clustered around the algae when there was enough light to support photosynthesis but not otherwise. Engelmann explained this clustering as the result of chemotactic movement to the high oxygen conditions around the algae.

It has long been known that algae naturally leak organic matter. There is a range of estimates for the rate at which this occurs. Recent measurements ranged from 5 to 30%, but were typically about 10%, of net photosynthesis (Mague *et al.*, 1980). For a given specific leakage rate (expressed as fraction of algal mass leaked per day), the concentration around an alga is a function of cell size. If the specific growth rate of algal cells is typically $1.0\,day^{-1}$, then typical specific leakage rates could be $0.1\,day^{-1}$.

Combining molecular diffusion and bacterial chemotaxis

I have developed a computer model which simulates the behavior of marine bacteria around an algal cell. It uses the model of Brown & Berg (1974) to describe behavior and a molecular diffusion model to describe the concentration around a cell (G. A. Jackson, unpublished data). With this system, I have tested factors which allow chemotactic orientation as proposed by Azam & Ammerman (1984).

In the presence of a very leaky, large cell, a bacterium can dance its way closer (Fig. 1). The average movement of a large number of bacteria initially uniformly distributed around the alga is more deliberate (Fig. 2). The rate of this net movement is a convenient way of comparing the effects of different conditions. For an alga of radius $10\,\mu m$, increased specific leakage rates cause increased drift rates toward the alga (Fig. 3); for an alga of radius $2.5\,\mu m$, there is no drift toward the alga for specific leakage rates as high as $2\,day^{-1}$. This suggests that there is a size below which chemotactic behavior cannot be used to sense another cell. Additional evidence for this comes from more extensive simulations of the system (G. A. Jackson, unpublished data).

An important determinant of bacterial sensitivity is the half-saturation binding constant, K_D, of the protein which acts as the chemical detector on the bacterial surface. The results discussed thus far have used the value of $0.1\,mM$ determined by Brown & Berg (1974) for *E. coli* movement towards aspartate. Increasing the value of K_D

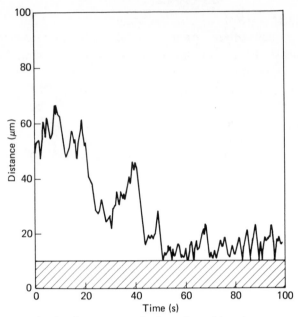

Fig. 1. Computer-simulated movement of a bacterium with a chemosensory response to a leaking alga. The alga has a radius of $10\,\mu m$ and is leaking 100% of its carbon content per day. The random nature of the bacterial movement causes its position to vary but its chemotactic response gives it a net movement toward the alga. Distance is from algal center. Shaded portion represents the alga.

by a factor of 1000, to $0.1\,\mu M$, does allow a bacterium to detect smaller sources (Fig. 4), but there is still a limit to the organism size that can be detected. Another constraint comes from the presence of a background concentration of the sensed substance (Fig. 5). Notice that the presence of a background concentration works to inhibit the effectiveness of lower values of K_D smaller than a critical value.

Importance of size and leakage rate

There are two important aspects to a chemotactic interaction: the sensing bacterium and the leaking alga. As the leaking alga becomes smaller, the chemical signal around it becomes smaller for two reasons. The first is that a smaller alga leaks less material to the water because is has less material available to leak. Because the biomass of a phytoplankter increases approximately as the square of its radius (Mullin, Sloan & Eppley, 1966), an algal cell with twice the radius of a smaller one will have approximately 4 times the biomass. If both have the same specific growth rate and the same fraction of this

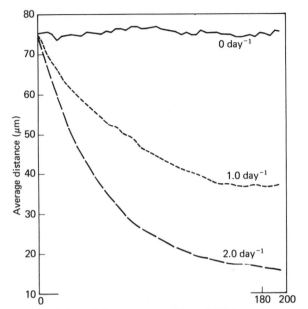

Fig. 2. Computer simulation of the average position of 800 bacteria around an algal cell with radius of $10\,\mu m$ and leakage rates of 0, 1 and $2\,day^{-1}$ as a function of time. The bacteria are initially uniformly distributed in a sphere that has a radius of $100\,\mu m$ around the algal center (10 times that of the algal cell). Distance is from algal center.

growth leaking, the larger will leak approximately 4 times as much material. There is a smaller concentration signal for a bacterium to sense around the smaller alga. The second reason is that the concentration of the leaked material decreases faster with increasing distance from a small cell than from a large one. For the case of two algal cells, one with radius A and the other with radius $2A$, the concentration at distance A from the surface is half its surface concentration for the smaller cell but two-thirds its surface concentration for the larger cell (eqn 2). Thus, the distance over which a cell projects a signal is also a function of its size. This effect has been compensated for in these results by simulating bacterial behavior between the alga and a distance 10 times its radius. Because the volumes affected differ, two ecosystems with different-sized algae have different ecological interactions.

The ability of a bacterium to find an algal source through chemotaxis is constrained. A sensory protein with a smaller half-saturation concentration can increase the bacterial sensitivity to a weak source. However, the background concentration of the sensed substance acts to mask the signal. This fact was used by Paerl & Gallucci

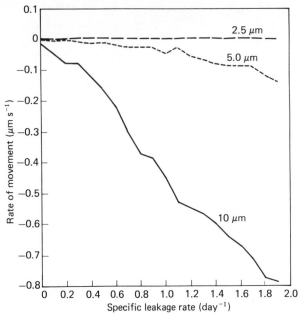

Fig. 3. The average rate at which 800 bacteria move relative to the alga as a function of leakage rate for the first 50 s of a computer simulation. This is essentially the initial slopes of the lines in Fig. 2. Negative values indicate that the average distance is decreasing as the bacteria approach the algal cell. The situations for algae of diameter 10, 5 and 2.5 µm are shown.

(1985) to demonstrate the chemotactic attraction of bacteria to nitro-gen-fixing algae. The statistical nature of molecular reactions at low concentrations places the ultimate constraint on detection. Berg & Purcell (1977) have noted that the small number of molecules available to react at the bacterial surface makes the relation between number of bound sites and average concentration a statistical one. A bacterial cell can overcome the large statistical noise associated with sensing small numbers of molecules by effectively averaging any sensory signals over time, but this is achieved by slowing the response time. Slower response time makes a bacterium more sus-ceptible to changes in position and direction caused by thermal Brown-ian motion and rotation. These factors place absolute constraints on the sensitivity with which bacteria can detect small sources.

The potential role of Brownian rotation suggests that large size is an important property of chemotactic bacteria. Berg (1983) has calculated that the root mean square (RMS) rotation caused by ther-mal Brownian rotation for *E. coli* is 30° in 1 s. This rate is similar to the angular deviation observed during *E. coli* runs, i.e. 27° in

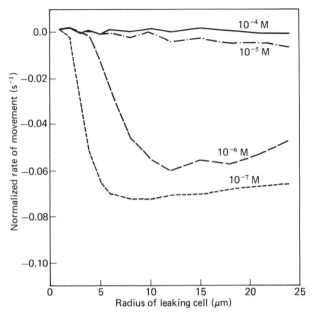

Fig. 4. Effect of algal size on bacterial sensitivity. Shown is the average normalized rate of movement for 800 bacteria as a function of leakage rate for the first 50 s of a simulation. Rates are normalized by the algal radius. The sensed molecule leakage rate is assumed to be 1.7% of the total cell carbon per day, calculated assuming that 10% of the total cell carbon is leaked per day and that half of this is in the form of a 3C molecule (such as an amino acid) detected by the bacteria. Half-saturation binding constants for these molecules are 10^{-4}, 10^{-5}, 10^{-6} and 10^{-7} M.

1 s. Because this rate is strongly dependent on size and is independent of behavioral responses, a bacterium with half the radius (0.5 μm) has the same RMS angular deviation in 0.12 s, and a bacterium with a radius of 0.25 μm has the same in 16 ms. The higher angular deviations make it more difficult for smaller organisms to maintain a direction during a run. Without a consistent heading, a cell cannot bias its runs in the direction of improving conditions. As a result, chemotaxis should be more effective with larger bacteria.

Brownian rotation could present a problem for marine bacteria because of their small size. Typical radii are on the order of 0.25 μm (e.g. Azam & Hodson, 1977; Fuhrman, 1981). Chemotactic behavior should be less of a problem for marine bacteria which are large or more elongate in shape.

The results discussed here imply that there is a minimum size of alga that bacteria can detect and around which they can maintain position. The calculations shown here suggest that this size is about 2 μm.

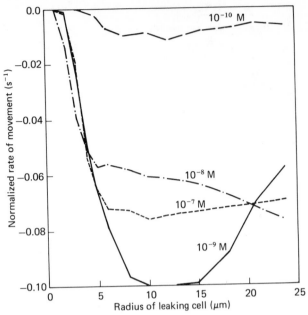

Fig. 5. Effect of algal size on bacterial sensitivity in the presence of a background concentration of the sensed substance of 10nM. Half-saturation binding constants for cases shown are 10^{-7}, 10^{-8}, 10^{-9} and 10^{-10} M. Other conditions are as in Fig. 4.

SIGNIFICANCE OF DIFFUSION AND CHEMOTAXIS IN NUTRIENT UTILIZATION

Mitchell, Okubo & Fuhrman (1985) considered the problem of bacterial chemotaxis to leaky algal cells by making several scaling arguments. They concluded that the microzone of high concentration around an algal cell with a radius of 5 μm would be too small to be very effective. The arguments presented here show this is the cell size at which algal cells tend to become undetectable by chemotaxis. Mitchell *et al.* also noted that algal cells tend to fall at rates faster than bacteria can swim but that this fall is slowed at density gradients. They concluded that the most likely zone in which chemotaxis would be important would be at a steep density gradient. The present study has not addressed the effect of bacterial settling. Algal motion may, in fact, limit the ability of bacteria to be chemotactically attracted to them. However, the hydrodynamics of sinking particles is complicated and should be explicitly considered in a study of chemotactic behavior.

The nature of bacterial chemotaxis offers an insight into the nature of interactions between organisms. The inability to use chemical cues to find organisms below a particular size should not be unique to bacteria but should be common to all organisms trying to sense other organisms smaller than that size. Chemotaxis is an important means allowing one organism to find another. It is important for feeding, for example, by allowing a copepod to detect edible food particles in the presence of inedible plastic spheres of the same size (Price & Paffenhöfer, 1986). It enables two zooplankters to find each other and mate (Katona, 1973). It could allow bacteria to find leaky food sources (Azam & Ammerman, 1984). It does not allow any of these if the particle to be sensed is too small.

That small particles, less than about 2 μm radius, are not detectable from a distance through chemotactic, visual, or any other sensory means suggests certain ways in which microorganisms in this size range interact. Interactions must involve physical contact, although there may be chemical aspects to this touching. Aggregation of small microorganisms must involve physical contact or association with a larger particle/organism releasing a detectable chemical signal. Any feeding on small microorganisms should be quite mechanical.

Observed microorganism feeding patterns are in accord with these deductions. Fenchel (1984) has measured the feeding rates of various bacterivores. His analysis of the feeding processes relies on simple physical models with no active prey detection.

Similarly, bacteria of the genus *Bdellovibrio* have been observed to prey on other aquatic bacteria (Varon & Shilo, 1980). A bdellovibrio finds its prey by making a straight run which is interrupted only by its striking another organism, such as its prey. The absence of adequate chemical cues from the small prey organisms makes this an efficient way to feed. If a bdellovibrio were to use a Brownian walk to find its prey, it would not only have no chemical cues but it would also spend considerable time re-searching the same volumes of water. The uninterrupted run is the reasonable response to the lack of chemical cues from bacterial-sized prey.

McManus & Fuhrman (1986) have found that some bacterivores feed on artificial spheres at the same rates as they feed on bacteria. The implication is that bacterivores are unable to distinguish between the two particle types and certainly unable to find the bacteria with any extra skill.

Chemosensing has been established as an important part of copepod feeding behavior, from both laboratory studies (Friedman & Strickler, 1975; Price & Paffenhöfer, 1986) and mathematical simulation studies (Andrews, 1983). There is a lower limit to the particle size on which copepods feed. For example, Frost (1972, 1977) reported that *Calanus pacificus* does not feed on particles smaller than about 14 μm diameter. Mullin (1980) has suggested that the lower limit on crustacean feeding is about 5 μm diameter. In contrast, salps filter food particles as small as bacteria with diameters of about 1 μm (Madin, 1974; Harbison & McAlister, 1979). Salp feeding involves the passive filtration of particle-containing water through a non-sensate mucus web. Rubenstein & Koehl (1977) have suggested that the lower limit of copepod filtration is determined by the nature of the filtration process and the low removal efficiency of particles in this size range. The inefficiency of small particle detection by chemosensing suggests another explanation, that is that copepods cannot detect bioparticles smaller than about 2.5 μm radius. While the details of chemosignal processing must be different in a larger, multicelled organism than in a bacterium, the problem of insufficient signal must occur for the copepod trying to find a small cell. The fact that these theoretical studies of bacterial chemotaxis yield an algal cell of a minimum detectable size that is similar to the minimum crustacean food size suggests that this limit is set by the inability of chemosensing to detect small organisms.

Despite pulse uptake by phytoplankton and the chemotactic orientation of bacteria, a large amount of released material is not taken up in the immediate vicinity of its release. This material must be taken up within the oceanic system if it is not to accumulate there. If there are cells which rely on localized uptake to survive, then there must also be cells which rely on the diffuse sources for their nutrient needs. The possible role of the localized uptake invoked in the pulse and chemotactic hypotheses does not eliminate the importance of more steady state uptake but rather adds a new method of survival through specialization. Localized uptake could thus provide more niches in plankton ecosystems.

CONCLUSION

The arguments in this chapter have emphasized the determination of conditions under which different types of microorganism interactions could occur in planktonic ecosystems. The difficulty of doing experiments on the very small scales involved places a premium on knowing how to relate scenarios, experiments, and environmental

CHEMICAL PROPERTIES OF AQUATIC ENVIRONMENTS 231

situations. The results developed here imply that pulses of nutrients are most important in high densities of large algae. Optimal conditions for chemosensory interactions include large leaking cells and low background concentrations of the sensed molecule. High abundance of cells would increase the relative volume influenced by the leakers. The large and small sizes and low and high densities which yield different results are those that are found within naturally occurring conditions. As such, they should be controlled in experimental studies and their importances acknowledged when trying to extrapolate from one situation to another.

ACKNOWLEDGEMENTS

This work was done in collaboration with E. Stewart, who assisted with the computer programming. It was funded by ONR Contract N00014-85-K-0473.

REFERENCES

ANDREWS, J. C. (1983). Deformation of the active space in the low Reynolds number feeding current of calanoid copepods. *Canadian Journal of Fisheries and Aquatic Science*, **40**, 1293–302.

AZAM, F. & AMMERMAN, J. W. (1984). Cycling of organic matter by bacterioplankton in pelagic marine ecosystems: microenvironmental considerations. In *Flows of Energy and Material in Marine Ecosystems*, ed. M. J. R. Fasham, pp. 345–60. New York, Plenum Press.

AZAM, F. & HODSON, R. E. (1977). Size distribution and activity of marine microheterotrophs. *Limnology and Oceanography*, **22**, 492–501.

BEERS, J. R., REID, F. M. H. & STEWART, G. L. (1975). Microplankton of the North Pacific Central Gyre: population structure and abundance, June 1973. *Internationale Revue der Gesamten Hydrobiologie*, **60**, 607–38.

BERG, H. C. (1983). *Random Walks in Biology*. Princeton, N.J., Princeton University Press.

BERG, H. C. & BROWN, D. A. A. (1972). Chemotaxis in *Escherichia coli* analyzed by three-dimensional tracking. *Nature*, **239**, 500–4.

BERG, H. C., MANSON, M. D. & CONLEY, M. P. (1982). Dynamics and energetics of flagellar rotation in bacteria. *Symposia of the Society for Experimental Biology*, **35**, 1–31.

BERG, H. C. & PURCELL, E. M. (1977). Physics of chemoreception. *Biophysics Journal*, **20**, 193–219.

BLOCK, S. M., SEGALL, J. E. & BERG, H. C. (1982). Impulse responses in bacterial chemotaxis. *Cell*, **31**, 2115–226.

BLOCK, S. M., SEGALL, J. E. & BERG, H. C. (1983). Adaptation kinetics in bacterial chemotaxis. *Journal of Bacteriology*, **154**, 312–23.

BROWN, D. A. & BERG, H. C. (1974). Temporal stimulation of chemotaxis in *Escherichia coli*. *Proceedings of the National Academy of Sciences of the United States of America*, **71**, 1388–92.

CARSLAW, H. S. & JAEGER, J. C. (1959). *Conduction of Heat in Solids*, 2nd edn. Oxford, Oxford University Press.

CRANK, J. (1956). *The Mathematics of Diffusion*. Oxford, Clarendon Press.

CURRIE, D. J. (1984*a*). Phytoplankton growth and the microscale nutrient patch hypothesis. *Journal of Plankton Research*, **6**, 591–9.

CURRIE, D. J. (1984*b*). Microscale nutrient patches: do they matter to the phytoplankton? *Limnology and Oceanography*, **29**, 211–14.

ENGELMANN, T. W. (1894). Die Erscheinungsweise der Sauerstoff ausscheidung chromophyllhaltiger Zellen im Licht bei Anwendung der Bacterienmethode. *Pfluegers Archiv fuer die Gesamte Physiologie des Menschen und der Tiere*, **57**, 375–86.

EPPLEY, R. W., HARRISON, W. G., CHISHOLM, S. W. & STEWART, E. F. (1977). Particulate organic matter in surface waters off Southern California and its relationship to phytoplankton. *Journal of Marine Research*, **35**, 671–96.

FENCHEL, T. (1984). Suspended marine bacteria as a food source. In *Flows of Energy and Material in Marine Ecosystems*, ed. M. J. R. Fasham, pp. 301–15. New York, Plenum Press.

FITT, W. K. (1985). Chemosensory responses of the symbiotic dinoflagellate *Symbiodinium microadriatica* (Dinophycae). *Journal of Phycology*, **21**, 62–7.

FRIEDMAN, M. M. & STRICKLER, J. R. (1975). Chemoreceptors and feeding in calanoid copepods (Arthropoda: Crustacea). *Proceedings of the National Academy of Sciences of the United States of America*, **72**, 4185–8.

FROST, B. W. (1972). Effects of size and concentration of food particles on the feeding behavior of the marine planktonic copepod *Calanus pacificus*. *Limnology and Oceanography*, **17**, 805–15.

FROST, B. W. (1977). Feeding behaviour of *Calanus pacificus* in mixtures of food particles. *Limnology and Oceanography*, **22**, 472–91.

FUHRMAN, J. A. (1981). Influence of method on the apparent size distribution of bacterioplankton cells: epifluorescence microscopy compared to scanning electron microscopy. *Marine Ecology – Progress Series*, **5**, 103–6.

GAVIS, J. (1976). Munk and Riley revisited: nutrient diffusion transport & rates of phytoplankton growth. *Journal of Marine Research*, **34**, 161–79.

GOLDMAN, J. C. (1984). Oceanic nutrient cycles. In *Flows of Energy and Material in Marine Ecosystems*, ed. M. J. R. Fasham, pp. 137–70. New York, Plenum Press.

HARBISON, G. R. & MCALISTER, V. L. (1979). The filter-feeding rates and particle retention efficiencies of three species of *Cyclosalpa* (Tunicata, Thaliacea). *Limnology and Oceanography*, **24**, 875–92.

HUTCHINSON, G. E. (1957). *A Treatise on Limnology*, 2 Vols. New York, John Wiley.

JACKSON, G. A. (1980). Phytoplankton growth and zooplankton grazing in oligotrophic oceans. *Nature*, **284**, 439–41.

KATONA, S. K. (1973). Evidence for sex pheromones in planktonic copepods. *Limnology and Oceanography*, **18**, 574–83.

KOSHLAND, D. E., JR (1979). A model regulatory system: bacterial chemotaxis. *Physiological Reviews*, **59**, 811–62.

LAWS, E. A., REDALJE, D. G., HAAS, L. W., BIENFANG, P. K., EPPLEY, R. W., HARRISON, W. G., KARL, D. M. & MARRA, J. (1984). High phytoplankton growth and production rates in oligotrophic Hawaiian coastal waters. *Limnology and Oceanography*, **29**, 1161–9.

LEHMAN, J. T. & SCAVIA, D. (1982*a*). Microscale patchiness of nutrients in plankton communities. *Science*, **216**, 729–30.

LEHMAN, J. T. & SCAVIA, D. (1982*b*). Microscale nutrient patches produced by zooplankton. *Proceedings of the National Academy of Sciences of the United States of America*, **79**, 5001–5.

LI, W. K. W., SUBBA RAO, D. V., HARRISON, W. G., SMITH, J. C., CULLEN, J. J., IRWIN, B. & PLATT, T. (1983). Autotrophic picoplankton in the tropical ocean. *Science*, **219**, 292–5.

MCCARTHY, J. J. & GOLDMAN, J. C. (1979). Nitrogen nutrition of marine phytoplankton in nutrient depleted waters. *Science*, **203**, 670–2.

MACNAB, R. M. & HAN, D. P. (1983). Asynchronous switching of flagellar motors on a single bacterial cell. *Cell*, **32**, 109–17.

MCMANUS, G. B. & FUHRMAN, J. A. (1986). Bacterivory in seawater studied with the use of inert fluorescent particles. *Limnology and Oceanography*, **31**, 420–6.

MADIN, L. P. (1974). Field observations on the feeding behavior of salps (Tunicata: Thaliacea). *Marine Biology*, **25**, 143–7.

MAGUE, T. H., FRIBERG, E., HUGHES, D. J. & MORRIS, I. (1980). Extracellular release of carbon by marine phytoplankton; a physiological approach. *Limnology and Oceanography*, **25**, 262–79.

MALONE, T. C. (1980). Algal size. In *The Physiological Ecology of Phytoplankton*, ed. I. Morris, pp. 433–63. Berkeley, University of California Press.

MITCHELL, J. G., OKUBO, A. & FUHRMAN, J. A. (1985). Microzones surrounding phytoplankton form the basis for a stratified microbial ecosystem. *Nature*, **316**, 58–9.

MULLIN, M. M. (1980). Interactions between marine zooplankton and suspended particles. In *Particulates in Water*, ed. M. C. Kavanaugh & J. Leckie, pp. 233–41. Advances in Chemistry Series, no. 189. Washington, D.C., American Chemical Society.

MULLIN, M. M., SLOAN, P. R. & EPPLEY, R. W. (1966). Relationship between carbon content, cell volume, and area in phytoplankton. *Limnology and Oceanography*, **11**, 307–11.

MUNK, W. H. & RILEY, G. A. (1952). Absorption of nutrients by aquatic plants. *Journal of Marine Research*, **11**, 215–40.

PAERL, H. W. & GALLUCCI, K. K. (1985). Role of chemotaxis in establishing a specific nitrogen-fixing cyanobacterial–bacterial association. *Science*, **277**, 647–9.

PRICE, H. J. & PAFFENHÖFER, G.-P. (1986). Capture of small cells by the copepod *Eucalanus elongatus*. *Limnology and Oceanography*, **31**, 189–94.

RUBENSTEIN, D. I. & KOEHL, M. A. R. (1977). The mechanism of filter feeding: some theoretical considerations. *American Naturalist*, **111**, 981–94.

SCAVIA, D., FAHNENSTIEL, G. L., DAVIS, J. A. & KREIS, R. G. JR (1984). Small-scale nutrient patchiness: some consequences and a new encounter mechanism. *Limnology and Oceanography*, **29**, 785–93.

SMAYDA, T. J. (1970). The suspension and sinking of phytoplankton in the sea. *Oceanography and Marine Biology Annual Review*, **8**, 353–414.

SOURNIA, A. (1982). Form and function in marine phytoplankton. *Biological Review*, **57**, 347–94.

SPERO, H. J. (1985). Chemosensory capabilities in the phagotrophic dinoflagellate *Gymnodinium fungiforme*. *Journal of Phycology*, **21**, 181–4.

VARON, M. & SHILO, M. (1980). Ecology of aquatic bdellovibrios. *Advances in Aquatic Microbiology*, **2**, 1–48.

WILLIAMS, P. J. L. & MUIR, L. R. (1981). Diffusion as a constraint on the biological importance of microzones in the sea. *Ecohydrodynamics*, ed. J. C. J. Nihoul, pp. 209–18. Amsterdam, Elsevier.

DIVERSITY IN FRESHWATER MICROBIOLOGY

J. GWYNFRYN JONES

Freshwater Biological Association, The Ferry House, Ambleside, Cumbria LA22 0LP, UK

It is not the purpose of this contribution to examine microbial diversity as expressed in the form of the many indices currently in use. Such notions are only of relevance if the community can be broken down into its component parts. Since we can only isolate, readily, approximately 0.25% of the bacteria seen in freshwater and sediments (Jones, 1977) there is little hope of assessing the diversity of a microbial community at present. Whether such a description of diversity in relation to population stability is of any relevance in microbial ecology is questioned by Brock in the opening chapter of this volume.

The diversity considered in this chapter is found on several levels. First, there is diversity within the environment itself, brought about by the establishment of physical and chemical gradients and by the imposition of physical force on these (turbulence, for example), which manifests itself in a fascinating collection of microbial communities. Our understanding of these communities is probably more advanced with reference to their function than their structure. It is possible to determine the flow of carbon to methane and carbon dioxide, but we know very little about the microbes responsible for this in the field. There are other bacterial populations which we observe in lakes, particularly those associated with the metalimnion, whose metabolic function remains a complete mystery. The presence in freshwater systems of solid and liquid phases and their interfaces ensures the diversity of form which we observed. But we can only guess at the metabolic adaptation which is necessary for a benthic (hydrophobic? specialist?) microbe as it converts to a planktonic (hydrophilic? generalist?) existence as environmental conditions change.

The development of anoxia in the hypolimnion of a eutrophic lake ensures that the microbiologist will encounter almost all the metabolic diversity possible within a single temperate field site. The subject of interactions between anaerobes in freshwaters has been reviewed extensively in recent years (Nedwell, 1984; Jones, 1985a), and therefore this chapter will attempt to concentrate more on diversity of the community than on processes.

Secondly, the freshwater system also provides the microbiologist with a diversity of opportunity, not only to study complex communities, but also to manipulate those communities so that a better understanding of controlling factors may be obtained. Such studies are not possible in the test tube, the chemostat, or even in laboratory microcosms, but are essential if we are to extend microbial ecology into a predictive science.

Finally, at the end of this chapter I shall return very briefly to the vexed question of microbial community diversity, but on this occasion to examine how the modern techniques of the molecular biologist might help the ecologist obtain a better understanding of the range of bacterial 'taxa' which might be encountered. The word 'taxa' was chosen after some deliberation. Perhaps one important barrier might be crossed if the ecologist, the geneticist and the systematist could agree on a working definition of the taxonomic unit which is most appropriate in the study of bacteria in the natural environment.

THE STRATIFIED EUTROPHIC LAKE

There is little need to dwell on the process of stratification in lakes except to say that it results in a series of gradients, both physical and chemical, which are sharper than those encountered in other aquatic habitats. The gradients may occur over a few micrometres in a sediment particle, over a few millimetres of depth in the bulk sediment itself, and over several decimetres or metres in the water column. The productivity of the lake, and its catchment, ensures the establishment of the gradients, and the morphometry and prevailing weather conditions govern their stability during the summer. The net result is the development of many microbial communities, whose population densities are far greater than those found in, for example, open oceanic water. Stratified lakes may, therefore, provide excellent model systems for the study of aquatic microbial populations and the processes which they mediate.

As a gross oversimplification, the stratified lake (Fig. 1) provides us with at least six major microbial communities for study. The plankton and the benthos may both be divided into three distinct components.

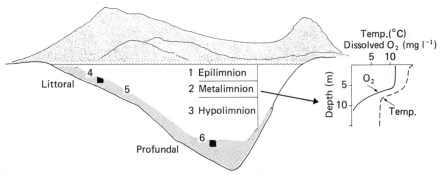

Fig. 1. Cross-section of a stratified eutrophic lake, identifying the sites of six major microbial communities (for details see text).

(1) The epilimnetic plankton has probably been the most intensively studied community in limnology. Not only are the major primary producers, the phytoplankton, identifiable under the microscope, but the community structure and its productivity provide sensitive indicators of the degree of eutrophication or changes in catchment land use. We are slowly moving towards an understanding of the interactions between bacteria and the phytoplankton both during photosynthesis (Jones, 1982; Jones, Simon & Cunningham, 1983; Jensen, 1985) and during decomposition of the algae (Cole, Likens & Hobbie, 1984). Similarly, the sources of dissolved organic matter in lakes have, to some degree, been identified (Cole, McDowell & Likens, 1984), yet we know very little about the majority of the bacteria present. It is likely that many of these bacteria are oligotrophs (see Poindexter, this volume) expressing considerable metabolic diversity, in that they are often generalists capable of utilizing a wide range of substrates (Sepers, 1981). Although techniques such as incorporation of tritiated thymidine or measurement of the frequency of dividing cells may provide some measure of the productivity of the bacterial population, it is certainly true that freshwater microbiologists are far behind their marine colleagues in determining the importance of microflagellate grazing in this community (Azam et al., 1983).

(2) The metalimnetic planktonic populations are often less stable than those in the epilimnion. They develop on physical (temperature) and chemical (O_2, NH_4^+, $Fe(II)$, CH_4, etc.) gradients, which are themselves subject to change mediated by turbulence

and the development of anoxia in the hypolimnion. In calm weather bacterial populations may be tightly stratified over depth intervals as small as a few decimetres, and the position of an individual component population is usually governed by its requirement for (or tolerance of) oxygen and the reduced chemical species of the hypolimnion (Rudd & Hamilton, 1978). Most of the bacteria found in this zone (many of them are morphologically distinct) remain to be isolated and characterized; the dominant forms depend on the underlying geology and sediment chemistry. For example, in many of the shallow more eutrophic water bodies of the English Lake District, the redox couple Fe(III)/Fe(II) dominates events in the hypolimnion (Davison *et al.*, 1981) and therefore the bacteria found in the metalimnion are usually those reputed to be involved in the iron cycle (Jones, 1981 and 1986). In some Cheshire Meres, on the other hand, the hypolimnia contain little Fe(II) and the major reduced species is Mn(II). Under these circumstances manganese-depositing bacteria, including the elusive morphotype '*Metallogenium*' (Fig. 2a), develop in profusion (Tipping, Jones & Woof, 1985). It is difficult to determine precisely where on the gradients the bacteria grow, thus hindering laboratory simulation in media and enrichment cultures, and the position of the population may also change frequently. It is not uncommon to determine the exact depth of a population and the conditions on the gradient in the morning to return in the afternoon (for some serious sampling) to find the population shifted by internal seiche or other water movements. The metalimnetic community is often complex and may also include components of the zooplankton, e.g. *Ceriodaphnia* (Smyly, 1974), whose vertical depth range may be constrained by their temperature and oxygen tolerances.

(3) The hypolimnetic plankton is also stratified but usually in response to light (rather than temperature) and the reduced chemical species. Encapsulated bacteria may be found at the top of the hypolimnion, many of which have been implicated in the iron cycle (e.g. *Ochrobium* sp., Fig. 2b) although the evidence for this is not good. Immediately below these, populations of photosynthetic bacteria may be observed, the composition of which may be controlled by the underwater light climate (Montesinos *et al.*, 1983). Until recently the anoxic hypolimnion had been considered largely as the domain of the bacteria, but observations by Finlay and co-workers (see Field Studies, below) have

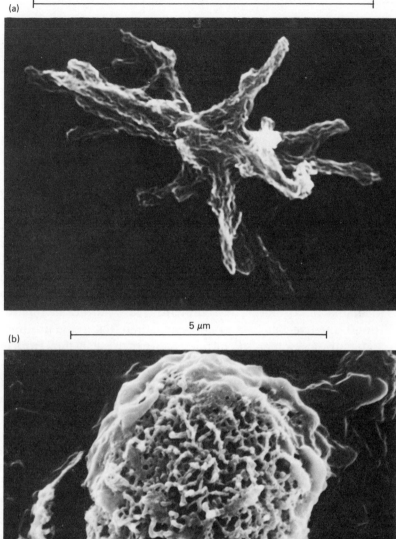

10 μm

(a)

5 μm

(b)

Fig. 2. 'Bacteria' commonly found in the metalimnion and the upper hypolimnion of stratified eutrophic lakes. (a) *Metallogenium*, a particulate form of manganese, the organism responsible for which remains to be isolated and cultured with consistency. (Photograph reproduced with kind permission of E. Tipping, F.B.A., and D. Thompson, University of Bristol.) (b) *Ochrobium*, an encapsulated bacterium, reported to deposit ferric iron and classified with the chemolithotrophs. Careful sampling and culture shows that it is an obligate anaerobe (Jones, 1981).

shown that ciliated protozoans form a significant part of the community and can play a significant part in geochemical cycling in this zone.

(4) The epilimnetic benthos, i.e. those communities which develop in the littoral sediments, has received little attention compared with the anoxic profundal benthos, possibly because turbulence results in greater surface sediment movement and instability (Sweerts, Rudd & Kelly, 1986). The littoral site, although it contains a smaller total bacterial population (Jones, 1980), is worthy of further study because of the biological and chemical interactions which can occur there. One particularly striking example is the difference in the nitrogen cycle observed at littoral and profundal sites, the former accounting for an approximately three-fold greater loss of nitrogen via denitrification (Jones & Simon, 1981). This can be attributed to several features of the littoral sediments of which the following are considered to be the most important. The sediments are oxidized all the year round, the overlying water is warmer than that in the profundal zone during summer and nitrate is always present, thus ensuring a flux into the sediment. The oxidized sediments bind any ammonia produced by mineralization processes, although this binding is not tight (the ammonia is KCl-extractable) and therefore the ammonia is likely to be available to the larger and more active population of nitrifying bacteria in the littoral sediments (Hall, in press). The nitrate produced is readily denitrified in the anaerobic microsites found in the larger sediment particles of this zone (Jones, 1979a). Similar microsites have since been shown to play a significant part in microbial metabolism in a wide variety of habitats (Paerl, 1984, 1985). Superimposed on this is a greater population of benthic invertebrates in the littoral zone; the enhanced bioturbation accelerates the coupled processes of nitrification and denitrification (Chatarpaul, Robinson & Kaushik, 1980) with attendant increased losses of nitrogen gas (Jones, 1985b). Diel shifts in the concentration of oxygen in the interstitial water as a result of photosynthesis by the benthic algae will also affect the process.

(5) The metalimnetic benthos has been largely ignored, even though it is situated in a zone which is well placed to receive supplies of both reduced species (from the hypolimnion) as well as oxygen and nitrate from the epilimnion. When examined, the metalimnetic sediment was shown to contain a large and active microbial

population (Jones, 1979b). This was usually dominated by filamentous forms, both cyanobacteria as well as non-photosynthe-
tic taxa, echoing the point raised by Brock in his introductory
chapter, that many microbial communities are dominated by
filamentous rather than unicellular bacteria. This is certainly true
of the benthos of more eutrophic lakes, where filamentous bacteria may account for up to 50% of the biomass (Godinho-
Orlandi & Jones (1981), as well as estuarine and near-shore
sediments (Jørgensen, 1977). In smaller water bodies such as
woodland ponds the population densities are even greater and
the range of taxa wider. These organisms have been described
for decades, based on morphological features (Skuja, 1956), but
many remain to be characterized. Of the few forms which have
been isolated, some show a high degree of nutritional specialization, for example the obligate amino acid-utilizing filamentous
glider isolated by Maiden & Jones (1984). Further characterization of this organism by 16S ribosomal RNA analysis provided
something of a surprise, in that the apparently Gram-negative
glider was most closely related to the genus *Bacillus* (Clausen,
Jones & Stackebrandt, 1985), a fact also reflected in its membrane fatty acids (Nichols *et al.*, in press). Clearly, an increased
effort is required if we are to understand the role of these organisms in the benthic community.

(6) The hypolimnetic benthos has, in recent years, been the zone
studied most intensively by microbial ecologists. This has reflected the considerable interest in the phenomenon of interspecies hydrogen transfer and its role in the terminal metabolism
of organic carbon; much of this information has been reviewed
by Mah (1982), Jones (1985a) and Ward & Winfrey (1985).
Rather than deal with the processes, this section will consider
briefly the populations involved. In freshwaters, which are frequently sulphate-limited, the sulphate-reducing bacteria are
often considered as net hydrogen donors to the methanogens.
However, in lakes such as those encountered in the English Lake
District the input of organic carbon, in the form of decaying
algal remains, is often insufficient to guarantee stable population
structures and interactions towards the end of the summer period. Changes in populations and their metabolic relationships
are frquently encountered. These may include transfer of hydrogen from sulphate reducer to methanogen (Jones, Simon &
Gardener, 1982), competition between sulphate reducer and

methanogen for acetate (Jones & Simon, 1984) and metabolic interaction between methanogens and autotrophic acetogens (Jones & Simon, 1985). With the exception of the confirmed presence of *Desulfotomaculum* spp., the organisms responsible have not been isolated.

This brief overview has described six very different communities, and then only as they might be seen during the period of summer stratification. During isothermal conditions in winter, several changes might occur. Certain planktonic forms may enter the benthos, either as resting stages or as active vegetative cells. Filamentous bacteria may be found in the profundal zone, from which they are largely absent during summer anoxia. The diversity and variability are enormous. If detailed transects of a single lake are taken, the variability in simple measures of microbial numbers, biomass and activity is as great on a single sampling occasion over the lake as a whole as is variability over a whole season at one site (Jones & Simon, 1980). More detailed analysis of spatial distribution in the processes of geochemical cycling (Klug, *et al.* 1980) revealed a similar degree of heterogeneity. The few attempts to study the problem in detail show us how patchy the planktonic populations and sediment characteristics can be (Fig. 3*a*, *b*). The work by Heaney (1976) and George & Heaney (1978) demonstrates the effect of wind and other forces on the planktonic community of Esthwaite Water whereas Hilton & Gibbs (1984/85) have examined the effects of erosion, transport and accumulation processes on the sediment chemistry of that lake (Fig. 3*c*–*g*). Clearly, differences in the distribution of variables such as carbon, nitrogen, iron and manganese cannot fail to influence the microbial communities. The question remains: in the light of such variability how is microbial ecology to transform itself into a predictive science? The answers lie not in the laboratory but in some extremely hard field work. This will be necessary if general principles developed in simpler model systems are to be validated. In Britain at least, ecologists will soon be called upon to predict the effect of a massive (50–60%) loss in our agricultural land, a result of overproduction of food in Western Europe. This land could well be turned over to forestry at much the same time as our older commercial forests are being cropped. What will be the effect of such large-scale afforestation and deforestation, for example, on the nitrogen cycle in the catchments and their receiving water bodies?

The above is a highly selective review of diversity in some fresh-water microbial communities. Some have been ignored completely, largely as a result of my ignorance of them. The neuston, the community of the air–water interface, is known to be quite different from that of the underlying water (Norkrans, 1977) but, as far as I am aware, its significance in terms of the population or the processes mediated has yet to be determined. It is more likely that a stable community would develop in smaller more sheltered water bodies such as ponds (Frølund, 1977) and under such circumstances it is possible to envisage its effect on several transport processes across the air–water interface. The microbial populations associated with macrophytes have been examined in more detail and their role in carbon turnover in the community has been estimated (Mann, 1972; Ramsay & Fry, 1976; Wetzel & Penhale, 1979). Macrophyte stands are likely to be of greater importance in smaller, shallower water bodies, and certainly in rivers (Baker & Orr, 1986). Probably the most stable freshwater microbial community, that of aquifers and subsurface waters, is also the most difficult to study either by direct observation of the population or measurement of metabolic activity. Removal by sampling is very likely to alter both, and therefore success in characterizing the community is most likely to be achieved through the analysis of biomarkers (Smith *et al.*, 1986). The use of lipid and other biomarkers, and the resolution achieved, is covered in more detail in this volume by Parkes.

FIELD STUDIES

One of the purposes of this symposium was to show that field-led microbial ecology could tell us something new about microbes and their communities. Such studies are just as important as those in the laboratory, where attempts are made to enunciate general principles from events which occur in the test tube, chemostat or some other experimental system. Clearly, both are required to provide the complete picture as is emphasized in the section on integrated field/laboratory studies in Brock's introduction to the symposium.

A fine example of the success achieved by this approach has been the work of Finlay and his colleagues on the ciliated protozoon *Loxodes* (Finlay & Fenchel, 1986). This started as a purely ecological investigation of *Loxodes* spp. in a hypereutrophic pond. The populations were seen to migrate, although not entirely, out of the sediment

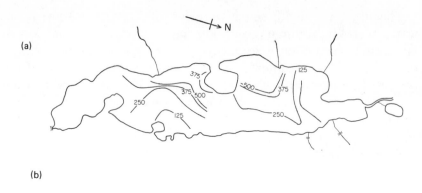

(a)

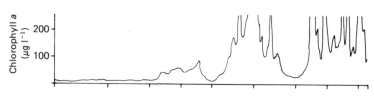

(b)

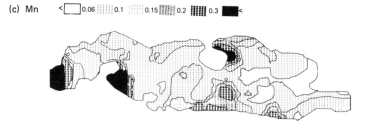

(c) Mn

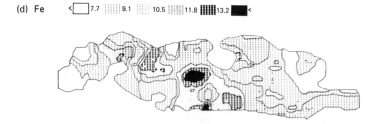

(d) Fe

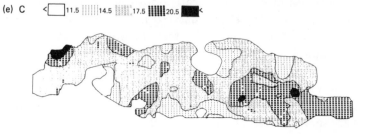

(e) C

(f) N $<\square$ 0.6 1.1 1.6 2.1 $\blacksquare<$

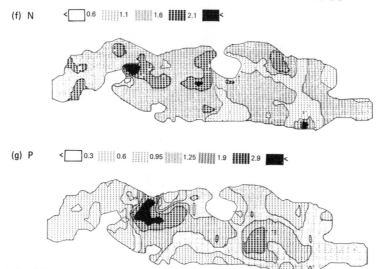

(g) P $<\square$ 0.3 0.6 0.95 1.25 1.9 2.9 $\blacksquare<$

Fig. 3. Spatial heterogeneity in the phytoplankton and the sediment characteristics of Esthwaite Water, English Lake District. (a) Horizontal distribution of the dinoflagellate *Ceratium hirundinella* in the 0–5 m depth zone of the epilimnion; the numbers are cells ml^{-1} obtained from random horizontal samples (taken from Heaney, 1976). Reprinted with permission from Freshwater Biology © copyright 1976 Blackwell Scientific Publications. (b) Continuous recording of chlorophyll fluorescence on a transect of the long axis of Esthwaite Water (taken from George & Heaney, 1978). Reprinted with permission from the British Ecological Society. (c–g) Distributions of manganese (c), iron (d), carbon (e), nitrogen (f) and phosphorus (g) in the sediments. The units of concentration are % MnO of mineral matter at 1100 °C, % Fe$_2$O$_3$ of mineral matter at 1100 °C, C and N as % of dry solid, % P$_2$O$_5$ of mineral matter at 1100 °C (taken from Hilton & Gibbs, 1984/85). Reprinted with permission from Chemical Geology © copyright 1984 Elsevier Science Publishers B.V.

into the water column (Finlay, 1981, 1982) during the development of summer anoxia in the hypolimnion (Fig. 4). However, the ciliates did not move into the aerobic epilimnion: what, then, governed their distribution? The *Loxodes* spp. were seen to position themselves in water containing little or no detectable oxygen (µmol l^{-1}) and was associated with an elevated concentration of nitrite (Fig. 5). Furthermore, the ciliates were shown to possess a respiratory nitrate reductase (the first demonstration in a eukaryote) and as the organisms switched from aerobic to anaerobic metabolism the number of mitochondria doubled, presumably to cope with the decreased energy yield (*c.* 50% in bacteria) associated with conversion to nitrate respiration (Finlay, Span & Harman, 1983; Finlay, 1985). The question of how the organism maintained itself at this depth, apparently perceiving the higher oxygen tension above, remained unanswered. Laboratory studies showed that *Loxodes* swam vertically upwards in the absence of oxygen and downwards in the presence of high oxygen concentrations. At low concentrations, random

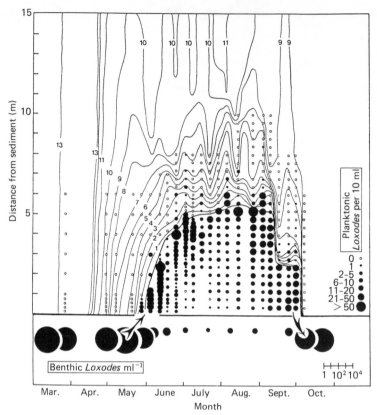

Fig. 4. The distribution and abundance of *Loxodes* spp. in the water column and sediment of Esthwaite Water. Contour lines show oxygen concentration in mg l⁻¹. Note the migration of most (but not all) of the ciliates out of the sediment and their concentration in zones of low oxygen concentration. (Taken from Finlay & Fenchel, 1986). Reprinted with permission from the Freshwater Biological Association.

reorientation increased until the organisms were apparently situated in water containing an optimal concentration of oxygen (Fenchel & Finlay, 1984). This response to gravity is unusual, possibly unique, in Protozoa. If *Loxodes* perceived gravity, then it was likely that it possessed a mechanoreceptor. Such organelles usually contain mineral concretions, the movement of which is detected by sensory cells. The Müller vesicles, mineral bodies containing barium sulphate enclosed in vacuoles, found along the dorsal rim of *Loxodes* appeared to be prime candidates for this task (Fig. 6). Sectioning showed that the vesicles were intimately connected to the row of cilia on the left side of the cell, and Fig. 6 shows how orientation of the organism alters the position of barium sulphate granule and the net effect of this on movement. At an ecological level this means

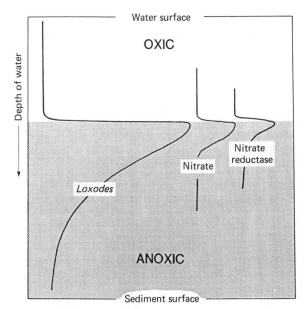

Fig. 5. Vertical distribution of *Loxodes* spp., nitrate concentration, and the nitrate reductase activity of ciliate-sized particles in the water column of Priest Pot. The ciliates were separated from bacteria and other smaller microorganisms with a 30 μm sieve. Microscopic examination confirmed that the fraction retained contained insignificant quantities of material, other than *Loxodes*, which could contribute to the activity measured. (Adapted from Finlay, Span & Harman, 1983).

that the ciliate accounts for almost all the particulate barium in the lake (Fig. 7) but the significance of this in the overall biogeochemical cycle of the element is, as yet, poorly understood (Finlay, Hetherington & Davison, 1983). Geotaxis is, however, not the only means of orientation available to *Loxodes*, and this has been shown by placing organisms in vessels where the oxygen gradient is in a horizontal, rather than a vertical, plane. Under such circumstances reduction in random motility as optimal conditions are approached results in a population distribution very similar to that obtained in the field and in vertical laboratory cultures, except that it takes much longer to reach equilibrium. But why should the optimum environment be depleted in oxygen? *Loxodes* produces both superoxide dismutase and catalase and therefore oxygen toxicity would appear to be an unlikely candidate. There was also no evidence of repulsion due to the presence of the potentially toxic radicals O_2^- and H_2O_2 in the lake water (Finlay, Fenchel & Gardener, 1986). If it was dissolved oxygen itself which repelled *Loxodes*, then how was this achieved? Subsequently *Loxodes* was observed to react, under

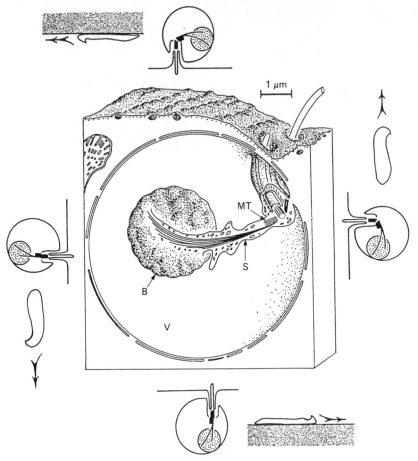

Fig. 6. The Müller vesicle of *Loxodes*. The Müller body (B), a barium sulphate concretion, is attached to the wall of the vacuole (V) by a stalk (S). The stalk contains nine microtubules (MT) anchored in a non-ciliated basal body adjacent to one containing a cilium. The possible positions of the barium granule and resulting orientation of the ciliate are also shown. (Adapted from Fenchel & Finlay, 1986).

certain conditions, to light. Under anoxic conditions the cells were insensitive but this was not so in the presence of oxygen. Organisms near the oxic–anoxic boundary responded to light by increased tumbling, followed by increased swimming velocity and pathlength and finally by positive geotaxis. These organisms remained in anoxic water until the source of light was removed. (Cells exposed to high light and low oxygen tensions, or darkness and high oxygen concentrations eventually explode.) Light and oxygen therefore interact in the effect they produce and in such a way that would be of considerable importance in the natural environment. The minimum swim-

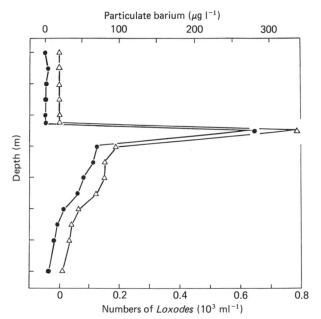

Fig. 7. Vertical distribution of *Loxodes* spp. ($\triangle$) and particulate barium (●) in the water column of Priest Pot in July 1982 (adapted from Finlay, Hetherington & Davison, 1983).

ming velocity of *Loxodes* occurred at progressively lower oxygen tensions as light levels increased, and only in anoxic water at light levels above $5\,W\,m^{-2}$. Escape from light could be detected at levels less than $10\,W\,m^{-2}$ and it was of particular interest to note that light levels greater than the critical 5–$10\,W\,m^{-2}$ could penetrate the lake water to depths where the pO_2 was *c.* 5% of saturation, i.e. depths within a few centimetres of the upper limit of *Loxodes* distribution. It is clear that, on occasions, light and oxygen interact to exclude *Loxodes* from oxygenated water. The photosensitive response of the ciliate is mediated by a pigment, possibly a flavin found in granules embedded in the cell membrane. The pigment absorbs principally at the blue end of the spectrum which corresponds to the action spectrum for escape in oxygenated lake water (Finlay & Fenchel, in press). The fact that cyanide blocked both oxygen and light perception in *Loxodes* (Fenchel & Finlay, 1986) suggested that the receptors for oxygen and light are identical. Although there is no firm evidence, the most favoured candidate as the active agent (in response to light as well as high oxygen tension in the dark) is superoxide.

The autecological and physiological studies described above, stimulated initially by field observations, have shown that ciliated protozoans are capable of metabolic adaptability hitherto ascribed only

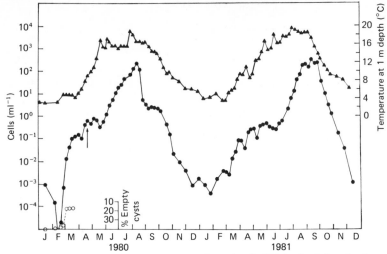

Fig. 8. Change in concentration of *Ceratium hirundinella* (●) in the 0–5 m depth zone of the epilimnion of Esthwaite Water and in water temperature (▲) with time. Populations increase rapidly in late winter when the water temperature is low and at the same time the number of empty cysts in the sediment increases (○). (Taken from Heaney, Chapman & Morrison (1983) and reproduced by permission of the British Phycological Society.)

to the prokaryotes. In addition they have demonstrated a fascinating range of mechanisms to ensure their location in optimal growth conditions.

Studies on the dinoflagellate *Ceratium hirundinella* (Heaney & Talling, 1980) have also provided insights into mechanisms of survival as well as optimization of resource utilization. Numbers of *Ceratium hirundinella* in the water column decline sharply during September and October of each year because of mass encystment of the population (Fig. 8) and sedimentation of the cysts into the sediment. The benthic cysts then provide the major inoculum for the planktonic population the following spring, a rapid increase being observed as the water temperature increases from 3 to 5 °C (Heaney, Chapman & Morrison, 1983). This is followed by a period of exponential growth but this results in a population increase which is considerably slower than that due to recruitment from the sediment. If, for some reason, the planktonic population fails to encyst in a given year (e.g. due to early mortality of the vegetative cells) recruitment in the following year is not diminished because of the stock of cysts, produced in earlier years, which remain in the sediment. One outcome of such a detailed study of these cysts has been the suggestion that the population consisted of not one but two species of *Ceratium*. Laboratory studies with experimentally stratified water columns (Heaney & Furnass, 1980) showed that diel vertical migrations of the vegetative cells of *Ceratium* differed with cells in the exponential and those in the stationary phases of growth, the latter showing a greater correlation with the light regime. At higher surface irradiance cells avoided the surface waters and population maxima were found in the metalimnetic zone, their further downward migration being restricted by thermal stratification. The diel migrations were later shown to depend also on the nutrient status of the cells (Heaney & Eppley, 1981). Nitrogen depletion resulted in downward migration earlier in the day, resulting in more complete penetration of the metalimnion into more nutrient rich water. This mixture of field and laboratory studies has also been used successfully to show that phytoplankton chemical composition close to the Redfield ratio $(C:N:P, 106:16:1)$ does not necessarily imply growth at near maximal rate in natural water as proposed by Goldman, McCarthy & Peavey (1979). The Droop Cell-Quota model implies that such ratios are possible in algae subject to other limitations such as light, and this was shown to be so in continuous culture experiments as well as from observations on natural phytoplankton (Tett, Heaney & Droop, 1985).

The diversity of strategies for survival and growth which algae exhibit is considerable. Parasitism of planktonic algae by protozoans and fungi has been recognized as being of importance in population control (Canter & Lund, 1968; Canter & Heaney, 1984) but it required much painstaking observational work to recognize the hypersensitive reaction of some algae to their chytrid parasites (Canter & Jaworski, 1979). In this study, about 4000 hypersensitively killed cells of *Asterionella*, each bearing only a single zoospore of the chytrid *Rhizophydium planktonicum*, were examined. In the hypersensitive reaction the host alga is killed by a single infective parasite. Both host and parasite die within 24 h of infection. This might be considered as a defence mechanism which removes parasites from the algal population as a whole.

Field studies on the whole phytoplankton community, as opposed to the population of a single species, have much to teach us about the survival strategies of algae. The use of large experimental enclosures and manipulation of both chemical and physical conditions, has provided new insights into algal periodicity (Reynolds, 1982; Reynolds, Wiseman & Clarke, 1984) and permitted the development of conceptual models relating morphology and survival strategy. Three basic strategies may be recognized: (1) the opportunist which is capable of diverting a significant part of its effort into reproduction; (2) the stress-tolerant form, capable of surviving deficiencies and grazing; and (3) the disturbance-tolerant form which can cope with rapid transitions in climate such as in light and temperature. Reynolds (1984) identifies 14 species assemblages which characterize the five limnologically distinct British lake systems and is able to accommodate most of these in a hypothetical three-dimensional matrix bounded by axes defining physical stability, the concentration of nitrogen and phosphorus, and the ratio of nitrogen to phosphorus. The application of such conceptual models and the identification of algal assemblages and their succession is of considerable value to the Water Industry in the management of reservoirs.

The use of the same large experimental enclosures also permitted the comparison of algal community carbon-specific growth rates (as measured by $^{14}CO_2$ uptake) with observed rates of cellular increase (Reynolds, Harris & Gouldney, 1985). Over a 24 h period net retention of the radiolabelled carbon represented 23–82% of the carbon fixed. The potential growth rates exceeded those observed in the field (after correction for *in situ* grazing and sinking) by a factor of between 1 and 30; this discrepancy was considered to be best explained in terms of physiological voiding of excess carbon (i.e. by respiration, photorespiration and excretion) before new cells are formed.

Experimental manipulation of natural, as opposed to enclosed, communities can illustrate the degree of fine tuning in certain freshwaters. Peterson et al. (1985) continuously enriched an Arctic river with phosphate, raising the concentration from ambient levels of 1–4 μg 1^{-1} to 10 μg 1^{-1} of phosphate-phosphorus. The result was an immediate growth of attached algae for more than 10 km downstream, suggesting that the system was phosphorus-limited. These authors were also able to quantify changes in heterotrophic (bacterial) and autotrophic (algal) processes and attempted to determine how these influenced the aquatic insects (Table 1). The addition of such a small quantity of phosphorus produced a shift in the major source of energy, from organic matter entering the stream to photosynthetic fixation within the stream community (i.e., a shift from heterotrophy to autotrophy). This, in turn, resulted in an increase in the size and developmental stage of some dominant aquatic insects. Although limited in its scope and duration, experiments such

Table 1. *The effect of the addition of 10 µg l⁻¹ of phosphate phosphorus to an Arctic tundra river. Data from Peterson* et al. *(1985)*

	Site in relation to point of addition	
	Upstream	Downstream
Concentration of chlorophyll *a* on rocks in riffles ($\mu g\,cm^{-2}$)	1–2	5–22
Respiratory heat production of the epilithic community ($\mu W\,cm^{-2}$)	0.5	1.4
Light[a]	0.5	1.6
Dark	0.4	0.4
[^{14}C]-acetate incorporation into the epilithic community ($dpm\,cm^{-2}\,h^{-1}$)		
Light	280	900
Dark	150	250
Direct counts of bacteria ($10^6\,cm^{-2}$)		
Light	4.2	9.2
Dark	2.6	2.6
Lipid phosphate ($nmol\,cm^{-2}$)		
Light	160	1551
Dark	69	92
Mean size of the blackfly *Prosimulium* sp. (mm)	3.4	4.0

[a]Values for light and dark refer to communities which developed in transparent and opaque plastic tubes.

as this provide an understanding of factors controlling production in stream communities which would not be obtained from laboratory studies.

It will not have escaped the notice of some readers that, even though my own research field has been that of the ecology of bacteria and the processes which they mediate, the examples of field studies I have chosen have been largely concerned with eukaryotic microorganisms. The reason is that protozoologists, algologists and mycologists have a distinct advantage over the bacterial ecologists. With a few exceptions, they are able to examine natural assemblages of their microorganisms under the microscope, identify them to species level, and make sensible statements about the community structure and its response to environmental variables. In other words, they are able to study microbial ecology. The emphasis on eukaryotes and field studies was also intended to shift the balance slightly in the publications of the Society in the general area of microbial ecology.

DIVERSITY IN THE BACTERIAL COMMUNITY

In the absence of an ability to describe natural communities bacterial ecologists have, over the past two decades, resorted to various measures of community activity or processes. It may be true that we have reached a better understanding of these processes and may be able to provide more reliable estimates of rates in the field. However, it is also true that the applications of some favoured concepts such as substrate affinity (Button, 1986) and competitiveness (Gottschal, 1985) are currently being questioned, particularly in the presence of transients or perturbations. In other words, there is a long-overdue move towards the study of non-steady state (i.e. natural) systems (Harris, 1985).

It is ironic that biotechnology, i.e. the proposed release of genetically engineered organisms, may, at one and the same time, provide both the impetus and the methodology to study the diversity of natural microbial communities. If we cannot define the natural community, if we are unable to isolate the majority of the bacteria present, how can we assess the transfer and survival of recombinant DNA in the environment? Apart from a much more imaginative approach to the isolation of bacteria, a better assessment of community diversity (at least in relatively simple communities or extreme environments) may be obtained from analysis of 5S ribosomal RNA sequences (Stahl et al., 1985; Pace et al., 1986). Measurements of population composition may become more reliable through the use of modern immunofluorescent techniques and the application of gene probes. Perhaps, eventually, the bacterial ecologist may be able to examine some of the factors which contribute to the diversity of a community, for example the dispersal, the survival and the ability of an organism to colonize a particular habitat. The role of an organism in interactions (including competition, mutualism and exploitation) will contribute to its fitness for a particular habitat. This information will also be required before release of a genetically engineered organism can be contemplated. However, such studies cannot be confined to laboratory model systems. Unless the diversity and variability encountered in the field are taken into account, then models designed to simulate the microbial community will be of little value (Harris, 1985).

ACKNOWLEDGEMENTS

I am grateful to the many colleagues at the Freshwater Biological Association for their invaluable help in the preparation of this paper. Much of the research reported here was funded by the Natural Environment Research Council and the Department of the Environment. I wish to thank Mrs Joyce Hawksford who typed the script with incredible efficiency and at very short notice.

REFERENCES

AZAM, F., FENCHEL, T., FIELD, J. G., GRAY, J. S., MEYER-REIL, L. A. & THINGSTAD, F. (1983). The ecological role of water-column microbes in the sea. *Marine Ecology Progress Series*, **10**, 257–63.

BAKER, J. H. & ORR, D. R. (1986). Distribution of epiphytic bacteria on freshwater plants. *Journal of Ecology*, **74**, 155–65.

BUTTON, D. K. (1986). Affinity of organisms for substrate. *Limnology and Oceanography*, **31**, 453–6.

CANTER, H. M. & HEANEY, S. I. (1984). Observations on zoosporic fungi of *Ceratium* spp. in lakes of the English Lake District; importance for phytoplankton population dynamics. *New Phytologist*, **97**, 601–12.

CANTER, H. M. & JAWORSKI, G. H. M. (1979). The occurrence of a hypersensitive reaction in the planktonic diatom *Asterionella formosa* Hassall parasitized by the chytrid *Rhizophydium planktonicum* Canter emend., in culture. *New Phytologist*, **82**, 187–206.

CANTER, H. M. & LUND, J. W. G. (1968). The importance of protozoa in controlling the abundance of planktonic algae in lakes. *Proceedings of the Linnaean Society of London*, **179**, 203–19.

CHATARPAUL, L., ROBINSON, J. B. & KAUSHIK, N. K. (1980). Effects of tubificid worms on denitrification and nitrification in stream sediment. *Canadian Journal of Fisheries and Aquatic Sciences*, **37**, 656–63.

CLAUSEN, V., JONES, J. G. & STACKEBRANDT, E. (1985). 16S ribosomal RNA analysis of *Filibacter limicola* indicates a close relationship to the genus *Bacillus*. *Journal of General Microbiology*, **131**, 2659–63.

COLE, J. J., LIKENS, G. E. & HOBBIE, J. E. (1984). Decomposition of planktonic algae in an oligotrophic lake. *Oikos*, **42**, 257–66.

COLE, J. J., MCDOWELL, W. H. & LIKENS, G. E. (1984). Sources and molecular weight of 'dissolved' organic carbon in an oligotrophic lake. *Oikos*, **42**, 1–9.

DAVISON, W., HEANEY, S. I., TALLING, J. F. & RIGG, E. (1981). Seasonal transformations and movements of iron in a productive English lake with deep-water anoxia. *Schweizerische Zeitschrift für Hydrologie*, **42**, 196–224.

FENCHEL, T. & FINLAY, B. J. (1984). Geotaxis in the ciliated protozoon *Loxodes*. *Journal of Experimental Biology*, **110**, 17–33.

FENCHEL, T. & FINLAY, B. J. (1986). Photobehaviour of the ciliated protozoon *Loxodes*: taxic, transient and kinetic responses in the presence and absence of oxygen. *Journal of Protozoology*, **33**, 139–45.

FINLAY, B. J. (1981). Oxygen availability and seasonal migrations of ciliated Protozoa in a freshwater lake. *Journal of General Microbiology*, **123**, 173–8.

FINLAY, B. J. (1982). Effects of seasonal anoxia on the community of benthic ciliated Protozoa in a productive lake. *Archiv für Protistenkunde*, **125**, 215–22.

FINLAY, B. J. (1985). Nitrate respiration by Protozoa (*Loxodes* spp.) in the hypolimnetic nitrite maximum of a productive freshwater pond. *Freshwater Biology*, **15**, 333–46.

FINLAY, B. J. & FENCHEL, T. (1986). Physiological ecology of the ciliated protozoon *Loxodes*. *Report of the Freshwater Biological Association*, **54**, 73–96.

FINLAY, B. J. & FENCHEL, T. (in press). Photosensitivity in the ciliated protozoon *Loxodes*: pigment granules, absorbance and action spectra, evidence for a flavin photoreceptor and ecological significance. *Journal of Protozoology*.

FINLAY, B. J., FENCHEL, T. & GARDENER, S. (1986). Oxygen perception and oxygen toxicity in the freshwater ciliated protozoon *Loxodes*. *Journal of Protozoology*, **33**, 157–65.

FINLAY, B. J., HETHERINGTON, N. B. & DAVISON, W. (1983). Active biological participation in lacustrine barium chemistry. *Geochimica Cosmochimica Acta*, **47**, 1325–9.

FINLAY, B. J., SPAN, A. & HARMAN, J. M. P. (1983). Nitrate respiration in primitive eukaryotes. *Nature*, **303**, 333–6.

FRØLUND, A. (1977). A seasonal variation of the neuston of a small pond. *Saertryk af Botanisk Tidsskrift*, **72**, 45–56.

GEORGE, D. G. & HEANEY, S. I. (1978). Factors influencing the spatial distribution of phytoplankton in a small productive lake. *Journal of Ecology*, **66**, 133–55.

GODINHO-ORLANDI, M. J. L. & JONES, J. G. (1981). Filamentous bacteria in sediments of lakes of differing degrees of enrichment. *Journal of General Microbiology*, **123**, 81–90.

GOLDMAN, J. C., MCCARTHY, J. J. & PEAVEY, D. G. (1979). Growth rate influence on the chemical composition of phytoplankton in oceanic waters. *Nature, London*, **279**, 210–15.

GOTTSCHAL, J. C. (1985). Some reflections on microbial competitiveness among heterotrophic bacteria. *Antonie van Leeuwenhoek*, **51**, 473–94.

HALL, G. H. (in press). Nitrification in lakes. In *Nitrification*, Society for General Microbiology Special Publication, ed. J. I. Prosser.

HARRIS, G. P. (1985). The answer lies in the nesting behaviour. *Freshwater Biology*, **15**, 375–80.

HEANEY, S. I. (1976). Temporal and spatial distribution of the dinoflagellate *Ceratium hirundinella* O. F. Muller within a small productive lake. *Freshwater Biology*, **6**, 531–42.

HEANEY, S. I., CHAPMAN, D. V. & MORRISON, H. R. (1983). The role of the cyst stage in the seasonal growth of the dinoflagellate *Ceratium hirundinella* within a small productive lake. *British Phycological Journal*, **18**, 47–59.

HEANEY, S. I. & EPPLEY, R. W. (1981). Light, temperature and nitrogen as interacting factors affecting diel vertical migrations of dinoflagellates in culture. *Journal of Plankton Research*, **3**, 331–44.

HEANEY, S. I. & FURNASS, T. I. (1980). Laboratory models of diel vertical migration in the dinoflagellate *Ceratium hirundinella*. *Freshwater Biology*, **10**, 167–70.

HEANEY, S. I. & TALLING, J. F. (1980). *Ceratium hirundinella* – ecology of a complex, mobile, and successful plant. *Report of the Freshwater Biological Association*, **48**, 27–40.

HILTON, J. & GIBBS, M. M. (1984/85). The horizontal distribution of major elements and organic matter in the sediment of Esthwaite Water, England. *Chemical Geology*, **47**, 57–83.

JENSEN, L. M. (1985). Characterization of native bacteria and their utilization of algal extracellular products by a mixed-substrate kinetic model. *Oikos*, **45**, 311–22.

JONES, A. K. (1982). The interactions of algae and bacteria. In *Microbial Interactions and Communities*, vol. 1., ed. A. T. Bull & J. G. Slater, pp. 189–247. London, Academic Press.

JONES, J. G. (1977). The effect of environmental factors on estimated viable and total populations of planktonic bacteria in lakes and experimental enclosures. *Freshwater Biology*, **7**, 67–91.

JONES, J. G. (1979*a*). Microbial nitrate reduction in freshwater sediments. *Journal of General Microbiology*, **115**, 27–35.

JONES, J. G. (1979*b*). Microbial activity in lake sediments with particular reference to electrode potential gradients. *Journal of General Microbiology*, **115**, 19–26.

JONES, J. G. (1980). Some differences in the microbiology of profundal and littoral lake sediments. *Journal of General Microbiology*, **117**, 285–92.

JONES, J. G. (1981). The population ecology of iron bacteria (genus *Ochrobium*) in a stratified eutrophic lake. *Journal of General Microbiology*, **125**, 85–93.

JONES, J. G. (1985*a*). Microbes and microbial processes in sediments. *Philosophical Transactions of the Royal Society, Series A*, **315**, 3–17.

JONES, J. G. (1985*b*). Denitrification in freshwaters. In *Denitrification in the Nitrogen Cycle*, ed. H. L. Golterman, pp. 225–39. New York, Plenum Publishing Corporation.

JONES, J. G. (1986). Iron transformations by freshwater bacteria. In *Advances in Microbial Ecology*, vol. 9, ed. K. C. Marshall, pp. 149–85. New York, Plenum Press.

JONES, J. G. & SIMON, B. M. (1980). Variability in microbiological data from a stratified eutrophic lake. *Journal of Applied Bacteriology*, **49**, 127–35.

JONES, J. G. & SIMON, B. M. (1981). Differences in microbial decomposition processes in profundal and littoral lake sediments, with particular reference to the nitrogen cycle. *Journal of General Microbiology*, **123**, 297–312.

JONES, J. G. & SIMON, B. M. (1984). The presence and activity of *Desulfotomaculum* spp. in sulphate limited freshwater sediments. *FEMS Microbiology Letters*, **21**, 47–50.

JONES, J. G. & SIMON, B. M. (1985). Interactions of acetogens and methanogens in anaerobic freshwater sediments. *Applied and Environmental Microbiology*, **49**, 944–8.

JONES, J. G., SIMON, B. M. & CUNNINGHAM, C. (1983). Bacterial uptake of algal extracellular products: an experimental approach. *Journal of Applied Bacteriology*, **54**, 355–65.

JONES, J. G., SIMON, B. M. & GARDENER, S. (1982). Factors affecting methanogenesis and associated anaerobic processes in the sediments of a stratified eutrophic lake. *Journal of General Microbiology*, **128**, 1–11.

JØRGENSEN, B. B. (1977). Bacterial sulfate reduction within reduced microniches of oxidized marine sediments. *Marine Biology*, **41**, 4–17.

KLUG, M. G., KING, G. M., SMITH, R. L. & LOVLEY, D. R. (1980). Comparative aspects of anaerobic microbial metabolism in sediments along a transect of a lake basin, p. 94. Abstracts of the Second International Symposium on Microbial Ecology, University of Warwick, U.K.

MAH, R. A. (1982). Methanogenesis and methanogenic partnerships. *Philosophical Transactions on the Royal Society, Series B*, **297**, 599–616.

MAIDEN, M. F. J. & JONES, J. G. (1984). A new filamentous, gliding bacterium, *Filibacter limicola* gen. nov. sp. nov. from lake sediment. *Journal of General Microbiology*, **130**, 2943–59.

MANN, K. G. (1972). Macrophyte production and detritus food chains in coastal waters. *Memorie dell'Istituto Italiano di Idrobiologia*, **29**, supplement, 353–83.

MONTESINOS, E., GUERRERO, R., ABELLA, C. & ESTEVE, I. (1983). Ecology and physiology of the competitition for light between *Chlorobium limicola* and *Chlorobium phaeobacteroides* in natural habitats. *Applied and Environmental Microbiology*, **46**, 1007–16.

NEDWELL, D. B. (1984). The input and mineralization of organic carbon in anaerobic aquatic sediments. In *Advances in Microbial Ecology*, vol. 7, ed. K. C. Marshall, pp. 93–131. New York, Plenum Press.

NICHOLS, P., STULP, B. K., JONES, J. G. & WHITE, D. C. (in press). Comparison of fatty acid content and DNA homology of the filamentous gliding bacteria *Vitreoscilla, Flexibacter*, and *Filibacter. Archives of Microbiology*.

NORKRANS, B. (1977). Surface microlayers in aquatic environments. In *Advances in Microbial Ecology*, vol. 4, ed M. Alexander, pp. 51–85. New York, Plenum Press.

PACE, N. R., STAHL, D. A., LANE, D. J. & OLSEN, G. J. (1986). The analysis of natural microbial populations by ribosomal RNA sequences. In *Advances in Microbial Ecology*, vol. 9, ed. K. C. Marshall, pp. 1–55. New York, Plenum Press.

PAERL, H. W. (1984). Alteration of microbial metabolic activities in association with detritus. *Bulletin of Marine Science*, **35**, 393–408.

PAERL, H. W. (1985). Microzone formation: its role in the enhancement of aquatic N_2 fixation. *Limnology and Oceanography*, **30**, 1246–52.

PETERSON, B. J., HOBBIE, J. E., HERSHEY, A. E., LOCK, M. A., FORD, T. E., VESTAL, J. R., MCKINLEY, V. L., HULLAR, M. A. J., MILLER, M. C., VENTULLO, R. M. & VOLK, G. S. (1985). Transformation of a tundra river from heterotrophy to autotrophy by addition of phosphorus. *Science*, **229**, 1383–6.

RAMSAY, A. J. & FRY, J. C. (1976). Response of epiphytic bacteria to the treatment of two aquatic macrophytes with the herbicide paraquat. *Water Research*, **10**, 453–9.

REYNOLDS, C. S. (1982). Phytoplankton periodicity: its motivation, mechanisms and manipulation. *Report of the Freshwater Biological Association*, **50**, 60–75.

REYNOLDS, C. S. (1984). Phytoplankton periodicity: the interactions of form, function and environmental variability. *Freshwater Biology*, **14**, 111–42.

REYNOLDS, C. S., HARRIS, G. P. & GOULDNEY, D. N. (1985). Comparison of carbon-specific growth rates and rates of cellular increase of phytoplankton in large limnetic enclosures. *Journal of Plankton Research*, **7**, 791–820.

REYNOLDS, C. S., WISEMAN, S. W. & CLARKE, M. J. O. (1984). Growth- and loss-rate responses of phytoplankton to intermittent artificial mixing and their potential application to the control of planktonic algal biomass. *Journal of Applied Ecology*, **21**, 11–39.

RUDD, J. W. M. & HAMILTON, R. D. (1978). Methane cycling in a eutrophic shield lake and its effects on whole lake metabolism. *Limnology and Oceanography*, **23**, 337–48.

SEPERS, A. B. J. (1981). Diversity of ammonifying bacteria. *Hydrobiologia*, **83**, 343–50.

SKUJA, H. (1956). Taxonomische und biologische Studien über das Phytoplankton schwedischer Binnengewässer. *Nova Acta Regiae Societatis Scientiarum Upsaliensis, Seria IV*, **16**, 1–404.

SMITH, G. A., NICKELS, J. S., KERGER, B. D., DAVIS, J. D., COLLINS, S. A., WILSON, J. T., MCNABB, J. F. & WHITE, D. C. (1986). Quantitative characterization of microbial biomass and community structure in subsurface material: a prokaryotic consortium responsive to organic contamination. *Canadian Journal of Microbiology*, **32**, 104–11.

SMYLY, W. J. P. (1974). Vertical distribution and abundance of *Ceriodaphnia quadrangula* (O. F. Müller) (Crustacea, Cladocera). *Freshwater Biology*, **4**, 257–66.

STAHL, D. A., LANE, D. J., OLSEN, G. J. & PACE, N. R. (1985). Characterization of a Yellowstone hot spring microbial community by 5S rRNA sequences. *Applied and Environmental Microbiology*, **49**, 1379–84.

SWEERTS, J. P., RUDD, J. W. M. & KELLY, C. A. (1986). Metabolic activities in flocculant surface sediments and underlying sandy littoral sediments. *Limnology and Oceanography*, **31**, 330–8.

TETT, P., HEANEY, S. I. & DROOP, M. R. (1985). The Redfield ratio and phytoplankton growth rate. *Journal of the Marine Biological Association of the United Kingdom*, **65**, 487–504.

TIPPING, E., JONES, J. G. & WOOF, C. (1985). Lacustrine manganese oxides: Mn oxidation states and relationships to 'Mn depositing bacteria'. *Archiv für Hydrobiologie*, **105**, 161–75.

WARD, D. M. & WINFREY, M. R. (1985). Interactions between methanogenic and sulfate-reducing bacteria in sediments. *Advances in Aquatic Microbiology*, **3**, 141–80.

WETZEL, R. G. & PENHALE, P. A. (1979). Transport of carbon and excretion of dissolved organic carbon by leaves and roots/rhizomes in seagrasses and their epiphytes. *Aquatic Botany*, **6**, 149–58.

BACTERIAL UTILIZATION OF ORGANIC MATTER IN THE SEA

FAROOQ AZAM AND BYUNG C. CHO

Scripps Institution of Oceanography, University of California, San Diego, La Jolla, California 92093, USA

INTRODUCTION

Until recently it was thought that bacteria in the sea played only a trivial role in organic fluxes; most of the primary production was believed to be utilized by the herbivores and passed on to the organisms comprising the 'grazing food chain' (Steele, 1974). This view had been perpetuated due to a lack of methods to quantify the *in situ* bacterial processes, and a feeling that seawater was too dilute to support significant bacterial growth. This notion has been challenged by results from new methods for measuring bacterial biomass and *in situ* growth rates. It has now been shown that: (1) bacterial abundance had been underestimated by an order of magnitude in the past (Hobbie, Daley & Jasper, 1977); (2) bacterial production rate is sufficiently rapid to require one-third to one-half of the primary productivity to support it (Hagström *et al.*, 1979; Fuhrman & Azam, 1980, 1982); and (3) bacteria are avidly preyed upon by a variety of protozoa such that growth and death of bacteria are tightly coupled processes (Hollibaugh, Fuhrman & Azam, 1980; Fenchel, 1982). In an ecosystem context, then, these new findings portray bacteria as a dynamic metabolic and trophic component representing a major pathway for matter and energy flux in the food web (Williams, 1981; Azam *et al.*, 1983).

If the above findings are correct then, somehow, bacteria manage to be highly effective competitors for organic matter in the sea. How do they do it? What physiological and biochemical strategies enable bacteria to be such effective competitors with other microbes and macrobes for organic matter utilization? Importantly, how can we reconcile this new view of bacterial activity with the long-standing perception that the ocean is an impoverished milieu for bacterial growth and a challenging environment for bacterial persistence? Either the recent information on the magnitude of the bacterial processes is wrong (i.e. greatly overestimates the magnitude of bacterial processes), or bacteria possess adaptations for growth in the sea which are not discerned in laboratory model systems (pure and mixed cultures studied in batch or the chemostat).

In this chapter we will attempt to develop the argument that marine bacteria employ a variety of biochemical and behavioural strategies for competing with other organisms for utilization of both dissolved organic matter (DOM) and particulate organic matter (POM). We will argue that the DOM 'soup' is not too dilute for bacterial persistence on it; it is the bacteria that strive to keep it at such great dilution that other organisms can not use it effectively. This is achieved by adaptations which allow a tight coupling between production of utilizable DOM in the environment and DOM uptake by bacteria.

THE IMPORTANCE OF AN ECOSYSTEM CONTEXT

Two conceptually distinct approaches have been used to study bacteria–organic matter interactions in the sea. The first approach employs pure cultures of bacteria and a defined organic matter regime in homogeneous growth medium. Generally, a chemostat is used as the model system. The implicit rationale for the approach is that the natural system is too complex to be tractable. The second approach is the 'ecosystem approach'. Interestingly, the adherents of this approach also invoke the extreme complexity of the natural system as rationale, arguing that the bacteria–organic matter interactions in the sea are so complex that the chemostat approach can not yield useful information. There are some fundamental difficulties with the laboratory approach (Jannasch, 1974). Mechanisms of DOM production can not be duplicated in the model systems. We do not even know what mechanisms are important. In a chemostat it is not possible to mimic the competition between bacteria and macrobes for organic matter utilization, and this is a major drawback. It is therefore essential to use an ecosystem approach in order to study how bacteria perform in the sea.

THE ENVIRONMENT OF MARINE BACTERIA

A challenge in studying bacteria–organic matter interactions is that the organic matter pool in seawater is chemically complex, physically heterogeneous, and variable in time and space. Also, since bacteria live in microenvironments, it is essential that bacterial performance be considered in the context of the organic matter regime in the microenvironment. DOM in seawater consists of literally hundreds of organic compounds, many of which are bacterial nutrients. These include sugars, amino acids, carboxylic acids, purines, pyrimidines, etc. Regulatory molecules such as cyclic AMP (cAMP), vitamins and antibiotics are also present. The concentrations of only a few compounds have been determined. They are vanishingly low, in pico-molar to nanomolar range (Williams, 1986). POM in the euphotic zone is a mixture of living plankton and detritus, roughly in equal proportions. POM concentration is highly variable, but often it is on the order of 100 μg C l^{-1} in the photic zone.

If all plankton, detritus and solute molecules were randomly distributed, then each microlitre of surface seawater will roughly contain one algal cell, 1000 bacteria, 1–10 cyanobacteria, one detritus particle, and a heterotrophic microflagellate. Ciliates and macrozooplankton will be present in this μl for a small fraction of time. Each bacterium will be of the order of 1000 μm from the nearest phytoplankton or detrital particle. The μl will contain a large variety of compounds at 10^9 to 10^{12} molecules. In terms of carbon, there is about 1 ng DOC per μl of which about 50 pg is utilizable by bacteria. This translates into a per-bacterium complement of 50 fg utilizable carbon. Each marine bacterium contains roughly 20 fg C (Bratbak, 1985; Bjørnsen, 1986); thus, in the absence of new DOM inputs, bacteria can double one or two times (Ammerman et al., 1984). This crude picture is intended to give some idea of the composition of the unstructured microenvironment of the bacterium. In reality it is possible that the microenvironment may be structured.

MECHANISMS OF DOM PRODUCTION

A central issue in bacteria–organic matter coupling is how utilizable DOM is produced in the bacterial microenvironment. This knowledge is critical to an understanding of the nutrient–growth relationships of bacteria, and bacterial performance in the ocean environment. Only limited information is available on the relative importance of various sources and mechanisms of utilizable DOM production.

Phytoplankton exudation

This is one of the most extensively studied and yet one of the least understood topics in marine microbiology. Numerous laboratory and field studies have measured exudation of specific molecular species or total exudates (Larsson & Hagström, 1979; Mague et al., 1979; Sharp, 1977). A variety of molecular species is released. However, exudation as a percentage of total photoassimilated carbon is highly variable. In field measurements, 0–70% of the fixed carbon has been reported to be exuded. This great variability is often attributed to methodological problems. Fragile algae may break during filtration to separate the cells from the medium, thus leading to overestimates of exudation. In field samples, bacteria take up and respire some exudates. Algal cell breakage during herbivore grazing ('sloppy feeding') leads to DOM release which can not be distinguished from exudation. Respiration of ingested carbon by herbivores may lead to underestimation of exudation.

Although the extent of exudation is uncertain, careful measurements using axenic cultures of algae show that some release of DOM is an integral part of phytoplankton physiology (Mague et al., 1979). Exudation may represent an overflow of photoassimilated carbon during periods when carbon fixation exceeds biomass production, thus, exudation of polysaccharides increases during nitrogen and phosphorus limitation (Mykelstad, 1977). However, exudation continues during the dark period (Mague et al., 1979), which is unlikely if it were merely an overflow of photoassimilated carbon. Alternatively (or additionally), significant losses of metabolic pools may occur by diffusion due to the large surface to volume ratios of microalgae and great concentration gradients across the cell membrane (T. K. Bjornsen, unpublished). Exudation could result in nutrient rich microzones around the algal cell (discussed below).

'Sloppy-feeding'

The earlier models of the marine food web assumed that phytoplankton is eaten by the herbivores without any loss of cell contents during grazing (e.g. Steele, 1974). There is evidence now that significant loss of algal cell contents can occur during handling and grazing by the herbivores (Lampert, 1978; Copping & Lorenzen, 1980; Eppley et al., 1981). The quantitative significance of 'sloppy feeding' as a mechanism for DOM production probably depends on the types of algae and herbivores involved. Systematic studies to determine this have not yet been done. Large chains of diatoms are likely to disintegrate during grazing and thus spill DOM. Fragile phytoflagellates may burst during handling by the herbivore, but this hypothesis has not yet been tested.

Autolysis

Nutrient limitation and other physicochemical stresses may cause death and autolysis of photoautotrophs. It is not known if non-predatory death of phytoplankton is a significant process in the sea. However, the autolysis of algal cells could result in the solubilization of a much greater fraction of the alga than sloppy feeding. Autolytic enzymes could, in a time-dependent manner, hydrolyse the structural proteins, polysaccharides, nucleic acids, lipids, etc. The gradual release of DOM also means that some components of DOM could elicit chemotactic response in heterotrophic bacteria in the microenvironment. Bacteria may attach to the autolysing alga, or remain swimming in close proximity to it in order to take up the released nutrients from a high concentration microzone (discussed later). Therefore carbon flow from an autolysing alga to heterotrophic bacteria may be an efficient process.

Excretion by microzooplankton and macrozooplankton

Most studies of zooplankton excretion have been concerned with the release of the mineral nutrients phosphate and ammonia, or the release of organic matter as faecal pellets. A substantial fraction of the ingested phytoplankton carbon is egested as faecal pellets. Copping & Lorenzen (1980) fed the zooplankton *Calanus pacificus* on uniformly [^{14}C]-labelled diatom *Thallasiosira fluviatilis*. They found that only a small fraction of the label was released as DOM. Metazooan faecal pellets may contain active digestive enzymes derived from the animal, and these enzymes may continue to hydrolyse the POM in the faecal pellet to DOM. Jacobsen & Azam (1985) found that faecal pellets become readily colonized by bacteria. These bacteria may use DOM produced from POM hydrolysis by animal digestive enzymes or by exoenzymes of the colonizing bacteria. Solubilization of large particles such as metazooan faecal pellets is a slow process (Jacobsen & Azam, 1985). Coprophagus animals may ingest faecal pellets before significant solubilization occurs. It is possible, however, that microzooplankton faecal material, because of its small size and large surface to volume ratio, is colonized and solubilized rapidly. The rejected food vacuoles of protozoa and the associated POM and DOM might be important sources of DOM production in oligotrophic oceans, where primary production is often dominated by cyanobacteria of the genus *Synechococcus* and other small photoautotrophs which are mainly grazed by microzooplankton (Welschmeyer & Lorenzen, 1985; Beers, Reid & Stewart, 1982; Platt, Subba Rao & Irwin, 1983).

POM as a source of DOM

As discussed, POM hydrolysis may be an important mechanism of DOM production. An interesting issue is whether bacteria hydrolyse POM in a controlled fashion and take up most of the hydrolysis products, or whether much of the hydrolysate diffuses into the environment as DOM. Jacobsen & Azam (1985) found that bacteria colonizing [^{14}C]-labelled faecal pellets of C. pacificus released into the seawater two to three times as much [^{14}C] as they assimilated. If bacteria actively compete with POM-consuming animals for organic matter, then a useful strategy might be to solubilize POM rapidly before an animal ingests the particle, even though much of the DOM produced may be lost to the colonizing bacteria due to molecular diffusion. The ecological consequences of this hypothetical scenario are discussed below.

Implications for bacteria–organic matter coupling

The modes and sources of DOM production have implications for bacterial adaptations for utilization of organic matter.

(i) All DOM sources (phytoplankton, zooplankton, other bacteria, POM) are particulate. Thus, DOM is produced at discrete loci, and DOM gradients must exist in the vicinity of the sources. Behavioural adaptations for sensing and responding to such gradients may therefore be important in bacteria–organic matter coupling.

(ii) A distinction should be made between episodic and sustained inputs of DOM. 'Sloppy-feeding' and animal excretion can be considered episodic, whereas algal exudation and autolysis, and POM hydrolysis would occur in a relatively sustained manner. One would expect that marine bacteria have strategies for effective nutrient uptake from short-lived DOM pulses as well as from sustained inputs.

(iii) Although the DOM pool in seawater is substantial (about 1.5 mg carbon per litre in surface waters) only about 5% of it is utilizable by bacteria (UDOM; Ammerman et al., 1984). New inputs of UDOM must sustain bacterial nutrient levels, suggesting a tight coupling between input and utilization.

(iv) Only a small fraction of UDOM can be directly taken up by bacteria; a major fraction consists of compounds such as proteins, polysaccharides, nucleic acids, which must first be hydrolysed to monomers or oligomers. Some small molecules also require prior hydrolysis (for example, nucleotides). Such molecules can be viewed as sources of bacterial nutrients, since they (like POM) must first be 'digested'.

STRATEGIES FOR ORGANIC MATTER UTILIZATION

The above discussion highlights the importance of an ecosystem context in considering the nutrient regime encountered by bacteria in their natural environment. Bacteria maintain the directly utilizable nutrients at extremely low concentrations, and this suggests strategies for a tight coupling between production and utilization of UDOM. Below, we discuss bacterial strategies for nutrient uptake *in situ*.

Uptake of directly utilizable compounds

High affinity uptake systems
Amino acids and sugars are present at 10^{-10} to 10^{-8} M (Williams, 1986), and they are taken up by extremely high affinity membrane transport system (k_m 1 to few nM for most amino acids and sugars (Azam & Hodson, 1981; Nissen, Nissen & Azam, 1984; Hagström *et al.*, 1984). Thus, free-living bacteria can effectively utilize extremely dilute DOM components from seawater.

High and low affinity uptake systems
Natural assemblages of marine bacteria exhibit high affinity low capacity (low flow) as well as low affinity high capacity (high flow) transport systems. Azam & Hodson (1981) found multiple transport kinetics (K_m range 10^{-8} to 10^{-4} M) for D-glucose uptake in natural planktonic assemblages. They reasoned that the high flow systems indicate the presence of substrate-enriched microzones in the bacterium's microenvironment. Nissen *et al.* (1984) found that D-glucose uptake in a marine bacterial isolate LNB-155 is mediated by a multiphasic transport system which changes its K_m (range 10^{-8} to 10^{-3} M) in an all-or-none fashion at critical substrate concentrations in seawater. Multiple and multiphasic transport systems should provide metabolic flexibility for a bacterium which experiences extremely low bulk-phase substrate concentration but also high concentration

gradients in the vicinity of a DOM source. A combination of high and low flow systems should allow the bacteria to enhance their rate of uptake over a broad concentration range.

Hagström et al. (1984) asked whether multiple transport systems persist in absence of particulate sources of DOM. Mixed assemblages of marine bacteria were grown in continuous cultures fed with particle-free unenriched seawater. After 30 generations, the assemblage still showed multiple uptake kinetics (K_m 10^{-9} to 10^{-7} M) for l-leucine uptake. The high K_m systems are either constitutive, or protein hydrolysis at bacterial surface can maintain high l-leucine concentrations to keep the high flow systems in an induced state. Considering the cost of maintaining a transport system, it is likely that free-living marine bacteria frequently encounter high l-leucine concentrations in their microenvironments.

Surge uptake
Marine bacteria might take advantage of episodic inputs of DOM by expressing very high uptake activity for brief periods (seconds to minutes). This strategy has been suggested for ammonium uptake by phytoplankton to take advantage of short-lived ammonium plumes produced by herbivore excretion (Goldman, McCarthy & Peavey, 1979). Surge uptake must involve the maintenance of excess transport capacity, which can be expressed instantaneously in response to high substrate inputs. The cost of maintaining 'excess uptake capacity' may be justified for a single critical growth-limiting substance (such as ammonium for phytoplankton), but not for a large variety of bacterial nutrients in the sea. However, surge uptake for one or few substrates might exist in different species in bacterial assemblages. Short-term uptake measurements (seconds to minutes) in natural assemblages need to be done to address this question.

Bacterial utilization of molecules which require modification prior to uptake

Free-living bacteria are faced with an intriguing challenge in utilizing those molecules which must be hydrolysed before uptake. It would be energetically prohibitive to release enzymes into solution in order to maintain significant activity in solution; bacteria account for only 10^{-7} of the volume in seawater. Below we discuss strategies which appear to be used by marine bacteria for utilization of dissolved polymers and some organophosphorus compounds.

Polymer utilization

Dissolved proteins and polysaccharides are quantitatively important nutrients for bacterial growth. The dissolved free amino acid (DFAA) pool which is directly utilizable by bacteria, is only 5–10% (w/v) of the dissolved combined amino acids (mainly protein and peptides). Therefore, it is important to know how the dissolved protein and peptides are utilized by free-living bacteria. Hollibaugh & Azam (1983) used tracer additions of haemoglobin-[^{14}C] or bovine serum albumin-[^{125}I] to seawater samples to measure protein degradation rates. They found rapid protein hydrolysis with pool turnover times on the order of a day. Significantly, protease activity was largely associated with bacterial cell surface (Hollibaugh & Azam, 1983; Sommville & Billen, 1983) and little or no activity was found in solution (<0.2 μm fraction). The location of protease activity on the bacterial cell is unknown; however, it is probably not located in the periplasmic space, since the dissolved protein molecules are too large to penetrate the outer membrane. The activity is probably located exterior to the outer membrane. If proteins were hydrolysed on cell surface, does much of the hydrolysate diffuse into seawater? Or do bacteria have strategies to maximize the hydrolysate uptake? Hollibaugh & Azam (1983) found indirect evidence of hydrolysis–uptake coupling. They found that in seawater samples in Southern California coastal waters the dissolved protein and l-leucine pool turnover rates were comparable (about 3% h^{-1}) although the protein pool was 20–50 greater than the amino acids pool. If all l-leucine produced by protein hydrolysis was channelled via the dissolved l-leucine pool, then l-leucine pool turnover rate would have to be much faster than the protein turnover rate. Thus protein hydrolysis and the uptake of the hydrolysate might be tightly coupled processes.

Dissolved oligosaccharides and polysaccharides are also major sources of nutrition for bacteria in the sea (Hagström et al., 1984). Hoppe (1983) and Sommville (1984) found that exoglucosidase activity in seawater samples is not present in solution (<0.2 μm filtrates) but is associated with the bacterial size fraction of plankton. The enzyme appeared to be associated with the bacterial cell surface. A consistent theme thus seems to be emerging. The enzymes responsible for polymer digestion are not secreted into the environment; they act while on cell surface. Further, the hydrolysis products may partly be utilized by the bacterium performing the hydrolysis. Although other polymers such as nucleic acids have not been studied in this respect, we predict that they are also hydrolysed by cell-associated hydrolyases and that some degree of hydrolysis–uptake coupling will be found.

Utilization of nucleotides

Nucleotides and many other organophosphorus compounds normally do not traverse the bacterial cell membrane; the phosphate must be cleaved by hydrolysis. It was found recently that 5′-nucleotides are rapidly hydrolysed by a bacterial periplasmic or membrane bound enzyme (Ammerman & Azam, 1985). This enzyme may account for a large fraction of phosphorus regeneration in the sea. Most of the 5′-nucleotidase activity was cell-associated rather than free in seawater. Using $[\gamma^{32}P]$ ATP as substrate, it could be shown that there was a strong (but not absolute) coupling between phosphate production and uptake. The liberated phosphate was 10–40 times more likely to be taken up than the phosphate in the bulk phase seawater. Interestingly, the hydrolysis–uptake coupling became tighter at low phosphate concentrations in bulk-phase seawater. When phosphate was depleted to unmeasurable ($<10^{-7}$ M) levels, 50% of liberated phosphate was taken up. This value was 15% in the presence of 1 μM phosphate, and the uptake could be eliminated by the addition of 1 mM phosphate. Thus, nucleotide hydrolysis and uptake is also consistent with the theme of tight coupling discussed above.

Implications for bacteria–organic matter coupling

The observations that bacteria 'digest' DOM with cell surface-associated hydrolyses and with some degree of hydrolysis–uptake coupling make perfect teleological sense. The bacterial biovolume is 10^{-7} of the environmental volume; it would be energetically prohibitive to maintain significant exoenzyme activity in the macroenvironment as well as to take up the hydrolysate after it has diffused to a great dilution. It appears that, for the free-living marine bacteria, the optimum strategy is to perform digestion on the cell surface or in the periplasmic space. The production of hydrolysis products in close vicinity to the bacterium may also mean that high substrate concentrations could be created and sustained in close proximity to the cell surface. Were hydrolysis to occur within the periplasmic space of the bacterium, the cell membrane transport systems could be exposed to high intraperiplasmic concentration of the hydrolysis product. This mode of DOM utilization may be a central theme in bacteria–organic matter interactions.

SIGNIFICANCE OF BACTERIAL BEHAVIOUR IN BACTERIA—ORGANIC MATTER INTERACTIONS

Non-random distribution of bacteria

We have so far discussed the utilization of DOM components reaching the cell surface by diffusion. Bacteria could enhance nutrient uptake by being in close vicinity of the source of UDOM. This, however, would require bacteria to be motile and chemotactic towards one or more components of UDOM.

Most isolates of marine bacteria are motile (Zobell, 1946; Rheinheimer, 1971). J. W. Ammerman (unpublished) directly observed motility in concentrates of natural marine assemblage. He found a fraction (not quantifiable by his procedure) of the assemblage to be motile with swimming speeds of the order of 20–40 μm s^{-1} (similar to *E. coli*). He also found a marine *Bdellovibrio* strain to be highly motile. B. C. Cho & F. Azam (unpublished), using a capillary assay with [^{35}S]-labelled bacteria, found that 5–30% of bacteria in concentrates of natural assemblages were motile.

Bell & Mitchell (1972) found that isolates of marine bacteria were chemoattracted by algal exudates. They proposed that, around algal cells, there is a region ('phycosphere') which is enriched in organic matter to which bacteria may be attracted. Chet & Mitchell (1976) found that bacteria from enrichment cultures grown on casein or albumin were attracted by the protein on which they were selected. Wellman & Paerl (1981) isolated a bacterium from tar balls in the Atlantic and found this isolate to be chemoattracted by Kuwait crude oil. In addition to chemoattraction, several instances of negative chemotaxis of marine bacteria have been reported. Using marine isolates, Young & Mitchell (1973) demonstrated negative chemotaxis by marine bacteria to hydrocarbons and heavy metals. B. C. Cho & F. Azam (unpublished) found that concentrates of marine natural assemblages of bacteria showed negative chemotaxis to mercuric chloride. Sieburth (1968) suggested that algae may secrete substances which inhibit bacterial association with algae. Thus, observations on both marine bacterial isolates and natural assemblages of bacteria support the idea that marine bacteria are capable of chemotaxis and motility. Marine bacteria may thus distribute non-randomly in responses to the concentration gradients in their microenvironments to maximize DOM uptake (Azam & Ammerman, 1984).

Sustained sources of DOM could attract bacteria present within reasonable distances. Bacteria in the surface waters are on the order of 1000 μm from the nearest algal cell, and this distance can be traversed within minutes (Azam & Ammerman, 1984). One might hypothesize therefore that free-living bacteria could, by chemotaxis and motility, locate and cluster around DOM sources. The operation of multiple and multiphasic transport systems would result in increasing nutrient flux as the bacterium moves up the concentration gradient.

Chemical cues

It is possible that chemical signals are involved in bacteria–algae and bacteria–bacteria interactions. The sensing of decomposing particles could involve both nutrient gradients and signals from the colonizing bacteria.

Algae–bacteria interactions

Bacteria are attracted by algal exudates (Bell & Mitchell, 1972) but the identity of the attractant(s) has not been established. Sieburth (1968) suggested that bacterial repellants might also be secreted by algae, and this could explain why healthy algae are generally free of attached bacteria whereas dead algae become heavily colonized. Smith & Higgens (1978) suggested the existence of a metabolic feed back between algae and bacteria. Such a feedback could regulate the rate and even the molecular composition of the exudates. Joiris et al. (1982) showed that DOM release by phytoplankton as a percentage of photosynthetically fixed carbon varied inversely with the concentration of nitrogen which was the limiting nutrient. Myklestad (1977) showed, in cultures of the diatom Chaetoceros affinis, that high amounts of polysaccharides were excreted during phosphorus-limited growth.

Algal exudation and its utilization by bacteria may be the basis of a commensal relationship, wherein algae provide organic matter for bacterial growth and bacteria in turn provide mineralized phosphorus and nitrogen which may be growth-limiting nutrients for algae. We suggest the following hypothetical scenario as a basis for algae–bacteria interaction. Phosphorus- or nitrogen-limited phytoplankton excrete carbon-rich compounds including polysaccharides and small molecular weight bacterial nutrients. Motile and chemotactic bacteria are attracted to the exuding alga, so bacterial abundance near the alga is enhanced. Since there are substantial pools of dissolved organic phosphorus (DOP) and dissolved organic nitrogen (DON) in surface seawater (Williams, 1986) the increased abundance of bacteria results in a high rate of N and P mineralization in the microenvironment of the algal cell. Bacterial metabolism near the algal cell would be upshifted because of the availability of high concentration of carbon rich nutrient. This may increase per cell mineralization of N and P as well.

Rosso and F. Azam (unpublished) found that cell-surface protease activity per cell shows strong direct correlation with the growth rates of bacteria in seawater. Thus, upshifted bacteria may show enhanced N mineralization by protein utilization. Ammerman & Azam (1985) showed that bacterial (periplasmic?) 5'-nucleotidase can satisfy 50–100% of the algal P requirements in euphotic zone Southern California coastal waters. Phosphorus mineralization by this cell-surface enzyme is particularly interesting in that the process is extracellular. Ammerman & Azam (1985) showed that at μM phosphate concentrations, 85% of the hydrolysed P is released into the environment. Thus, once clustered bacteria create a high P microzone around the algal cell, most of the hydrolysed P is released by the hydrolysing bacteria which sustains the high P concentration within the microenvironment.

It is not clear whether a microalga could, by exudation, sustain a high concentration of a diffusible molecule in its microenvironment. High surface to volume ratio in nanoplankton would result in rapid diffusion. Mitchell, Okubo & Fuhrman (1985) and Jackson (this volume) argue that small algal cells could not sustain extensive enriched microzones and therefore would not be surrounded by bacterial clusters. Larger alga, on the other hand, might sink too fast for cluster formation. Most of the photosynthesis in the sea is due to $<10\,\mu$m algae. Could then the algae in the sea be surrounded by bacterial clusters?

We suggest that a large fraction of exudation is in the form of polysaccharaides which remain associated with algal surface. We envision that small diffusible molecules (including informational molecules such as cAMP; below) serve to attract bacteria to this region of highly hydrated polysaccharide. Bacteria may attach to the periphery of polysaccharide zone, hydrolyse the polysaccharide chains, and utilize the sugar moieties thus produced. The polyionic nature of algal exopolysaccharides (Myklestad, 1977) could bind the remineralized phosphate and ammonium within the polysaccharide zone.

This scenario could result in a tight coupling of exudate flux and mineralized nutrients between bacteria and algae. The exudates are not allowed to diffuse freely in the macroenvironment, and are made available mainly to those bacteria which are metabolically active (and hence producing mineralized plant nutrients) within the microenvironment of the alga. The extent of exudation could be regulated in accordance with the N, P, status of the algae, thus providing a feedback relationship between bacteria and algae.

It is possible that informational molecules are also exchanged between bacteria and algae. It is interesting in this context that cAMP was found to be excreted by freshwater algae (Franko & Wetzel, 1980, 1981); marine algae have not been examined in this respect). Ammerman & Azam (1981) used a radioimmunoassay to measure cAMP concentration in seawater. They found 1–30 pM cAMP in California coastal waters, and cAMP concentration showed a diel pattern similar to some bacterial nutrient of algal origin. They found that bacteria in seawater samples took up cAMP by an extremely high affinity active transport system (K_m was few pM) which was highly specific for cyclic nucleotides. Among the cyclic nucleotides, cAMP was the best substrate for this transport system. This transport system was very effective in cAMP uptake from the environmental concentrations, and it could double the intracellular concentration of cAMP in marine bacteria in minutes to hours.

Ammerman & Azam (1981) speculated that cAMP might act as a metabolic cue in bacteria–algae interaction. They argued that the kinetic characteristics of the cAMP transport system indicate a regulatory rather than a nutritional role for cAMP. This transport system shows a near-absolute specificity for cyclic nucleotides and a preference for cAMP. AMP, which occurs at about 100 times greater concentration in seawater, is not a substrate for the cyclic nucleotide transport system. It is improbable that the cyclic nucleotide transport system evolved to take up cAMP as a nutrient, because this compound is so dilute compared with AMP into which it would be converted before further metabolism. It is tempting to speculate that cAMP is used as a metabolic cue between algae and bacteria. It could serve as a unifying signal of the metabolic and nutritional status of the microenvironment.

Bacteria–POM interactions
Bacteria solubilize POM with exoenzymes and then utilize the resulting DOM, so the DOM concentrations in the particle microenviron-

ment could be quite high. Therefore, it is often thought that POM and surfaces coated with DOM provide a nutrient-rich microenvironment for bacterial growth. Recently, however, several laboratories have found that attached bacteria do not necessarily grow faster than the free-living bacteria (Fletcher, 1984; Hoppe, 1984; Kirchman, 1983). B. C. Cho & F. Azam (unpublished) found in ocean profiles in Santa Monica Basin that, per cell, tritiated thymidine incorporation rates for free-living and attached bacteria were comparable, indicating that free-living and attached bacteria had comparable growth rates. How can we reconcile these observations with the view that particles are sites of nutrient enrichment and rapid growth?

In the sea, bacteria must compete with a great variety of particle-eaters. For each size-class of POM that bacteria colonize there are animals capable of ingesting the particle, often within hours of particle production. A strategy for bacteria could be to rapidly convert POM to DOM even if a large fraction of DOM diffuses into the seawater. This 'hypersolubilization' could be achieved by overproduction and secretion of hydrolytic enzymes into the particle microenvironment. Whereas free-living bacteria do not release the hydrolytic enzymes, we suggest that, upon attachment to a particle, bacteria do release enzymes into the environment of the particle. Stepped-up exoenzyme synthesis may involve cell enlargement and this is observed in attached bacteria. The colonized particle will thus become a source of DOM which the free-living bacteria could utilize.

We speculate that the progeny of the colonizing bacteria is released into free-living state and stays within the enriched microzone around the hydrolysing particle. We further envision a 'sorting out' of bacterial strains within the microenvironment by the secretion of bacteriocins or other allelopathic molecules by the colonizing species. This would result in colonization of the space around the particle by the particle colonizing species, followed by enhanced growth of the free-living progeny, and rapid hydrolysis of the particle. This strategy would also save the progeny from being ingested by the particle-eaters when the partially spent particle is eaten. Jacobsen & Azam (1985) found that colonized faecal pellets of C. pacificus become baby-machines. After an initial phase of rapid colonization the progeny was released in free-living state. Colonized POM can thus become a source of both bacteria and DOM in the microenvironment. The slow growth of attached bacteria might be due to energy requirement for exoenzyme hyperproduction.

BACTERIAL GROWTH IN THE SEA AND ITS REGULATION BY ENVIRONMENTAL CONDITIONS

Growth is the main goal of bacterial performance. Until recently, however, there were no methods for measuring *in situ* growth rates. Two methods have recently been developed. One method is based on measurement of the frequency of dividing cells (FDC) and an empirically determined relationship between growth rate and FDC (Hagström *et al.*, 1979). A second method measures tritiated thymidine incorporation into bacterial DNA and computes from it the bacterial growth rate by using a theoretically calculated and empirically calibrated conversion factor (Fuhrman & Azam, 1980, 1982). Importantly, these two methods are specific for bacteria in the presence of the normal biota of marine environments. This methodology, therefore, allows one to address questions about *in situ* bacterial growth and to study, by natural and experimental perturbations, how bacterial growth is regulated in the natural environment. Bacterial enumeration by epifluorescence microscopy now allows reliable estimates of natural abundances needed for computing the specific growth rates.

Bacterial production can also be interpreted in terms of overall carbon flux into heterotrophic bacteria. Thus, even though it is not possible to measure *in situ* uptake of the myriad DOM components, the rate of bacterial production can serve as a cumulative measure of carbon flux into bacteria. This flux can then be considered in the context of overall carbon and energy fluxes in the ecosystem to see if bacteria play a significant role. A number of such studies have now been done. In most environments, 30–50% of the primary production is needed to support the measured bacterial production. These results have led to a new conception of the pelagic marine food web which emphasizes a bacteria–protozoa 'loop' ('microbial loop'; Azam *et al.*, 1983) as a major pathway for matter and energy flow in the food web.

The patterns of bacterial production in large areas of the sea have now been determined to find out what environmental factors control bacterial growth in the sea. Of particular interest is the relationship between nutrient supply and bacterial growth *in situ*. In an extensive study in the Southern California Bight, Fuhrman & Azam (1982) found that bacterial specific growth rate, μ, was strongly correlated with chlorophyll *a* (a measure of algal abundance) in seawater. Cho & Azam (1986), working in the oligotrophic north Pacific gyre, found a strong correlation between μ and phytoplankton primary production. These results are consistent with the notion that bacteria derive their nutrients mainly from phytoplankton, but that the exact nature and mechanism of transfer of organic matter from algae to bacteria may vary (probably with the type of algae, their physiological status, and the dominant herbivores present in the environment).

Growth versus dormancy

It is often thought that the most bacteria in natural marine assemblages are dormant because of nutrient starvation (Stevenson, 1978). This perception of bacterial growth is not based on *in situ* measurements but rather on extrapolations from chemostat cultures (Jannasch, 1968, 1974). Measurements of *in situ* growth rates generally yield assemblage mean generation times of 0.5–2 days in the photic zone in coastal and open ocean waters. Is this growth so slow that marine bacteria should be considered dormant? This is not just a semantic issue. Clearly, the performance of bacteria in the sea should be considered in relation to the nutrient status of the natural environment and not with laboratory cultures. Perhaps the question should be: Given the rate of organic matter input (primary production), how well do heterotrophic bacteria compete with other organisms in the environment?

Bacterial biomass in the photic zone is 10–50% of the total plankton biomass, and bacteria manage to take up a very substantial fraction of the photosynthetically fixed carbon to maintain a large standing stock (Fuhrman & Azam, 1982). It is not known whether the standing stock of bacteria in the photic zone (generally about $10^6\,ml^{-1}$) is controlled by predation or by nutrient limitation. However, the large bacterial standing stocks can only grow slowly because of the ceiling imposed by primary production. Bacteria in the sea are subject to intense grazing pressure (Fenchel, 1982). It is difficult to see how large bacterial assemblage with a 1–2 day turnover time could maintain a large dormant component. Tritiated thymidine autoradiography shows that 30–80% of bacteria in coastal surface waters are growing (Tabor & Neihof, 1982; Fuhrman & Azam, 1982). These arguments do not support the notion that most bacteria in the sea are dormant. In any event, bacteria as an assemblage compete very effectively for organic matter in the ocean environment.

CONCLUSIONS

The question of the performance of marine bacteria in an ecosystem context has just begun to be addressed. The finding that heterotrophic bacteria are a major metabolic component in the sea has prompted interest in biochemistry of bacterial growth in the sea, and in the ecological role of bacteria. The extreme complexity of nutrient–growth relationship of bacteria in the sea means that we can not hope to mimic this relationship in a chemostat; we need *in situ* and single cell techniques. In terms of ecology of bacteria, the complex and transient nutrient patterns may represent varied niches for the coexistence of many species in a seemingly homogeneous environment. The study of interspecific competition in a nutritionally dynamic environment is a major intellectual challenge. A new conceptual framework and new techniques are needed to study the biochemistry of bacterial growth in an ecosystem context.

The findings about bacterial adaptations for organic matter utilization lead to the optimism that some predictable unifying concepts may emerge. The characteristics of nutrient 'digestion' and transport systems discussed above support this view. New concepts of the biochemical bases of microbe–environment interactions in the sea may also be relevant to other ecosystems including soil, animal, and human.

ACKNOWLEDGEMENT

This work was supported by National Science Foundation Grant OCE85-01363 to F. Azam. We thank Michelle Pontius-Brewer and R. W. Eppley for critically reading the manuscript for us.

REFERENCES

AMMERMAN, J. W. & AZAM, F. (1981). Dissolved cyclic adenosine monophosphate (cAMP) in the sea and uptake of cAMP by marine bacteria. *Marine Ecology Progress Series*, **5**, 85–9.

AMMERMAN, J. W. & AZAM, F. (1985). Bacterial 5′-nucleotidase in aquatic ecosystems: A novel mechanism of phosphorus regeneration. *Science*, **277**, 1338–40.

AMMERMAN, J. W., FUHRMAN, J. A., HAGSTRÖM, A. & AZAM, F. (1984). Bacterioplankton growth in seawater: I. Growth kinetics and cellular characteristics in seawater cultures. *Marine Ecology Progress Series*, **18**, 31–9.

AZAM, F. & AMMERMAN, J. W. (1984). Cycling of organic matter by bacterioplankton in pelagic marine ecosystems: Microenvironmental considerations. In *Flows of Energy and Materials in Marine Ecosystems*, ed. M. J. R. Fasham, pp. 345–360. New York: Plenum Publishing Company.

AZAM, F., FENCHEL, T., FIELD, J. G., GRAY, J. S., MEYER-REIL, L. A. & THINGSTAD, F. (1983). The ecological role of water column microbes in the sea. *Marine Ecology Progress Series*, **10**, 257–63.

AZAM, F. & HODSON, R. E. (1981). Multiphasic kinetics for D-glucose uptake by assemblages of natural marine bacteria. *Marine Ecology Progress Series*, **6**, 213–22.

BEERS, J. R., REID, F. M. H. & STEWART, G. L. (1982). Seasonal abundance of the microplankton population in the North Pacific Central Gyre. *Deep-Sea Research*, **29**, 227–45.

BELL, W. H. & MITCHELL, R. (1972). Chemotactic and growth responses of marine bacteria to algal extracellular products. *Biological Bulletin*, **143**, 265–77.

BJORNSEN, T. K. (1986). Automatic determination of bacterioplankton biomass by image analysis. *Applied and Environmental Microbiology*, **51**, 1199–204.

BRATBAK, G. (1985). Bacterial biovolume and biomass estimations. *Applied and Environmental Microbiology*, **49**, 1488–93.

CHET, I. & MITCHELL, R. (1976). An enrichment technique for isolation of marine chemotactic bacteria. *Microbial Ecology*, **3**, 75–8.

CHO, B. C. & AZAM, F. (1986). Bacterial secondary production and flux of organic matter in the North Pacific Central Gyre: PRPOOS Program Abstracts, p. 21. *American Society of Limnology and Oceanography*, Kingston, Rhode Island, USA.

COPPING, A. E. & LORENZEN, C. J. (1980). Carbon budget of a marine phytoplankton–herbivore system with carbon-14 as a tracer. *Limnology and Oceanography*, **25**, 873–82.

EPPLEY, R. W., HORRIGAN, S. G., FUHRMAN, J. A., BROOKS, E. R., PRICE, C. C. & SELLNER, K. (1981). Origins of dissolved organic matter in Southern California coastal waters: Experiments on the role of zooplankton. *Marine Ecology Progress Series*, **6**, 149–59.

FENCHEL, T. (1982). Ecology of heterotrophic microflagellates. IV. Quantitative occurrence and importance as bacterial consumers. *Marine Ecology Progress Series*, **9**, 35–42.

FLETCHER, M. (1984). Comparative physiology of attached and free-living bacteria. In *Microbial Adhesion and Aggregation*, ed. K. C. Marshall, pp. 223–32. Dahlem Konferenzen. Berlin: Springer–Verlag.

FRANKO, D. A. & WETZEL, R. G. (1980). Cyclic adenosine $-3':5'$-monophosphate: Production and extracellular release from green and blue–green algae. *Physiologia Plantarum*, **49**, 65–7.

FRANKO, D. A. & WETZEL, R. G. (1981). Dynamics of cellular and extracellular cAMP in *Anabaena flos* Aquae (cyanophyta): Intrinsic culture variability and correlation with metabolic variables. *Journal of Phycology*, **17**, 129–34.

FUHRMAN, J. A. & AZAM, F. (1980). Bacterioplankton secondary production estimates for coastal waters of British Columbia, Antarctica, and California. *Applied and Environmental Microbiology*, **39**, 1085–95.

FUHRMAN, J. A. & AZAM, F. (1982). Thymidine incorporation as a measure of heterotrophic bacterioplankton production in marine surface waters: Evaluation and field results. *Marine Biology*, **66**, 109–20.

GOLDMAN, J. C., McCARTHY, J. J. & PEAVEY, G. D. (1979). Growth rate influence on the chemical composition of phytoplankton in oceanic waters. *Nature*, **279**, 210–15.

HAGSTRÖM, A., AMMERMAN, J. W., HENRICHS, S. & AZAM, F. (1984). Bacterioplankton growth in seawater: II. Organic matter utilization during steady-state growth in seawater cultures. *Marine Ecology Progress Series*, **18**, 31–9.

HAGSTRÖM, A., LARSSON, U., HÖRSTEDT, P. & NORMARK, S. (1979). Frequency of dividing cells, a new approach to the determination of bacterial growth rates in aquatic environments. *Applied and Environmental Microbiology*, 37, 805–12.

HOBBIE, J. E., DALEY, R. J. & JASPER, S. (1977). Use of Nuclepore filters for counting bacteria by fluorescence microscopy. *Applied and Environmental Microbiology*, 33, 1225–8.

HOLLIBAUGH, J. T. & AZAM, F. (1983). Microbial degradation of dissolved proteins in seawater. *Limnology and Oceanography*, 28, 1104–16.

HOLLIBAUGH, J. T., FUHRMAN, J. A. & AZAM, F. (1980). Radioactively labeling of natural assemblages of bacterioplankton for use in trophic studies. *Limnology and Oceanography*, 25, 172–81.

HOPPE, H.-G. (1983). Significance of exoenzymatic activities in the ecology of brackish water: Measurements by means of methyl-umbelliferyl substrates. *Marine Ecology Progress Series*, 11, 299–308.

HOPPE, H.-G. (1984). Attachment of bacteria: Advantages and disadvantages for survival in the aquatic environment. In *Microbial Adhesion and Aggregation*, ed. K. C. Marshall. Dahlem. Konferenzen. Berlin: Springer–Verlag.

JACKSON, G. A. (1987). Physical and chemical properties of aquatic environments. Symposium on Ecology of microbial communities, 108th Ordinary Meeting of the Society for General Microbiology, University of St Andrews, 7–19 April 1987.

JACOBSEN, T. R. & AZAM, F. (1985). Role of bacteria in copepod fecal pellet decomposition: Colonization, growth rates and mineralization. *Bulletin of Marine Science*, 35, 495–502.

JANNASCH, H. W. (1969). Estimations of bacterial growth rates in natural waters. *Journal of Bacteriology*, 99, 156–60.

JANNASCH, H. W. (1974). Steady-state and the chemostats in ecology. *Limnology and Oceanography*, 19, 716–20.

JOIRIS, C., BILLEN, G., LANCELOT, C., DARO, M. H., MOMMAERTS, J. P., BERTELS, A., BOSSICARTA, M., NIJS, J. & HECQ, J. H. (1982). A budget of carbon cycling in the Belgian coastal zone: Relative roles of zooplankton, bacterioplankton and benthos in the utilization of primary production. *Netherlands Journal of Sea Research*, 16, 260–75.

KIRCHMAN, D. (1983). The production of bacteria attached to particles suspended in a freshwater pond. *Limnology and Oceanography*, 28, 858–72.

LAMPERT, W. (1978). Release of dissolved organic carbon by grazing zooplankton. *Limnology and Oceanography*, 23, 831–4.

LARSSON, U. & HAGSTRÖM, A. (1979). Phytoplankton exudate release as an energy source for the growth of pelagic bacteria. *Marine Biology*, 52, 199–206.

MAGUE, T. H., FRIBERG, E., HUGHES, D. H. & MORRIS, I. (1979). Extracellular release of carbon by marine phytoplankton: A physiological approach. *Limnology and Oceanography*, 25, 262–79.

MITCHELL, J. G., OKUBO, A. & FUHRMAN, J. A. (1985). Microzones surrounding phytoplankton form the basis for a stratified marine microbial ecosystem. *Nature*, 316, 58–9.

MYKLESTAD, S. (1977). Production of carbohydrates by marine planktonic diatoms. II. Influence of the N/P ratio, growth rates, and production of cellular and extracellular carbohydrates by *Chaetoceros Affines* var. Willei (Gran) Hustedt and *Skeletonema costatum* (Grev.) Cleve. *Journal of Experimental Marine Biology and Ecology*, 29, 161–79.

NISSEN, H., NISSEN, P. & AZAM, F. (1984). Multiphasic uptake of D-glucose by an oligotrophic marine bacterium. *Marine Ecology Progress Series*, 16, 155–60.

PLATT, T., SUBBA RAO, D. V. & IRWIN, B. (1983). Photosynthesis of picoplankton in the oligotrophic ocean. *Nature*, 301, 702–4.

RHEINHEIMER, G. (1971). 'Aquatic Microbiology' London, J. Wiley and Sons.

SIEBURTH, J. McN. (1968). Observations on planktonic bacteria in Narragansett Bay, Rhode Island: A resumé. Misaki. Marine Biological Institute, Kyoto University, 12, 49–64.

SHARP, J. H. (1977). Excretion of organic matter by marine phytoplankton. Limnology and Oceanography, 22, 381–99.

SMITH, D. F. & HIGGENS, H. W. (1978). An interspecific regulatory control of dissolved organic carbon production by phytoplankton and incorporation by microheterotrophs. In Microbial Ecology, eds. M. W. Loutit & J. A. R. Miles, pp. 34–9. Berlin: Springer–Verlag.

SOMMVILLE, M. (1984). Measurement and study of substrate specificity of exoglucosidase activity in eutrophic water. Applied and Environmental Microbiology, 48, 1181–5.

SOMMVILLE, M. & BILLEN, G. (1983). A method for determining exoproteolytic activity in natural waters. Limnology and Oceanography, 28, 190–3.

STEELE, J. H. (1974). The structure of marine ecosystems. Harvard University Press, Cambridge, Mass., 128 pp.

STEVENSON, L. H. (1978). A case for bacterial dormancy in aquatic systems. Microbial Ecology, 4, 127–33.

TABOR, P. S. & NEIHOF, R. A. (1982). Improved microautoradiographic method to determine individual microorganisms active in substrate uptake in natural waters. Applied and Environmental Microbiology, 44, 945–53.

WELLMAN, A. M. & PAERL, H. W. (1981). Rapid chemotaxis assay using radioactively labeled bacterial cells. Applied and Environmental Microbiology, 42, 216–21.

WELSCHMEYER, N. A. & LORENZEN, C. J. (1985). Chlorophyll budgets: Zooplankton grazing and phytoplankton growth in a temperate fjord and the central Pacific gyres. Limnology and Oceanography, 30, 1–21.

WILLIAMS, P. J. LEB. (1981). Incorporation of microheterotrophic processes into the classical paradigm of the planktonic food web, Fifteenth European Symposium on Marine Biology, Kiel, F. R. G., Kieler Meeresforsch., Sonderh., 5, 1–28.

WILLIAMS, P. M. (1986). Chemistry of the dissolved and particulate phases in the water column. In Plankton Dynamics of the Southern California Bight, ed. R. W. Eppley, pp. 53–83. Berlin: Springer–Verlag.

YOUNG, L. Y. & MITCHELL, R. (1973). Negative chemotaxis of marine bacteria to toxic chemicals. Applied and Environmental Microbiology, 25, 972–5.

ZOBELL, C. E. (1946). Marine Microbiology. A monograph on Hydrobacteriology. Waltham, Chronica Botanica Company.

BACTERIAL RESPONSES TO NUTRIENT LIMITATION

JEANNE S. POINDEXTER

*The Public Health Research Institute
of The City of New York, Inc.,
New York, NY 10016, USA*

Oligotrophy was considered in an earlier review (Poindexter, 1981) as an existence that alternated between fast and famine conditions, in contrast to the feast and famine conditions (Koch, 1971) characteristic of copiotrophy. This paper will focus principally on fasting, and examine bacterial strategies for growth under conditions in which an essential nutrient is available in such low supply that its availability limits the rate of bacterial growth. Famine (absence of nutrients), common to both types of existence, will not be dealt with here.

Field studies and models are amply represented among other presentations in this symposium. I have chosen to examine laboratory studies of monotypic populations subjected to prolonged nutrient limitation during growth. As an ecological approach, this is defensible in the sense that the laboratory pure culture is the most practical way of magnifying the microhabitat. From such studies, we can learn the capabilities of a kind of organism under a defined set of conditions. Most of the studies to be considered are continuous cultivation experiments. Single-cycle batch cultures are not suitable for the study of responses to nutrient limitation that allow continued reproduction because limitation is experienced by only one generation, viz., the one just prior to maximum stationary phase. Some adaptations may require more than one generation for their full development (Law & Button, 1977; Höfle, 1982, 1983; see below). Several reviews of the various uses of continuous cultures in microbial ecology are available (Veldkamp, 1970, 1976, 1977; Jannasch & Mateles, 1974; Harder, Kuenen & Matin, 1977; see also Matin, 1979; Tempest & Neijssel, 1981; Harder & Dijkhuizen, 1982, 1983).

The fundamental thesis presented is that specific physiological properties enable bacteria to exploit fasting (oligotrophic) conditions. Some bacteria seem able to express these properties following environmental changes, without a detectable change in genotype. Other bacteria depend on genetic change, so that only mutant progeny of the original populations survive prolonged periods of fasting. Investigation of such mutants and comparison of their properties with those of their unchanged parents is useful in evaluating the relevance of particular physiological properties that are present in wild-types of other groups. It is possible that those organisms that adapt to nutrient limitation through mutation are themselves descendants of populations in which phenotypic changes accomplished adaptation to fasting conditions. The techniques of enrichment and isolation used for the vast majority of cultured bacteria select strongly for those organisms able to multiply rapidly in the presence of abundant nutrients, i.e., copiotrophs, and may well select against bacteria with properties especially valuable under oligotrophic conditions.

A priori, it is to be expected that each property will exact a cost – metabolically, physiologically, or with respect to competitiveness – and that few, if any, species will be capable of adapting to every kind of nutrient limitation. This recognition has resolved Hutchinson's 'paradox of the phytoplankton'; that is, several species of phytoplankton can occur in what seems to be the same niche because minor fluctuations in environmental conditions, e.g. nutrient ratios, occur, favouring one organism then another, and even favouring species that would not be competitive under constant conditions but are adapted to fluctuating conditions (Tilman, Kilham & Kilham, 1982).

Of the major essential nutrients for heterotrophic bacteria (carbon, hydrogen, oxygen, nitrogen, phosphorus and sulphur), emphasis will be placed on carbon, nitrogen and phosphorus. Since bacteria cannot grow without these elements, the strategies to be examined are those that improve the capacity of the organism or the population to acquire and utilize them.

POSSIBLE ADAPTIVE RESPONSES TO NUTRIENT
LIMITATION

In a nutrient-limited environment, not all the scarce nutrients available will be in the same form; even the soluble carbon sources will be present as a mixture of compounds. It follows that an appropriate adaptation to fasting would be acceptance of a wider variety of carbon, nitrogen or phosphorus sources. Not only the enzymes of intermediary metabolism (both degradative and assimilatory), but also transport systems must become active in the uptake of whatever type of nutrient is available. The synthesis and secretion of hydrolytic enzymes to solubilize organic debris would also expand the utilizable resources (Harder & Dijkhuizen, 1983). Studies of chemotaxis in several non-enteric Gram-negative bacteria show that enhanced detection of resources may be another response to nutrient limitation and that derepressed positive chemotaxis improves acquisition of utilizable nutrients by directing cells towards slightly less oligotrophic regions.

Increased variety of catabolic enzymes as a response to low nutrient flux

This response has been reviewed elsewhere with attention focused largely on carbon resources (Harder & Dijkhuizen, 1982). In this discussion, examples will be cited that illustrate the ability of bacteria to increase their utilization of alternative sources of both carbon and nitrogen, even in the presence of preferred substrates, but only when the flux of limiting nutrient is low. In the majority of such studies, a reduction in nutrient flux has been achieved by reducing the flow rate of chemostat cultures. This practice unfortunately complicates interpretations because it reduces both nutrient flux and growth rate. Reduction of nutrient flux can also be achieved by reducing an individual nutrient concentration in the reservoir while maintaining the flow rate and therefore the dilution rate, or by dilution of the entire medium, but the yield of cell material is correspondingly reduced. A happy medium – and an illuminating approach – is to impose relative nutrient limitation by increasing the concentrations of all but one nutrient without reducing the flow rate. All three methods mimic changes that occur in natural habitats.

In a study with a marine corynebacterium, Law & Button (1977) varied nutrient flux. Nutrient ratios were varied but total organic content was kept constant, and each ratio was tested at different dilution rates. They found that the ability of the population to scavenge glucose was increased as glucose carbon was replaced by amino acid carbon. With glucose as sole carbon source, at a dilution rate (D) of 0.008 to $0.05\,h^{-1}$, more than $200\,\mu g\,l^{-1}$ of glucose remained unutilized. With the addition of one, two or twenty amino acids, residual glucose was progressively reduced, to a minimum of $0.3\,\mu g\,l^{-1}$ with 20 amino acids, at $D = 0.02\,h^{-1}$. Although the physiological explanation is not obvious, it is clear that affinity for a given substrate is not constant for a given organism; affinity is influenced by both growth rate and the ratios of the substrates available. Law and Button expressed surprise at the greatly increased capacity for glucose scavenging from low concentrations. However, their study may provide the microbiological equivalent of the fox and the last rabbit. Supposedly, the probability that the fox will be able to dine on the last rabbit in the neighborhood increases with the availability of alternative menu items.

In the majority of studies in which the reservoir medium was of constant composition and nutrient flux was varied by changing the flow rate, it has been observed that there is a threshold nutrient flux below which the diversity of catabolic activities increases sharply (Harder & Dijkhuizen, 1982). In contrast, the phenomenon of diauxie (sequential utilization of preferred, then alternative substrates) is characteristic of nutrient abundance. At present, it is assumed that repression of catabolic enzymes is the principal mechanism for diauxie, and it is generally the case that the preferred substrate of a diauxie pair will support faster growth. The implied corollary is that an inability to repress enzymes involved in the catabolism of substrates that support slower growth would be a disadvantage under conditions of nutrient abundance. There are, however, several conceivable situations in which lack of repression might be advantageous at high nutrient fluxes, particularly if abundance is only transient.

(1) If catabolism of the secondary substrate resulted in excretion of by-products inhibitory for potential competitors, the growth rate of those organisms might be retarded while both kinds of organisms consumed the preferred substrate. For example, in an unbuffered environment, continued oxidation of thiosulfate might improve the competitiveness of mixotrophic thiobacilli relative to organotrophs for an organic substrate. The same would be true for an organism that continued to oxidize one substrate incompletely, releasing an organic acid or an alcohol, while it and its competitor oxidized the preferred substrate completely.

(2) If catabolism of the second substrate effectively detoxified that material, allowing its accumulation to a higher concentration by repression of enzymes for its catabolism could interfere with growth even in the presence of an abundance of the preferred substrate. This disadvantage of repression was suggested by

Harder & Dijkhuisen (1982) to explain their observations of *Pseudomonas oxalaticus*. In continuous culture at $D < 0.2\,h^{-1}$, acetate and oxalate were used simultaneously. At $D \geqq 0.2\,h^{-1}$, oxalate utilization was repressed, but yield of dry weight per mole of acetate consumed was lower than that observed with acetate alone. Oxalate accumulated to >75mM when utilization was repressed, well above the 20mM threshold at which oxalate retards growth of this organism (Dijkhuisen & Harder, 1975).

(3) A third potential advantage of a lack of repression is the phenomenon of cometabolism. In our studies with *Caulobacter crescentus*, the practice of employing dual carbon sources was adopted several years ago (Poindexter, 1978; Poindexter & Eley, 1983). At the suggestion of B. Ely (personal communication), glutamate was added as a second carbon (and nitrogen) source to glucose–ammonium–salts media. The principal result of this addition was prevention of the pH drop typical of *C. crescentus* cultures in glucose minimal media caused by production of acetic acid (Riley & Kolodziej, 1976). Glutamate alone, as the sole source of carbon in ammonium-containing media, or as the sole source of both carbon and nitrogen, is utilized only slowly for growth. The doubling time is 17.0 h, and the yield is 215 µg dry weight mg^{-1} glutamate. However, in the presence of glucose (with or without ammonium), the glutamate is utilized as rapidly as the glucose, the doubling time of *Caulobacter* remaining unchanged at 2.7 h and the yield staying in proportion to the sum of glucose + glutamate carbon initially available (449 µg dry weight mg^{-1} glucose with glucose alone, and 667 µg [mg glucose + mg glutamate]$^{-1}$). Diauxie, observable in glucose + lactose media (Kurn, Shapiro & Agabian, 1977; Poindexter, unpublished), is not observable in glucose + glutamate media. It thus seems that failure to repress glutamate utilization in the presence of abundant glucose allows simultaneous utilization of both carbon sources at a rate substantially higher than glutamate can be utilized in the absence of glucose. The metabolic basis for this cometabolism has not been investigated, but it is not an isolated observation (Linton, Griffiths & Gregory, 1981; Hommes *et al.*, 1985).

(4) Lack of repression would also be an appropriate adaptation to brief fluctuations in nutrient availability. If a readily utilizable substrate suddenly becomes abundant, organisms that utilize it

exclusively by repression of other enzymes will be less well prepared for its disappearance, even though they were probably
more efficient in its utilization during its transient abundance.
Full and effective repression would increase lag time in utilization
of secondary substrates, during which time the re-adapting cells
would compete poorly with bacteria in which the relevant catabolic enzymes were constitutive or only mildly repressed, or only
inhibited but not repressed.

An obviously advantageous shift in catabolic enzyme synthesis
has been reported for *Hyphomicrobium* X (Meiberg, Bruinenberg
& Harder, 1980). This methylotroph can use dimethylamine (dma)
as its sole source of carbon and energy, relatively rapidly under
aerobic conditions and more slowly under anaerobic conditions by
dissimilatory reduction of nitrate. In batch cultures provided with
$O_2 + N_2$ mixtures resulting in dissolved oxygen tensions (DOT) of
>30 to 90 mm Hg, dma oxidation was catalyzed by dma mono-
oxygenase. At 3–30 mm Hg DOT, dma dehydrogenase was also
present. In a continuous culture at $D = 0.15\,h^{-1}$, derepression
of the dehydrogenase was observed at DOT $\leqq$ 15 mm Hg, and
a concomitant decrease in the oxygenase specific activity was
observed.

Subsequent continuous cultures provided with nitrate-containing
dma medium were studied from 0 to 70 mm Hg DOT. Nitrate and
nitrite reductase activities as well as the catabolic enzymes were
assayed. *Hyphomicrobium* X exhibited three conditions relative to
specific enzyme activities, each clearly appropriate to environmental
availability of oxygen. (1) At DOT > 15 mm Hg, only the mono-
oxygenase was detectable. (2) Below 15 mm Hg, but as long as some
oxygen was available in the atmosphere provided, both catabolic
enzymes were present, with dehydrogenase activity fully derepressed, oxygenase activity one-third its maximum specific activity,
and the dehydrogenase approximately 1.6 times as active as the oxygenase; specific activity of each reductase was inversely related to
DOT. (3) Under anaerobic conditions, the reductases were fully
derepressed, dehydrogenase level was unchanged, and oxygenase
activity was not detectable. Thus, as oxygen availability was decreased, the manner of dma catabolism shifted toward that appropriate to anaerobic conditions, but full commitment to catabolism
via the dehydrogenase did not occur until oxygen was not just scarce,
but absent. However, the specific DOT at which this many-faceted

shift occurred was also dependent on growth rate. At lower growth rates, derepression of nitrate reductase and of dma dehydrogenase was detectable at higher DOT; i.e. oxygen was less effective in repressing the synthesis of these enzymes in slower-growing cells. This implies that incompletely repressed, slower-growing populations in nature would be able to utilize either manner of dma catabolism over a wider range of DOT than would faster-growing cells.

It is apparent that heterotrophic bacteria in general – not just those that occur in oligotrophic habitats – are prepared to increase the diversity of substrates utilized for growth when confronted with a decrease in availability of a preferred substrate. Alternative pathways may become available when an environmental condition, e.g. oxygen tension, limits metabolic access to the carbon source. The most commonly elucidated mechanism is derepression of the enzymes required for catabolism; relief of catabolite inhibition has also been described (Baumberg, 1981). Less frequently reported is the synthesis of enzymes with greater affinity for substrate. Whether derepressed catabolic enzymes are less substrate-specific is generally unknown.

Increased variety of assimilatory enzymes as a response to low nutrient flux

A wide variety of free-living heterotrophic bacteria are known to utilize ammonia as the preferred nitrogen source. Under conditions of ammonia limitation, many such bacteria are able to increase their affinity for ammonia through derepression of the assimilatory system in which the task of ammonia assimilation is performed by glutamine synthetase (GS). This enzyme is repressed and inactivated when ammonia is in adequate supply by a regulatory mechanism that seems about as universal as GS occurrence (Tyler, 1978). Repression and inactivation occur even in bacteria that use GS as their sole means of ammonia assimilation and lack the lower-affinity, metabolically more efficient glutamate dehydrogenase (Ely, Amarasinghe & Bender, 1978). Among bacteria capable of dinitrogen assimilation, GS is generally the catalyst for assimilation into organic material of the ammonia resulting from dinitrogen fixation.

Besides ammonia scavenging, bacteria can turn to alternative nitrogen sources such as dinitrogen, nitrate, amino acids or urea. The most thoroughly studied alternative is dinitrogen, the principal atmospheric form of nitrogen. Air serves as a vast reservoir of nitrogen that can be scavenged when the preferred form becomes scarce, so that oligonitrotrophs, once adapted, are nitrogen-replete.

Non-symbiotic dinitrogen fixation is restricted to prokaryotes but occurs among bacteria of diverse types: strict aerobes such as *Azotobacter*, facultative anaerobes such as *Klebsiella*, strict anaerobes such as *Clostridium*, oxygenic photoautotrophic cyanobacteria and anoxygenic photoheterotrophic purple bacteria. In all of them, ammonia availability is known to repress, and usually also to inhibit, the dinitrogen-fixing systems. Dinitrogen fixation is a metabolically expensive means of nitrogen assimilation, and for each of these organisms, growth is slower when dinitrogen is used as sole nitrogen-source than when combined nitrogen is available in adequate supply. This discussion will focus on dinitrogen fixation in photoheterotrophic bacteria in which use of this alternative nitrogen source provides a means of escaping from competition for the widely preferred ammonia.

Many years ago, before the identity and regulatory mechanisms governing the two principal enzymes of nitrogenase systems (the iron-containing dinitrogenase reductase and the MoFe-containing protein dinitrogenase (Mortenson & Thorneley, 1979)) were known, Munson & Burris (1969) were attempting to prepare cell extracts of *Rhodospirillum rubrum* possessing nitrogenase activity. Extracts of cells grown under dinitrogen-fixing conditions in batch cultures were low and erratic in activity. Suspecting that small, unidentifiable and uncontrollable variations in cultural conditions were responsible, they turned to continuous cultivation as a means of stabilizing the environment during growth. They hypothesized that any conditions that resulted in internal (intracellular) nitrogen limitation would result in nitrogenase activity. To test this hypothesis, nitrogen limitation of continuous cultures was imposed in three ways: by provision of a yield-limiting supply of dinitrogen, or of glutamate or ammonia under an atmosphere of helium + carbon dioxide; malate was the carbon source. The highest and most consistent activity was found in extracts prepared from ammonia-limited populations, an observation that implicated derepression, rather than induction by dinitrogen, as the principal regulatory mechanism, and established dinitrogen fixation by these bacteria as a response to nitrogen limitation of growth. The form of the limited amount of nitrogen available was not the determining factor.

The nitrogenase of photoheterotrophic purple bacteria exhibits a second property, viz., hydrogen evolution, that, like dinitrogen fixation, is light-dependent. Hydrogen can also be consumed by these very versatile organisms in the light-dependent fixation of carbon dioxide. Comparison of hydrogen production and hydrogen consumption was the overall purpose of a study of *Rhodopseudomonas capsulata* by Hillmer & Gest (1977). In this study, the activities were studied with cells that had been grown in batch cultures in media containing glutamate or ammonia as nitrogen source, and lactate, pyruvate, glycerol, malate, fumarate or succinate as carbon source. The 'standard medium' contained lactate and glutamate. Ammonia availability influenced both hydrogen production and hydrogen consumption. When ammonia was in excess over carbon, hydrogen evolution was reduced by both repression and inhibition of nitrogenase, and hydrogen consumption in carbon dioxide fixation was stimulated. When ammonia was relatively scarce, nitrogenase activities (hydrogen production and dinitrogen fixation) were accelerated, and hydrogen consumption for carbon dioxide fixation (an expensive form of carbon scavenging) was reduced. Nitrogen 'scarcity' could be imposed by adding an excess carbon source without removing the nitrogen source. Lactate, malate and succinate, all readily utilizable carbon sources for this organism, were especially effective in promoting nitrogenase activity.

More recently, the effect of the C : N ratio on nitrogenase synthesis and activity has been used as a practical means of stimulating nitrogenase activity in *Rhodospirillum rubrum* (Hoover & Ludden, 1984). The *R. rubrum* strain employed required carbon dioxide when grown with glutamate as the sole carbon and nitrogen source; carbon dioxide was not required in media containing glutamate as nitrogen source and malate as an additional carbon source, or in media containing ammonia and malate. This observation and the finding that α-ketoglutarate was a poor carbon source for growth implied that glutamate was a poor carbon source for *R. rubrum*, and that cells provided with glutamate alone were relatively carbon-limited and nitrogen-adequate. In continuous culture with glutamate alone, excess nitrogen was excreted as ammonia. At $D = 0.034\,h^{-1}$ with glutamate at $7.8\,mM$, ammonia accumulated in the medium to a maximum of $200\,\mu M$ and nitrogenase activity was undetectable. When malate was added (directly to the culture as well as to the reservoir medium) to a concentration of $25\,mM$, medium ammonia decreased much faster than could be accounted for by dilution, i.e., it was consumed by the cells, to a constant minimum concentration of 20 μM. Nitrogenase activity rose to a maximum within less than half a volume change; i.e. it was promptly derepressed. This rise was paralleled by an increase in Fe-protein antigen, as detected by rocket immunoelectrophoresis. Three other activities were also assayed; two (malic dehydrogenase and glutamate dehydrogenase) did not change significantly, but glutamine synthetase (GS) activity increased approximately 10-fold as nitrogenase rose to its maximum, and doubled again by the end of the first volume change after malate addition. In contrast to Fe-protein antigen, however, GS antigen increased only two-fold, indicating that activation was the principal mechanism of change in GS activity, and that derepression was a lesser mechanism.

To examine these events further, batch cultures in glutamate-only medium were amended during exponential phase by the addition of one of several other potential carbon sources. Five of these (acetate, β-hydroxybutyrate, succinate, malate and oxaloacetate) increased the final yield of cells by approximately 50%; in these cells, nitrogenase activity was 30–45 times higher than in the glutamate-alone culture (except only a 12-fold increase with oxaloacetate). Citrate, malonate and α-ketoglutarate reduced yield in the presence of glutamate and failed to stimulate nitrogenase activity. The authors concluded that the addition of 'amino acceptors' (including acetate, but not including α-ketoglutarate in these glutamate dehydrogenase-containing cells) served to remove the ammonia produced from glutamate and thereby relieved nitrogenase repression. However, in the light of the two earlier studies just described, it is more reasonable to regard this as another test of Munson and Burris' hypothesis, and to conclude that without a change in nitrogen availability, the internal activities of the cells were shifted from those appropriate during carbon limitation to those appropriate during nitrogen limitation, and their response was a shift to dinitrogen consumption.

The photoheterotrophic bacteria are remarkably well prepared to make this switch, not only by repression and derepression according to combined-nitrogen availability, but also according to the availability of reducing equivalents. Their nitrogenase system is capable of immediate responses (switch-on and switch-off) to fluctuations in both ammonia availability and light. Kanemoto & Ludden (1984) investigated the correlation between Fe-protein inactivation (by modification) and the loss of nitrogenase activity upon the addition of ammonia or the removal of light, and re-activation (by demodification) by illumination. In studies with continuous cultures of *R. rubrum*, they found that change in nitrogenase activity during these fluctuations paralleled reversible enzyme modification, and that the changes occurred more rapidly than Fe-protein turnover. Oxidants (including oxygen) caused a switch-off that was faster than enzyme modification and was not reversible. This implies that the ability of these organisms to scavenge nitrogen by dinitrogen fixation has limitations other than metabolic cost: it is oxygen-sensitive, so that dinitrogen fixation (like photosynthesis) will not occur during aerobic growth, during which the organisms must compete with other heterotrophs for combined nitrogen.

Similar accounts could be given for other dinitrogen-fixing bacteria. The developing story in the case of purple non-sulfur bacteria was selected for this discussion because it illustrates a point that will arise again in this paper, viz., limitation by a particular essential element is not definable by the absolute amount of flux of the element available, but by its availability relative to other resources. This principle applies not only to the relative availability of different substances, but also to the manner in which a single, mixed-resource substrate may be used. A single example is offered.

In the course of studies of amino acid transport in *Pseudomonas aeruginosa*, Kay & Gronlund (1969) determined the metabolic fate of amino acids taken up by cells previously subjected to a period of carbon or of nitrogen starvation. Amino acids were rapidly taken up by cells with either history and transiently accumulated in intracellular pools; a fraction of the ^{14}C label subsequently appeared in protein. However, the fate of the remainder of the ^{14}C differed in the two types of cells: the label was lost when the cells had been carbon-starved, presumably as $^{14}CO_2$, whereas nitrogen-starved cells excreted ^{14}C as pyruvate and α-ketoglutarate (from [^{14}C]-glutamate, -aspartate, -alanine and -proline), or as unidentified deaminated keto-products (from [^{14}C]-lysine, -isoleucine and -phenylalanine). Degradation did not occur when [^{14}C]-proline was offered to a mutant unable to use proline as a carbon source. Fifteen minutes after supplementation with 0.1% $(NH_4)_2SO_4$, the previously nitrogen-limited cell suspensions behaved like the carbon-starved cells, implying that the presence of ammonia either inhibited the activity of enzymes that degraded amino acids or excreted their deaminated products. In short, whether *Pseudomonas aeruginosa* utilized the amino acids as sources of carbon or of nitrogen depended on the internal C : N balance of the cells, a balance that could be altered rapidly by the appearance of exogenous ammonia.

Changes in transport activities as a response to low nutrient flux

The ability of heterotrophic bacteria to increase their uptake affinity for limiting nutrients is well documented, particularly for sugars and for phosphate (e.g. Harder & Dijkhuizen, 1984). As in the case of catabolic enzymes and as a contributing mechanism in N_2 fixation, derepression is the most commonly described regulatory mechanism. There is also a growing realization that the higher-affinity systems derepressed upon nutrient limitation are qualitatively different from lower affinity systems that operate during nutrient adequacy (Andrews & Lin, 1976). The former are more likely to involve binding proteins, to be lower in substrate specificity, and, in the case of phosphorus, to be co-regulated with other kinds of scavenging enzymes such as phospholipases and organophosphatases.

The present discussion, rather than describing such systems and their regulation, will focus on the ecophysiological question of the value of increased substrate affinity in the acquisition of nutrients present at levels not adequate to support growth of unadapted cells. Three questions form the background for this discussion. Can higher velocity systems also serve to improve nutrient scavenging? Does quantitatively improved scavenging demonstrably improve competitiveness? Does greater substrate affinity increase the susceptibility of the cell to adverse consequences of sudden abundance or imbalance of the nutrient supply, that could be attributed to unmanageable metabolite imbalance?

The majority of studies relevant to these questions have concerned intentionally or unintentionally selected mutants. Mutants appear as the consequence of one of two types of procedure: prolonged incubation under conditions of nutrient limitation, and selection for mutants resistant to inhibitory effects of single nutrients or to analogs of utilizable nutrients.

In some instances, the role of mutation was not established clearly, but seems a probable explanation for the events observed. One such example is the study of *Cytophaga johnsonae* by Höfle (1982, 1983). In the first report, glucose uptake capacity was determined for cells grown in batch and continuous cultures in which the nutrient flux was varied by varying the flow rate (and therefore dilution rate). The specific glucose uptake velocity (normalized to dry mass) of cells from batch culture rose during exponential growth to a maximum that was sustained during the last two generations of growth; at the onset of the maximum stationary phase, uptake velocity decreased suddenly and rapidly to an insignificant level. In a glucose-limited continuous culture ($D = 0.15\,h^{-1}$; μ_{max} in this medium was $0.2\,h^{-1}$), a steady state with respect to cell density and glucose uptake velocity was achieved within 20 and 40 h, respectively; steady state uptake velocity was approximately the same as the rate exhibited by late-exponential phase cells in batch culture ($c.\ 0.1$ nmole glucose $min^{-1}\,mg^{-1}$ (dry wt)). In marked contrast, when the chemostat was operated at $D = 0.03\,h^{-1}$, a steady state was not achieved until 200 h. During the first 200 h, glucose uptake rose almost continuously, finally reaching a constant velocity of 1.2 nmoles $min^{-1}\,mg^{-1}$ (dry wt), more than 10 times the velocity observed with fast-growing cells. Residual glucose concentrations in steady states at $D = 0.15\,h^{-1}$ and $0.03\,h^{-1}$ were $25\,mg\,l^{-1}$ and $0.3\,mg\,l^{-1}$, respectively, a difference consistent with the respective uptake capabilities of the cells. In the slower-growing population, full development of glucose uptake capacity required more than six volume changes, changing steadily with time from the onset of operation of the chemostat; i.e. it did not appear to result from mutation but from progressive adaptation of the population. Such adaptation could not have been observed in batch cultures, where nutrient limitation at the level imposed in the chemostat would occur for only a fraction of a generation.

Determination of the kinetic constants for batch-cultivated and slowly growing, continuously cultivated cells ($D = 0.012\,h^{-1}$) showed that the fully adapted, slowly growing cells had a greater affinity for glucose ($K_t = 1.55$ as opposed to $8.54\,\mu M$ for batch-grown cells) and a higher maximum velocity ($V_{max} = 11.49$ as opposed to 1.09 nmoles glucose $min^{-1}\,mg^{-1}$ (dry wt), respectively). These results imply the presence of two uptake systems with different affinities, and further comparisons revealed that the low affinity system was less sensitive to CCCP inhibition, resulted in a higher proportion of unmetabolized intracellular glucose after 10 s uptake, was not sensitive to competitive inhibition by 3-O-methylglucose, D-fructose, D-fucose, and D-galactose and was less sensitive to competitive inhibition by D-mannose, D-glucosamine, and 1-O-methylglucose. These differences are consistent with the general pattern of differences between high and low affinity uptake systems for carbohydrates (Andrews & Lin, 1976).

In the second report (Höfle, 1983), the chemostat was operated for up to 57 volume changes at $D = 0.03\,h^{-1}$ and 200 volume changes at $D = 0.15\,h^{-1}$. During prolonged incubation at both rates, three types of change occurred, occasionally and suddenly: glucose uptake velocity increased, residual glucose decreased, and viable counts decreased while turbidity remained constant. At each dilution rate, more than twenty volume changes were completed before the first set of these abrupt changes occurred. At $D = 0.15\,h^{-1}$, the second sudden rise in glucose uptake velocity was accompanied by a sharp, three-fold decrease in colony forming units (cfu) on the minimal medium, and the third rise was accompanied by a similar decrease in cfu on a complex medium. At $D = 0.03\,h^{-1}$, each of the two rises in uptake velocity was preceded by a rise in cfu on the minimal medium, then accompanied by a decrease ($>$ three-fold) in cfu on the minimum medium; the cfu on the complex medium decreased almost continuously from the time of onset of dilution until the first rise in glucose uptake velocity and remained relatively stable thereafter.

These observations are consistent with the idea that selection favoured descendants whose glucose uptake velocity was considerably higher (up to c. 10-fold, at each D) than that of their parent inoculum. These descendants were, however, much less capable (or incapable) of developing into visible colonies on either the reservoir (minimal) medium or a complex medium. Perhaps their greater uptake velocity increased their sensitivity to glucose and other organic compounds. Employment of the reservoir medium as a plating medium is not sufficient to ensure that the plating environment is similar to the growth environment within the vessel. Once a steady state is established, the bacteria within the vessel grow in a highly diluted form of the reservoir medium, at least with respect to the limiting nutrient. Thus, in Höfle's cultures, the reservoir medium would have been diluted 1.2×10^4-fold in the culture. It seems quite reasonable to interpret the apparent loss of viability of his cultures to their increased sensitivity to sudden nutrient abundance in the plating medium.

In these two studies, there is ample evidence that C. johnsonae adapted to glucose limitation by a phenotypic shift to a higher affinity, higher velocity uptake system, and was also capable of mutations that further increased glucose uptake velocity by each cell. The second mechanism, more than the first, resulted in increased sensitivity to organic material.

The leucine transport systems of *Escherichia coli* also vary in velocity, substrate affinity, repressibility, and substrate specificity (Rahmanian, Claus & Oxender, 1973; Guardiola *et al.*, 1974; Wood, 1975; Anderson, Quay & Oxender, 1976). There is a non-repressible, membrane-bound, low affinity system (LIV-II), as well as two high affinity, binding protein-associated systems that are repressed by leucine. One of the high affinity systems is leucine-specific (LST); the other (LIV-I) transports also isoleucine, valine, threonine, alanine, serine and homoserine (Wood, 1975). Wild-type cells grown in minimal medium without leucine are susceptible to growth inhibition by leucine, which causes a lengthened lag period after transfer to leucine-containing medium (see Quay, Dick & Oxender, 1977), but leucine-grown cells are not; the LIV-I system is repressed in the latter case. Mutants that are constitutively de-repressed (Rahmanian & Oxender, 1972) for this system are better scavengers of leucine for growth under leucine-limited conditions, but are more sensitive to inhibition by leucine and by leucine analogs (Quay *et al.*, 1975; Quay, Dick & Oxender, 1977). Growth inhibition is attributed to isoleucine limitation due to at least two effects of a sudden influx of leucine via the LIV-I system: feedback inhibition of isoleucine biosynthesis (Levinthal *et al.*, 1973), and rapid exchange of intracellular isoleucine with exogenous leucine (Quay, Dick & Oxender, 1977).

Growth of unadapted wild-type *E. coli* is also inhibited by valine, an inhibition that is reduced when isoleucine is available. Like leucine, valine exerts feedback inhibition on isoleucine biosynthesis, so that an excess of valine entering the cell can inhibit growth by starving the cell for isoleucine of endogenous origin. Valine-resistant mutants can be isolated in which resistance is attributable to changes in enzymes involved in isoleucine biosynthesis, but also in at least one case (Guardiola & Iaccarino, 1971) to a decreased velocity of valine uptake without a change in affinity for valine. Thus, like affinity changes, changes in velocity can affect sensitivity to substrates whose entry in excess can disturb internal metabolism.

In a study of the proton/substrate stoichiometry of lactate and alanine transport in *E. coli*, Collins *et al.* (1976) discovered that prolonged incubation of alanine-limited continuous cultures resulted in selection for mutants altered in this stoichiometry for alanine. Such mutants appeared within 10–20 volume changes; the basis for their favoured selection over the parent inoculum could not be attributed to changes in alanine transport constants. However, a third such mutant isolated after about 150 volume changes (in a separate continuous culture) exhibited a four-fold increase in the velocity of alanine uptake, while affinity was slightly reduced ($K_t = 13 \mu M$ as opposed to $7 \mu M$ for the parent). Although μ_{max} was not different from that of the parent, the change in velocity in the mutant was accompanied by a 20-fold decrease in K_s for growth, from $0.5 \mu M$ for the parent to $0.025 \mu M$ for the mutant. At the D and reservoir concentration of alanine employed ($0.1 h^{-1}$ and $27 mM$, respectively), the maximum instantaneous concentration upon entry of one drop into the 500 ml culture was $2.7 \mu M$; the instantaneous growth rate of the mutant at this concentration would have been approximately 1.37 times that of the parent, and this could have accounted for its selection. The relative effectiveness of the mutant in alanine removal from the medium was reflected in a steady state residual alanine concentration of $0.01–0.04 \mu M$ compared with $0.1–0.2 \mu M$ for the parent. Thus, a change in uptake velocity without an increase in substrate affinity clearly improved the ability of the mutant to acquire alanine from growth rate-limiting ambient concentrations and to favour its growth over that of the parent.

Such studies with *Salmonella* and *Escherichia* have established the toxicity of certain amino acids when suddenly abundant within the cell, and have elucidated some of the mechanisms by which that toxicity is mediated. Autotrophic thiobacilli have also been examined in detail with regard to their sensitivity to amino acids. These bacteria have long been thought of as organic material-shy. Although capable of assimilating organic material during autotrophic growth (with carbon dioxide and thiosulfate), they are inhibited by the addition of certain organic substances, notably amino acids (Kelly, 1969*a*, *b*, *c*; Johnson & Vishniac, 1970; Lu, Matin & Rittenberg, 1971). Their amino acid transport system exhibits relatively low substrate specificity (Kelly, 1969*b*).

As in the enteric bacteria, inhibition of growth of thiobacilli by a single amino acid can be relieved by simultaneous addition of another amino acid; relief for thiobacilli can, however, be brought about by the addition of any of several other amino acids, not just by those that are closely related biosynthetically. This implies that competitive inhibition of transport is responsible for relief, but effectiveness at a metabolic level has not been excluded (Lu, Matin & Rittenberg, 1971). The majority of histidine-tolerant mutants isolated in one study were also tolerant of phenylalanine and threonine; phenylalanine uptake by the one mutant that was tested was

negligible (Johnson & Vishniac, 1970). Thiosulfate oxidation by resting cells was unaffected by any of the growth-inhibiting amino acids.
Thus, inhibition of growth of thiobacilli by single amino acids seems
to be caused by interference with biosynthetic pathways rather than
by effects related to energy metabolism. The mechanisms may well
be similar to those elucidated in studies with enteric bacteria. The
greater sensitivity of the thiobacilli, in the sense that a majority
of protein amino acids exert such effects on this group, is best interpreted as a reflection of the lower specificity of their amino acid
transport systems.

Among oligotrophic heterotrophs, i.e. apparently organic-shy
organic utilizers, sensitivity to single amino acids has been reported
for *Asticcacaulis* (Larson & Pate, 1975), *Hyphomicrobium* (Hirsch,
1974; Moore, 1981), and *Caulobacter* (Ferber & Ely, 1982). As with
enteric bacteria and thiobacilli, inhibition results not from amino
acid mixtures but from the presence of single amino acids, and can
be relieved by the simultaneous presence of other amino acids. Only
in the last study, with *Caulobacter* (Ferber & Ely, 1982), were amino
acid-resistant mutants isolated. Each mutant of *C. crescentus* selected
for resistance to inhibition by any one of nine inhibitory amino acids
was resistant to all nine, and in addition acquired resistance to several
inhibitory amino acid analogs. Ability to utilize amino acids for
growth was also reduced in such mutants, and all of those tested
exhibited severely reduced uptake (three to seven times lower than
the parent strain) of a mixture of ^{14}C-labelled amino acids. Glucose
uptake was not altered. ^{14}C-labelled amino acid uptake by the parent
was competitively inhibited by alanine alone, but not significantly
by glutamate at comparable concentrations. Alanine alone could
relieve inhibition by any of the nine inhibitory amino acids, but
glutamate did not relieve inhibition in any case. In this study,
the relation of constitutive, relatively non-specific transport of
amino acids to amino acid sensitivity was clearly demonstrated.
Whether these systems (system?) are repressible has not been
reported.

Like the thiobacilli, wild-type caulobacters are sensitive to amino
acid imbalance, not amino acids *per se*, and the implication is that
metabolic disturbance due to a sudden excess of any one of several
amino acids inhibits growth. The proximal cause of such imbalance
is the lack of substrate specificity of transport. Thus, while versatile
bacteria have an improved capacity for nutrient scavenging, loss
of transport specificity also confers a greater sensitivity to nutrient

imbalance. If such systems are not repressible, a nutrient scavenger would be intolerant of unbalanced nutrient abundance.

When the pantry is empty, go shopping

For motile bacteria, there is another means of improving access to scarce nutrients, and that is chemotaxis. The development of a quantitative assay (Adler, 1973; Mesibov, Ordal & Adler, 1973) has proved adaptable with only minor modifications to a variety of bacteria. Most studies have employed enteric bacteria (*Escherichia* and *Salmonella*) or *Bacillus*, and the similarities and differences of chemotaxis in these two groups were reviewed recently, particularly with respect to biochemical events and components (Ordal & Nettleton, 1985). One similarity between these two groups is that tactic responses to amino acids are more dramatic than responses to sugars; in addition, responses to amino acids are typically constitutive (Mesibov & Adler, 1972; Van der Drift & de Jong, 1974; Ordal & Gibson, 1977), whereas some (*E. coli* (Adler, Hazelbauer & Dahl, 1973)) or most (*B. subtilis* (Ordal, Villani & Rosendahl, 1979)) positive responses to sugars are inducible, requiring previous growth in the presence of the sugar. There does not seem to be any evidence that chemotaxis in these organisms is repressible.

The question of repressibility is relevant to strategies for adaptation to nutrient limitation. Derepression could allow nutrient-limited cells to use some of their limited resources for detecting and moving towards utilizable nutrients, behaviour that, like dinitrogen fixation, is not so valuable during times of adequate nutrient supply. On the other hand, in natural oligotrophic habitats, gradients sufficiently steep to elicit chemotactic responses may occur so rarely that selection for chemotactic responses in oligotrophic populations may be relatively weak. This hypothesis is highly susceptible to disproof if and when evidence for chemotaxis in a diversity of oligotrophic bacteria is obtained.

Because the majority of motile oligotrophs are Gram-negative, polar-flagellate bacteria, the remainder of this discussion of chemotaxis will focus on studies of chemotaxis in polar flagellates. Such studies on their chemotaxis are far fewer than those with peritrichous bacteria, and the evidence regarding the responses of some groups is quite recent. Among the earliest studies were those employing *Bdellovibrio bacteriovorus* (Straley & Conti, 1974; LaMarre, Straley & Conti, 1977) in which only very weak responses were detected (with batch-grown cells). Only a few amino acids were attractive, the threshold concentrations eliciting responses were far higher than in enteric bacteria (10^{-4} M as opposed to $\leq 10^{-6}$ M), and bacterial accumulation relative to controls was only four- to five-fold, rather than 100- to 1000-fold as found for enteric bacteria and bacilli.

However, relative accumulations and threshold levels observed with *Pseudomonas aeruginosa* are similar to those seen with peritrichous bacteria (Moench & Konetzka, 1978; Moulton & Montie, 1979; Craven & Montie, 1985). This versatile bacterium, which occurs naturally in habitats ranging from clean waters to wounded animal tissues, responds chemotactically particularly to potential nitrogen sources (amino acids and ammonia). Responses to some but not all organic acids and sugars (including glucose) are inducible, whereas responses to potential nitrogen sources are usually constitutive. However, Craven & Montie (1985) found that responses to amino acids were markedly affected by the identity and quantity of the nitrogen source present during growth. The responses to serine or to arginine of cells grown in a rich complex broth (L-broth) were barely detectable, whereas cells grown in glucose–ammonium–salts medium responded strongly to these amino acids as attractants. In the same minimal medium with nitrate as the sole nitrogen source, growth was slower, and the responses of these nitrogen-limited cells was further enhanced. In contrast, the responses were practically abolished when the medium was supplemented with either glutamate (an excellent nitrogen source for *P. aeruginosa*) or casamino acids. The nitrate-grown cells were not attracted to glucose even when they had been grown with glucose as the sole carbon source. Craven & Montie (1985) concluded that 'chemotaxis towards amino acids certainly may be considered to be a mechanism of nitrogen acquisition'. Their studies have provided a clear example of the ability of nitrogen-limited cells to improve their capacity for locating and approaching potential nitrogen sources. That induction of the response to glucose did not occur in nitrogen-limited cells indicates that – like their photosynthetic relatives – these chemoheterotrophic pseudomonads probably derepress nitrogen acquisition properties according to their internal C : N ratio. This ratio may govern their shopping list as it does their amino acid transport (Kay & Gronlund, 1969).

A study of chemotaxis in *Spirochaeta aurantia* has been reported (Terracciano & Canale-Parola, 1984) that parallels the work of Munson & Burris (1969) discussed above. *S. aurantia* cells grown in batch cultures with glucose or xylose as sole carbon source responded positively to both sugars; response to each sugar was unaffected by the presence of the other, implying the presence of two independent chemoreceptors. Response to each sugar was increased markedly, in terms of both relative accumulation and threshold concentration, when the cells were prepared from continuous cultures. For each sugar, there was an optimal nutrient flux that resulted in maximum enhancement, although responses at all fluxes were still stronger than with batch-grown cells. The figures illustrating relative response plotted against dilution rate in this study provide yet another example of the dependence of bacterial adaptations on nutrient flux/growth rate. Adaptation was reduced at higher growth rates, and also at very low growth rates. Terracciano & Canale-Parola (1984) suggested that bacteria adapted for growth at low nutrient fluxes might not exhibit a decrease in responsiveness at very low dilution rates. Resolution of this question is a challenge to those whose experimental organisms are oligotrophs.

The symbiotic dinitrogen-fixing bacteria *Rhizobium* and *Bradyrhizobium* occupy similar niches, yet are distinguished by their maximum growth rates. In a comparative study of chemotaxis in *R. trifolii* (fast-growing, with peritrichous flagella) and *B. japonicum* (slower-growing, with sub-polar flagella), Parke, Rivelli & Ornston (1985) found that the slower-growing species was much more sensitive to, and active in, its chemotactic responses to a variety of aromatic compounds. Threshold concentrations for *B. japonicum* ranged from $<10^{-7}$ to 10^{-6} M, whereas the lowest threshold concentration exhibited for any aromatic compound by *R. trifolii* was 10^{-5} M. The lowest accumulation of *B. japonicum* was one-half the highest of *R. trifolii*, and its highest accumulation was 10 times greater than the highest of *R. trifolii*. The authors noted that 'the slow metabolism of *B. japonicum* is not reflected in sluggish chemotaxis'. If the relatively slow growth of bradyrhizobia reflects their having stepped aside for the faster-growing rhizobia, they may have done so in part by retaining or acquiring in the course of their evolution the ability to sit down first at the table.

Although evidence is still fragmentary, the hypothesis regarding the unlikelihood of selection for chemotaxis in oligotrophs appears quite vulnerable. This very brief examination of chemotaxis in polar flagellates encourages the hypothesis that chemotaxis has been and is an important strategy for the pursuit of nutrients in oligotrophic habitats.

NUTRIENT LIMITATION AS NUTRIENT IMBALANCE

In the majority of continuous culture studies, results of experiments on the effects of nutrient limitation, nutrient uptake, enzyme activity, chemotactic behaviour, and reserve storage are expressed in terms of dry weight, protein content or cell number. The composition of a population and the magnitude of its collective activities in a constant volume of liquid is considered a suitable unit only in the case of competition studies. However, if the continuous culture is regarded as a community occupying a given volume of environment, and the loss of community members to the effluent is taken as representative of natural attrition such as that due to predation, a monotypic culture can serve as an experimental model of a stable community. It can be subjected to environmental fluctuations that mimic changes that occur in natural habitats, and the responses of the community to such fluctuations can be studied. This approach should be regarded not so much as a means of predicting what will occur in a natural community, which is never contained and controlled in the manner possible with a laboratory culture, but as a means of examining predictions made from natural observations of microorganisms. This approach will be illustrated by reference to the ecology of a lake.

Spring, summer and autumn in a lake

In addition to diurnal cycles, predictable and recurring fluctuations in nutrient availability occur with the seasons in temperate zones. Considerable attention has been given to the influence of annual fluctuations in inorganic nutrients on algal populations in fresh water (Tilman, Kilham & Kilham, 1982). The determinative role of phosphorus availability has been described repeatedly for natural fluctuations and for fluctuations caused by human agency. Nitrogen availability typically influences species composition of phototrophic communities and, less often, the total algal crop. When the $N:P$ ratio is low, dinitrogen-fixing cyanobacteria are favoured, and nitrogen limitation is relieved by their activity (Megard, 1972). Major cations and sulfate do not limit algal population development, although Ca^{2+} availability may – like silicate, combined nitrogen, and phosphate availability – determine species composition of the community (Moss, 1972).

Although pH and temperature exert strong influences on both species composition and rate of photosynthetic activity per unit chlorophyll in algal communites, the size of the community of freshwater phytoplankton is determined most often by phosphorus availability. Using their own and others' accumulated data, Dillon & Rigler (1974) were able to predict the summer chlorophyll concentration in freshwater lakes of the Canadian Shield employing a single independent variable, viz. phosphorus made available during 'spring turnover'. The major causes of variation in the numerical relation between spring phosphorus and summer chlorophyll concentration were physical factors such as lake shape and flushing time. Once begun, however, the extent of the summer bloom of algae is typically phosphorus-limited. The phosphorus (and nitrogen, at least initially) are rapidly consumed by the algal community, and for most of the summer the phosphorus recycles rapidly through very short pathways from algae, to herbivores, to their excreta, which are then mineralized by decomposers (Lehman, 1980). Phosphorus-limited phototrophs are avid phosphorus scavengers, particularly certain widely distributed diatoms as well as some frequently encountered cyanobacteria.

Meanwhile, suspended in the water column or attached to submerged surfaces (including algal thalli) is a persistent population of heterotrophic bacteria. The majority of types of bacteria found year-round in freshwater aquatic habitats are not known to form dormant stages. They persist as vegetative cells, although undoubtedly their rates of metabolic activity fluctuate with the seasons. During the winter, their activity is presumably limited predominantly by physical factors such as low temperature, lowered solubility of nutrients and decreased diffusion rates. In the spring, as mineral nutrients are freed, there may be a period during which the heterotrophic bacteria are carbon-limited. However, once the algal community begins to increase, carbon becomes relatively abundant, as does nitrogen (from heterotrophic dinitrogen-fixation as well as cyanobacterial activity), and for most of the summer, the season of peak biological activity, the heterotrophs are phosphorus-limited. The principal inorganic phosphorus supply is the release of phosphorus from herbivore excreta by decomposers (predominantly heterotrophic bacteria) capable of digesting organic phosphates and releasing soluble, assimilable phosphorus. For this phosphorus, the heterotrophic bacteria must compete with the algal community.

In the autumn, as algal activity decreases, dead algal and herbivore material accumulates and is slowly degraded. Algal exudates and herbivore excreta are replaced by decomposition products from the necromass. However, the bulk of material decomposed consists of algal (and plant) cell walls which is high in carbon and low in phosphorus.

Thus, like the summer 'feast', the autumn 'harvest' provides a phosphorus-poor diet. Through both seasons, the nutrient supply is unbalanced for heterotrophic bacteria; it is a time of fasting, of neither feast nor famine, since only one food, phosphorus, is relatively unavailable. Only spring turnover provides phosphorus; phosphorus will not, as do sulfur and nitrogen, fall with the rain or arrive, as do carbon and nitrogen, on the wind. During the summer the ecosystem must operate with a fixed amount of phosphorus, and as the phosphorus is recycled within the system, the phototrophs and oligotrophic bacteria must compete for this limiting nutrient.

Simulated spring, summer and autumn in the chemostat

The principal purpose of the experiment described below was to elucidate the mechanisms by which *Caulobacter crescentus* adapts to a decrease in phosphorus relative to carbon. The role of morphogenesis and the adaptations that occur within the individual cell will be described elsewhere. In the present paper, the behaviour of a single-species, artificially cultivated community will be described. As the 'seasons' progressed, of all the parameters determined, only the genotype and the mode of reproduction remained unchanged; in all other ways, the population changed appropriately for community persistence during the summer and autumn periods of phosphorus fasting. Whether the changes occur in nature cannot be inferred from this study but it is clear that caulobacters are capable of appropriate adjustment.

The medium (Poindexter, 1978; Poindexter & Eley, 1983) contained two carbon sources (glucose and glutamate), NH_4Cl as the principal nitrogen source, phosphorus as orthophosphate, sulfur as sulfate, and other mineral nutrients as provided by Hutner's base (Cohen-Bazire, Sistrom & Stanier, 1957) without vitamins. At the carbon and phosphorus levels used, all other nutrients were in excess; at all glutamate levels used, more nitrogen was available from ammonia than from glutamate. It was particularly important that Ca^{2+} availability was adequate, since morphogenesis (Poindexter, 1984*a*, *b*), nutrient uptake capacity and synthesis and activity of

Table 1. *Reservoir concentrations of carbon and phosphorus during 'seasonal progression'*[a]

'Season'	μg carbon ml^{-1}	μg phosphorus ml^{-1}	Duration of 'season' (h)
Early spring	446	5.0	96
Late spring	549	5.0	96
Early summer	810	5.0	97
Late summer	981	5.0	84
Early autumn I[b]	451	2.1	203
Early autumn II[b]	390	2.1	71
Late autumn	567	2.1	108

[a] Concentrations were determined by assay of glucose and of inorganic phosphate at least twice during each 'season'. Dilution rate was 0.083 h^{-1}. (Data of Poindexter, unpublished.)
[b] Early 'autumn' was interrupted by a contaminant and even though it was present as only one per 1000 caulobacter cells, the apparatus was dismantled, sterilized, and reinoculated and 'autumn' re-established.

alkaline phosphatase (Poindexter, unpublished) are calcium-dependent in *C. crescentus*; a constant reservoir medium concentration of 0.5 mM calcium chloride provided a 3- to 7.5-fold molar excess of Ca^{2+} over phosphate. This medium has been employed for nine continuous cultures among which 30 phosphorus-, carbon-, or phosphorus plus carbon-limited equilibria have been achieved and characterized (Poindexter, unpublished). In phosphorus- and phosphorus plus carbon-limited steady state populations, the number of cells per unit volume (but not bacterial mass) was directly proportional to the reservoir concentration of phosphorus. In carbon- and carbon plus phosphorus-limited populations, the mass per unit volume at equilibrium (but not cell number) was directly proportional to the reservoir concentration of carbon. Both relations were independent of dilution rate within the limits tested.

The change of nutrient availabilities with the 'seasons' was mimicked by modifying the reservoir medium as shown in Table 1; temperature, aeration, agitation, and dilution rate were kept constant. The six successive 'seasons' were designed to have excess phosphorus relative to carbon in early spring, followed by increasing carbon availability as summer approached and progressed, then a loss of available carbon (reservoir glucose + glutamate) in the autumn simulating a decrease in phototrophic exudate. This decrease in carbon was accompanied by a decrease in phosphorus availability to mimic decreased herbivore feeding and excretion. In late autumn, carbon increased again, but phosphorus did not, mimicking the release of materials from decaying carbon-rich algal remains.

To determine how the caulobacter community fared through the simulated seasons, two kinds of assays were performed at 12 or 24 h intervals. The first set of assays measured the composition of the community in terms of cell number, biomass, and cell composition (protein, RNA, DNA) (Fig. 1). The second set of assays measured the activities of the community with respect to consumption, storage and capacity for uptake of carbon and phosphorus, and ability to scavenge organic phosphorus (Fig. 2). The results are interpreted as follows.

Only early spring provided an excess of phosphorus over carbon for the community. This was the only season during which inorganic polyphosphate was stored by the cells. The number of cells in the community was slightly less than was later produced with the same amount of phosphorus. During the next three seasons (late spring, early and late summer), cell number, limited by the unchanged phosphorus availability, did not significantly increase over the level achieved during late spring. However, through the spring and summer and into early autumn, the community managed to gather at least 97% of the available carbon, even though available carbon increased more than 220% while cell number increased by only 24%. The bulk of the excess carbon was stored as poly-β-hydroxybutyrate (PHB), which increased relative to cell protein more than 11-fold from spring to late summer and 15-fold by late autumn, to 0.98 μg PHB per μg protein. Since there is a clear correlation between carbon-reserve content and starvation survival of bacteria (Dawes, 1976), this community seemed well prepared for the famine of winter.

The only surprising change in macromolecular composition of the cells was the rise in RNA relative to protein. This was unexpected for two reasons: first, growth rate, regarded as the principal determinant of RNA content, (Nierlich, 1978) was constant; secondly, the rise occurred as relative phosphorus availability decreased. While a constant proportionality of about 0.20 μg RNA per μg protein was maintained through the spring, the proportion rose to 0.26 in the summer; in early autumn the proportion was back to about 0.20, but rose in late autumn to 0.24, a repetition of the summer response to an increased carbon excess over phosphorus but at lower concentrations of both nutrients. It is not clear how this could contribute to persistence during phosphorus fasting unless RNA is the main

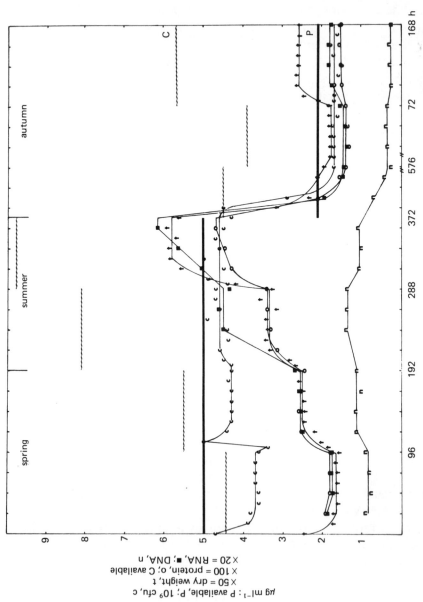

Fig. 1. Population composition during simulated seasonal progression in the chemostat. c: colony forming units. t: dry weight. o: protein. ■: RNA.

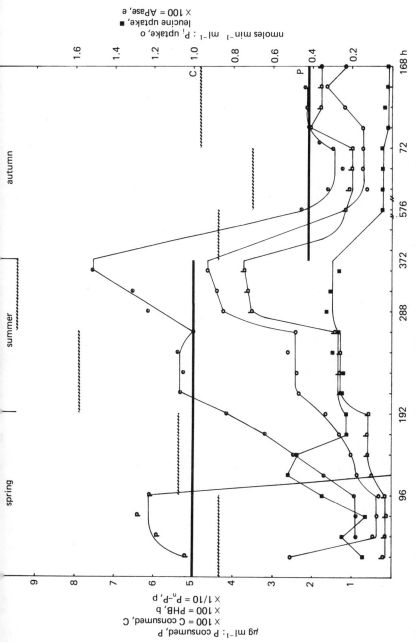

Fig. 2. Population activities during simulated seasonal progression in the chemostat. e: alkaline phosphatase activity, as nmoles *p*-nitrophenyl phosphate hydrolyzed. o: $^{32}P_i$ uptake. ■: [^{3}H]-leucine uptake. p: inorganic polyphosphate. b: poly-β-hydroxybutyric acid. wavy lines: C consumed. heavy straight lines: P consumed. See also legend to Fig. 1. (Data of Poindexter, unpublished.)

phosphorus reserve under conditions of carbon and nitrogen excess, while polyphosphate is the main phosphorus reserve for carbon-limited cells.

The nutrient scavenging activities of the population were sustained throughout the summer. Consumption of at least 97% of available glucose was maintained until autumn, when it decreased to 90% and, in late autumn, to 85%. Population capacity for uptake of the amino acid leucine (employed as an intended measure of constitutive uptake activity, this strain being prototrophic for leucine, which was not provided in the medium) increased at first (late spring) following increased availability of carbon, then decreased to a relatively constant level through the summer. The autumn decrease, however, was two-fold greater than the population decrease and there was a further two-fold decrease in late autumn. This occurred as glucose consumption dropped and suggested that the carbon-replete cells of the autumn had begun to reduce their intake of this long-in-excess element.

Phosphorus scavenging, in contrast, rose steadily: per cell, phosphate uptake increased 11-fold and alkaline phosphatase (AP) activity increased seven-fold by late summer. In late spring and late summer, both of these activities continued to rise from generation to generation, while all other parameters arrived at equilibrium levels. The autumn decrease in AP activity was proportional to the population decrease, although after the unfortunate need to reinoculate (see footnote [b], Table 1), the second early autumn population exhibited lower activity than the first. However, the maximum specific activity reached by the end of late autumn was comparable to that of midsummer. There was a prolonged period over which enzyme activity increased (late spring through late summer), so that development of maximum activity per cell was clearly not achieved within a generation, or even a few generations. The lower activity after reinoculation emphasizes the importance of prolonged incubation (and exclusion of contaminants) when determining responses to nutrient limitation. Continuous cultivation serves this purpose quite well, whereas batch cultivation provides too little time for elicitation of the full complex of responses.

The autumn decline in phosphate uptake activity was also greater than the population decline, but this occurred prior to reinoculation. The explanation is apparent in the details of the uptake constants exhibited by these populations, to be reported elsewhere (Poindexter, unpublished), but briefly summarized as follows.

C. crescentus, like many other bacteria, possesses two systems for phosphate uptake – a faster, lower affinity system ($K_t c. 25 \mu M$), and a slower, higher affinity system ($K_t c. 1 \mu M$). At an ambient phosphorus input of $5 \mu g\, ml^{-1}$, the lower affinity system predominated in the cells. When the phosphorus input was reduced to $2 \mu g\, ml^{-1}$ (or possibly when the Ca:P molar ratio rose from 3 to 7.5), only the higher affinity system was detectable. Thus, although the uptake velocity was lower in the autumn cells, their scavenging capacity and sensitivity to very low phosphorus concentrations was increased. Whether such a change occurred in leucine uptake could not be inferred because leucine uptake had become so slow that transport constant determinations were ambiguous. Presumably, in nature, autumn would coincide with decline of the principal phosphorus-scavenging community (algae and cyanobacteria), and so a shift to dependence on the slower system would not be a disadvantage. Even so, the velocity per cell could still be increased upon further increase in carbon during late autumn (Fig. 2).

This experiment has been presented in detail for several reasons. One is to suggest that some aspects of microbial ecology can be studied indoors. Since all natural habitats are heterogeneous in space and in time, and macrodeterminations of heterogeneous samples cannot characterize microhabitats, it is valid to magnify the microhabitat and to expand the monospecific subcommunity artificially in order to determine bacterial responses to environmental changes. It is thereby possible to isolate the changes; in this experiment, the 'seasons' were experimentally defined by changes in nutrient availability, while temperature and other changes that would have occurred in nature were not allowed. The puzzle pieces will fit together, and the shape of each will probably be retained even when it is removed from the whole.

A second (but not secondary) reason has been to describe the several appropriate responses of a widely distributed freshwater aquatic oligotroph, *Caulobacter*, to a sequence of nutritional stresses it must accommodate in nature. Four of the observations were, frankly, unexpected: the possibility that RNA can serve as a phosphorus reserve, the apparent loss of carbon-scavenging ability, the loss of the low affinity phosphate uptake system (providing, without mutation, populations that possessed only the high affinity system) and the seemingly indefinite increase in alkaline phosphatase activity. The last of these observations implies that these oligotrophs may participate in decomposition of insoluble organic material, a role that has not yet been investigated but which is consistent with detection of caulobacters in sewage sludge, leaf litter and other decomposer communities.

Whether the several responses of *C. crescentus* to relative phosphorus limitation can be elicited in response to other limiting nutrients has not been determined. However, while acceleration of stalk elongation is a major factor allowing caulobacter cells to increase their uptake of phosphate (Poindexter, unpublished), this change in morphology is not elicited by carbon limitation (in contrast to *Asticcacaulis biprosthecum*; Larson & Pate, 1975). Nitrogen-limited swarmers do not even proceed with loss of motility (Chiaverotti *et al.*, 1981; Poindexter, unpublished), the first step in stalk development. This implies that if these organisms respond to nitrogen limitation, the response may relate to motility and chemotaxis, not to changes in the stalk surface. Behavioural evidence of chemotaxis in *C. crescentus* has not been obtained, e.g. by microcapillary assay of differential migration in chemical gradients. Nevertheless, there is ample evidence of the presence in swarmer cells of the biochemical components of chemotaxis (Shapiro, 1985), and behavioral evidence has been obtained for *C. vibrioides* (Matveeva & Gromov, 1983). *C. vibrioides* responded strongly to 16 of 17 amino acids tested, with minimum initial concentrations across the gradients rising to only 10^{-7} M for cysteine, methionine, glutamate and glutamine, and relative accumulations of cells being several hundred- to 1000-fold. Whether the responses were inducible or repressible was not investigated.

SUMMARY

A selection of possible adaptive responses appropriate to exploitation of nutrient-limited environments by heterotrophic bacteria has been examined here. Each response has been found not only to occur but often to be coordinated with a repertoire of responses within a given species. The importance of nutrient ratios, not just absolute amounts or fluxes of nutrients, in the perception of nutrient limitation by bacteria has been illustrated repeatedly; both types of variation in nutrient availability occur in natural ecosystems. Differences between oligotrophic and copiotrophic bacteria seem to relate to the extent of the response repertoire and possibly less to the mechanisms by which responses are mediated. As expected, each response and each mechanism seem to confer some disadvantage along with an associated advantage, so that no one species is designed to respond appropriately to every environmental stress. If that were not so, the earth would know but one nutrient-scavenging species.

REFERENCES

ADLER, J. (1973). A method for measuring chemotaxis and use of the method to determine optimum conditions for chemotaxis by *Escherichia coli*. *Journal of General Microbiology*, **74**, 77–91.

ADLER, J., HAZELBAUER, G. L. & DAHL, M. M. (1973). Chemotaxis toward sugars in *Escherichia coli*. *Journal of Bacteriology*, **115**, 824–7.

ANDERSON, J. J., QUAY, S. C. & OXENDER, D. L. (1976). Mapping of two loci affecting the regulation of branched-chain amino acid transport in *Escherichia coli* K-12. *Journal of Bacteriology*, **126**, 80–90.

ANDREWS, K. J. & LIN, E. C. C. (1976). Selective advantages of various bacterial carbohydrate transport mechanisms. *Federation Proceedings*, **35**, 2185–9.

BAUMBERG, S. (1981). The evolution of metabolic regulation. In *Molecular and Cellular Aspects of Microbial Evolution*, Society for General Microbiology Symposium 32, ed. M. J. Carlile, J. F. Collins & B. E. B. Moseley, pp. 229–72. Cambridge, Cambridge University Press.

CHIAVEROTTI, T. A., PARKER, G., GALLANT, J. & AGABIAN, N. (1981). Conditions that trigger guanosine tetraphosphate accumulation in *Caulobacter crescentus*. *Journal of Bacteriology*, **145**, 1463–5.

COHEN-BAZIRE, G., SISTROM, W. R. & STANIER, R. Y. (1957). Kinetic studies of pigment synthesis by non-sulfur purple bacteria. *Journal of Cellular and Comparative Physiology*, **49**, 25–68.

COLLINS, S. H., JARVIS, A. W., LINDSAY, R. J. & HAMILTON, W. A. (1976). Proton movements coupled to lactate and alanine transport in *Escherichia coli*: isolation of mutants with altered stoichiometry in alanine transport. *Journal of Bacteriology*, **126**, 1232–44.

CRAVEN, R. & MONTIE, T. C. (1985). Regulation of *Pseudomonas aeruginosa* chemotaxis by the nitrogen source. *Journal of Bacteriology*, **164**, 544–9.

DAWES, E. A. (1976). Endogenous metabolism and the survival of starved prokaryotes. In *The Survival of Vegetative Microbes*, Society for General Microbiology, Symposium 26, ed. T. R. G. Gray & J. R. Postgate, pp. 19–53. Cambridge, Cambridge University Press.

DIJKHUIZEN, L. & HARDER, W. (1975). Substrate inhibition in *Pseudomonas oxalaticus* OX1: a kinetic study of growth inhibition by oxalate and formate using extended cultures. *Antonie van Leeuwenhoek Journal of Microbiology and Serology*, **41**, 135–46.

DILLON, P. J. & RIGLER, F. H. (1974). The phosphorus–chlorophyll relationship in lakes. *Limnology and Oceanography*, **19**, 767–73.

ELY, B., AMARASINGHE, A. B. C. & BENDER, R. A. (1978). Ammonia assimilation and glutamate formation in *Caulobacter crescentus*. *Journal of Bacteriology*, **133**, 225–30.

FERBER, D. M. & ELY, B. (1982). Resistance to amino acid inhibition in *Caulobacter crescentus*. *Molecular and General Genetics*, **187**, 446–52.

GUARDIOLA, J., DEFELICE, M., KLOPOTOWSKI, T. & IACCARINO, M. (1974). Multiplicity of isoleucine, leucine, and valine transport systems in *Escherichia coli* K-12. *Journal of Bacteriology*, **117**, 382–92.

GUARDIOLA, J. & IACCARINO, M. (1971). *Escherichia coli* K-12 mutants altered in the transport of branched-chain amino acids. *Journal of Bacteriology*, **108**, 1034–44.

HARDER, W. & DIJKHUIZEN, L. (1982). Strategies of mixed substrate utilization in microorganisms. *Philosophical Transactions of the Royal Society of London, Series B*, **297**, 459–80.

HARDER, W. & DIJKHUIZEN, L. (1983). Physiological responses to nutrient limitation. *Annual Review of Microbiology*, **37**, 1–23.

HARDER, W., DIJKHUIZEN, L. & VELDKAMP, H. (1984). Environmental regulation of microbial metabolism. In *The Microbe 1984, Part II, Prokaryotes and Eukaryotes*, Society for General Microbiology, Symposium 36, ed. D. P. Kelly & N. G. Carr, pp. 51–95. Cambridge, Cambridge University Press.

HARDER, W., KUENEN & MATIN (1977). Microbial selection in continuous culture. *Journal of Applied Bacteriology*, **43**, 1–24.

HILLMER, P. & GEST, H. (1977). H_2 metabolism in the photosynthetic bacterium *Rhodopseudomonas capsulata*: production and utilization of H_2 by resting cells. *Journal of Bacteriology*, **129**, 732–9.

HIRSCH, P. (1974). Budding bacteria. *Annual Reviews of Microbiology*, **28**, 391–444.

HÖFLE, M. G. (1982). Glucose uptake of *Cytophaga johnsonae* studied in batch and continuous culture. *Archives of Microbiology*, **133**, 289–94.

HÖFLE, M. G. (1983). Long-term changes in chemostat cultures of *Cytophaga johnsonae*. *Applied and Environmental Microbiology*, **46**, 1045–53.

HOMMES, R. W., VAN HELL, B., POSTMA, P. W., NEIJSSEL, O. M. & TEMPEST, D. W. (1985). The functional significance of glucose dehydrogenase in *Klebsiella aerogenes*. *Archives of Microbiology*, **143**, 163–8.

HOOVER, T. R. & LUDDEN, P. W. (1984). Derepression of nitrogenase by addition of malate to cultures of *Rhodospirillum rubrum* grown with glutamate as the carbon and nitrogen source. *Journal of Bacteriology*, **159**, 400–3.

JANNASCH & MATELES (1974). Experimental bacterial ecology studied in continuous culture. *Advances in Microbial Physiology*, **11**, 165–212.

JOHNSON, C. L. & VISHNIAC, W. (1970). Growth inhibition in *Thiobacillus neapolitanus* by histidine, methionine, phenylalanine, and threonine. *Journal of Bacteriology*, **104**, 1145–50.

KANEMOTO, R. H. & LUDDEN, P. W. (1984). Effect of ammonia, darkness, and phenazine methosulfate on whole-cell nitrogenase activity and Fe protein modification in *Rhodospirillum rubrum*. *Journal of Bacteriology*, **158**, 713–20.

KAY, A. A. & GRONLUND, A. F. (1969). Influence of carbon or nitrogen starvation on amino acid transport in *Pseudomonas aeruginosa*. *Journal of Bacteriology*, **100**, 276–82.

KELLY, D. P. (1969a). Regulation of chemoautotrophic metabolism. I. Toxicity of phenylalanine to *Thiobacilli*. *Archiv für Mikrobiologie*, **69**, 330–42.

KELLY, D. P. (1969b). Regulation of chemoautotrophic metabolism. II. Competition between amino acids for incorporation into *Thiobacillus*. *Archiv für Mikrobiologie*, **69**, 343–59.

KELLY, D. P. (1969c). Regulation of chemoautotrophic metabolism. III. DAHP synthetase in *Thiobacillus neapolitanus*. *Archiv für Mikrobiologie*, **69**, 360–9.

KOCH, A. L. (1971). The adaptive responses of *Escherichia coli* to a feast and famine existence. *Advances in Microbial Physiology*, **6**, 147–217.

KURN, N., SHAPIRO, L. & AGABIAN, N. (1977). Effect of carbon source and the role of cyclic adenosine 3′,5′-monophosphate on the *Caulobacter* cell cycle. *Journal of Bacteriology*, **131**, 951–9.

BACTERIAL RESPONSES TO NUTRIENT LIMITATION   315

LaMarre, A. G., Straley, S. C. & Conti, S. F. (1977). Chemotaxis toward amino acids by *Bdellovibrio bacteriovorus*. *Journal of Bacteriology*, **131**, 201–7.

Larson, R. J. & Pate, J. L. (1975). Growth and morphology of *Asticcacaulis biprosthecum* in defined media. *Archives of Microbiology*, **106**, 147–57.

Law, A. T. & Button, D. K. (1977). Multiple-carbon-source-limited growth kinetics of a marine coryneform bacterium. *Journal of Bacteriology*, **129**, 115–23.

Lehman, J. T. (1980). Release and cycling of nutrients between planktonic algae and herbivores. *Limnology and Oceanography*, **25**, 620–32.

Levinthal, M., Williams, L. S., Levinthal, M. & Umbarger, H. E. (1973). Role of threonine deaminase in the regulation of isoleucine and valine biosynthesis. *Nature, London, New Biology*, **246**, 65–8.

Linton, J. D., Griffiths, K. & Gregory, M. (1981). The effect of mixtures of glutamate and formate on the yield and respiration of a chemostat culture of *Beneckea natriegens*. *Archives of Microbiology*, **129**, 119–22.

Lu, M. C., Matin, A. & Rittenberg, S. C. (1971). Inhibition of growth of obligately chemolithotrophic thiobacilli by amino acids. *Archiv für Mikrobiologie*, **79**, 354–66.

Matin, A. (1979). Microbial regulatory mechanisms at low nutrient concentrations as studied in chemostat. In *Strategies of Microbial Life in Extreme Environments*, Life Science Research Report 13, ed. M. Shilo, pp. 323–39. Weinheim, Verlag Chemie.

Matveeva, M. A. & Gromov, B. V. (1983). Attraction of amino acids for motile cells of *Caulobacter vibrioides*. *Microbiology*, **52**, 357–9.

Megard, R. O. (1972). Phytoplankton, photosynthesis, and phosphorus in Lake Minnetonka, Minnesota. *Limnology and Oceanography*, **17**, 68–87.

Meiberg, J. B. M., Bruinenberg, P. M. & Harder, W. (1980). Effect of dissolved oxygen tension on the metabolism of methylated amines in *Hyphomicrobium* X in the absence and presence of nitrate: evidence for 'aerobic' denitrification. *Journal of General Microbiology*, **120**, 453–63.

Mesibov, R. & Adler, J. (1972). Chemotaxis towards amino acids in *Escherichia coli*. *Journal of Bacteriology*, **112**, 315–26.

Mesibov, R., Ordal, G. W. & Adler, J. (1973). The range of attractant concentrations for bacterial chemotaxis and the threshold and size of response over this range. *Journal of General Physiology*, **62**, 203–23.

Moench, T. T. & Konetzka, W. A. (1978). Chemotaxis in *Pseudomonas aeruginosa*. *Journal of Bacteriology*, **133**, 427–9.

Moore, R. L. (1981). *The Genera Hyphomicrobium, Pedomicrobium, Hyphomonas*. In *The Prokaryotes*, vol. 1, ed. M. P. Starr, H. Stolp, H. G. Trüper, A. Balows & H. G. Schlegel, pp. 480–7. Berlin, Springer-Verlag.

Mortenson, L. E. & Thorneley, R. N. F. (1979). Structure and function of nitrogenase. *Annual Reviews of Biochemistry*, **48**, 387–418.

Moss, B. (1972). The influence of environmental factors on the distribution of freshwater algae: an experimental study. I. Introduction and the influence of calcium concentration. *Journal of Ecology*, **60**, 917–32.

Moulton, R. C. & Montie, T. C. (1979). Chemotaxis by *Pseudomonas aeruginosa*. *Journal of Bacteriology*, **137**, 274–80.

Munson, T. O. & Burris, R. H. (1969). Nitrogen fixation by *Rhodospirillum rubrum* grown in nitrogen-limited continuous culture. *Journal of Bacteriology*, **97**, 1093–8.

Nierlich, D. P. (1978). Regulation of bacterial growth, RNA, and protein synthesis. *Annual Reviews of Microbiology*, **32**, 393–432.

Ordal, G. W. & Gibson, K. J. (1977). Chemotaxis toward amino acids by *Bacillus subtilis*. *Journal of Bacteriology*, **129**, 151–5.

ORDAL, G. W. & NETTLETON, D. O. (1985). Chemotaxis in *Bacillus subtilis*. In *The Molecular Biology of the Bacilli*, vol. 2, ed. D. Dubnau, pp. 53–72. New York, Academic Press.

ORDAL, G. W., VILLANI, D. P. & ROSENDAHL, M. S. (1979). Chemotaxis towards sugars by *Bacillus subtilis*. *Journal of General Microbiology*, 115, 167–72.

PARKE, D., RIVELLI, M. & ORNSTON, L. N. (1985). Chemotaxis to aromatic and hydroaromatic acids: comparison of *Bradyrhizobium japonicum* and *Rhizobium trifolii*. *Journal of Bacteriology*, 163, 417–22.

POINDEXTER, J. S. (1978). Selection for nonbuoyant morphological mutants of *Caulobacter crescentus*. *Journal of Bacteriology*, 135, 1141–5.

POINDEXTER, J. S. (1981). Oligotrophy. Fast and famine existence. In *Advances in Microbial Ecology*, vol. 5, ed. M. Alexander, pp. 63–89. New York, Plenum Publishing Corporation.

POINDEXTER, J. S. (1984*a*). The role of calcium in stalk development and in phosphate acquisition in *Caulobacter crescentus*. *Archives of Microbiology*, 138, 140–52.

POINDEXTER, J. S. (1984*b*). Role of prostheca development in oligotrophic aquatic bacteria. In *Current Perspectives in Microbial Ecology*, ed. M. J. Klug & C. A. Reddy, pp. 33–40. Washington, D.C., American Society for Microbiology.

POINDEXTER, J. S. & ELEY, L. F. (1983). Combined procedure for assays of poly-β-hydroxybutyric acid and inorganic polyphosphate. *Journal of Microbiological Methods*, 1, 1–17.

QUAY, S. C., DICK, T. E. & OXENDER, D. L. (1977). Role of transport systems in amino acid metabolism: leucine toxicity and the branched-chain amino acid transport systems. *Journal of Bacteriology*, 129, 1257–65.

QUAY, S. C., OXENDER, D. L., TSUYUMU, S. & UMBARGER, H. E. (1975). Separate regulation of transport and biosynthesis of leucine, isoleucine, and valine in bacteria. *Journal of Bacteriology*, 122, 994–1000.

RAHMANIAN, M., CLAUS, D. R. & OXENDER, D. L. (1973). Multiplicity of leucine transport systems in *Escherichia coli* K-12. *Journal of Bacteriology*, 116, 1258–66.

RAHMANIAN, M. & OXENDER, D. L. (1972). Derepressed leucine transport activity in *Escherichia coli*. *Journal of Supramolecular Structure*, 1, 55–9.

RILEY, R. G. & KOLODZIEJ, B. J. (1976). Pathway of glucose catabolism in *Caulobacter crescentus*. *Microbios*, 16, 219–26.

SHAPIRO, L. (1985). Generation of polarity during *Caulobacter* cell differentiation. *Annual Review of Cell Biology*, 1, 173–207.

STRALEY, S. C. & CONTI, S. F. (1974). Chemotaxis in *Bdellovibrio bacteriovorus*. *Journal of Bacteriology*, 120, 549–51.

TEMPEST, D. W. & NEIJSSEL, O. M. (1981). Metabolic compromises involved in the growth of microorganisms in nutrient-limited (chemostat) environments. In *Trends in the Biology of Fermentations for Fuels and Chemicals*, ed. A. Hollaender, pp. 335–56. New York, Plenum Publishing Corporation.

TERRACCIANO, J. S. & CANALE-PAROLA, E. (1984). Enhancement of chemotaxis in *Spirochaeta aurantia* grown under conditions of nutrient limitation. *Journal of Bacteriology*, 159, 173–8.

TILMAN, D. KILHAM, S. S. & KILHAM, P. (1982). Phytoplankton community ecology: the role of limiting nutrients. *Annual Review of Ecology and Systematics*, 13, 349–72.

TYLER, B. (1978). Regulation of the assimilation of nitrogen compounds. *Annual Reviews of Biochemistry*, 47, 1127–62.

VAN DER DRIFT, C. & DEJONG, M. H. (1974). Chemotaxis toward amino acids in *Bacillus subtilis*. *Archives of Microbiology*, 96, 83–92.

VELDKAMP, H. (1970). Enrichment cultures of prokaryotic organisms. In *Methods in Microbiology*, vol. 3A, ed. J. R. Norris and D. W. Ribbons, pp. 305–61. New York, Academic Press.

VELDKAMP, H. (1976). *Continuous Culture in Microbial Physiology and Ecology*. Bushey, U.K., Meadowfield Press.

VELDKAMP, H. (1977). Ecological studies with the chemostat. In *Advances in Microbial Ecology*, vol. 1, pp. 59–94. New York, Plenum Publishing Corporation.

WOOD, J. M. (1975). Leucine transport in *Escherichia coli*. The resolution of multiple transport systems and their coupling to metabolic energy. *Journal of Biological Chemistry*, **250**, 4477–85.

BACTERIAL PRODUCTION AT DEEP-SEA HYDROTHERMAL VENTS AND COLD SEEPS: EVIDENCE FOR CHEMOSYNTHETIC PRIMARY PRODUCTION

DAVID M. KARL

Department of Oceanography, University of Hawaii, Honolulu, HI 96822, USA

HISTORICAL PERSPECTIVE

It is well known that the outer surface of the earth is constructed of numerous, semi-rigid plates which together comprise the lithosphere. According to the theory of seafloor spreading (Hess, 1962), new oceanic lithosphere is continuously generated as specialized plate boundaries (spreading centers), which are almost exclusively restricted to ocean basins. Here, molten magma rises from the asthenosphere, cools and accretes to the existing seafloor, and replaces crustal materials displaced laterally and subducted during the spreading process.

Direct measurements of the conductive heat flux at oceanic spreading centers fall short of the expected value predicted from theoretical calculations (Lister, 1980; van Andel, 1981). This discrepancy in the mid-ocean ridge thermal budget led scientists to speculate that convection, in addition to conduction, may be an important process of heat dissipation at active spreading centers. In this model, bottom seawater ($\sim 2\,^{\circ}C$) penetrates deep into the permeable crust through a complex system of fissures. There it becomes heated and emerges as a warm, buoyant hydrothermal spring or vent. The existence of convective circulation of seawater through newly formed oceanic crust was strongly indicated by measurements of bottom water temperature anomalies at the Galapagos Rift (Williams *et al.*, 1974). It was later confirmed by remote sensing and sampling of the resultant hydrothermal plumes (Klinkhammer, Bender & Weiss, 1977; Lonsdale 1977; Weiss *et al.*, 1977; Bolger, Betzer & Gordon, 1978; Jenkins, Edmond & Corliss, 1978).

Deep-sea bottom photographs obtained at the Galapagos Rift using the unmanned Deep-Tow vehicle (Lonsdale, 1977) revealed oceanic basalts covered with hydrothermal mineral precipitates and an unexpected community of large benthic organisms. The dense colonies of suspension-feeding molluscs appeared to be intimately associated with the hydrothermal vents, and it was suggested that the high standing stock of macrobenthos resulted from localized increases of food near the hydrothermal plumes.

Typically, life in the deep sea is characterized by sparse populations of metabolically inactive organisms. This is believed to be the result of the extreme conditions imposed on them by food limitation. The availability of planktonic carbon, originally produced in the near-surface seawater through the process of photosynthesis, exhibits a characteristic decrease with increasing water depth. Generally, less than 5% of this photosynthetically produced carbon descends to the seafloor in waters greater than 2000 m (Suess, 1980). The existence of deep-sea hydrothermal vent ecosystems and cold-water seep analogs suggests alternative pathways of primary production to drive the benthic system.

It has been hypothesized that the internal heat of our planet, produced and maintained through the radioactive decay of long-lived isotopes of uranium, thorium and potassium, may represent an alternative energy source to incoming electromagnetic radiation. In this alternative ecosystem model, the intense heating of seawater and subsequent admixture with volcanic solutes and gases yields a liquid medium which could support bacterial primary production either by chemolithoautotrophy (Whittenbury & Kelly 1977; i.e. the reduction of inorganic carbon to carbohydrate at the expense of reduced inorganic compounds, also referred to in this text as chemosynthetic primary productivity by analogy to photosynthetic primary production) or by chemoorganotrophy (i.e. utilization of primordial methane (CH_4) and other abiotically produced organic compounds in the volcanic solutes). On the other hand, the diverse benthic life surrounding hydrothermal vents in the deep sea may originate from, and be sustained by, recycled carbon originally derived from photosynthetic sources. An experimental evaluation of the existence and quantitative significance of these two fundamentally different modes of bacterial nutrition is an objective of the highest priority.

In 1977, a manned-submersible expedition to the Galapagos Rift produced a most remarkable and serendipitous biological discovery. For the first time, hot springs on the seafloor and their associated microbial and macrofaunal communities were directly observed and sampled (Corliss et al., 1979). Subsequent research expeditions to geographically and geologically distinct regions of the seafloor have revealed the presence of similar deep-sea faunal communities.

While the validity of the primary productivity hypothesis has not been substantially challenged, neither has it been (in my opinion) rigorously tested or thoroughly established as fact. On the tenth anniversary of the discovery of the deep-sea hydrothermal vents it is perhaps appropriate to review critically the experimental evidence supporting the novel production hypothesis and, at the same time, to formulate a research prospectus for the next decade.

GEOGRAPHIC LOCATIONS AND DIVERSITY OF GEOLOGIC SETTINGS

Following the initial discovery of microorganisms and macrofauna at the Galapagos Rift, similar hydrothermal vent ecosystems have been discovered at numerous sites along the East Pacific Rise, on the Mid-Atlantic Ridge and at off-axis volcanic island arcs. More recently, animal communities thought to be analogous to the hydrothermal vent ecosystems have been discovered surrounding ambient temperature seeps in the Gulf of Mexico and at the subduction margins of major oceanic plates (Fig. 1). However, major differences might be anticipated between individual 'hot vents' habitats and between the 'hot vents' and 'cold seeps'. This is especially true with regard to the potential source(s) of carbon and energy required to sustain these otherwise morphologically similar communities. The large-scale geologic and small-scale geochemical settings determine, to a large extent, the diversity of potential habitats for the growth and reproduction of chemolithoautotrophic and chemo-organotrophic bacteria. Consequently, a thorough understanding of the chemistry of seawater–basalt interactions and a complete chemical description of the environments are essential for a rigorous evaluation of the chemosynthetic primary production hypothesis.

Hard lava oceanic ridge hydrothermal vents

As a group, these systems are all located at active oceanic spreading centers and consequently have numerous shared characteristics. The circulation of seawater through mid-ocean ridge crustal materials is responsible not only for dissipating a significant amount of heat but also for mediating important water–rock interactions. These interactions constitute a major exchange of ions between the ocean and the igneous basement and provide a mechanism for the transport of selected chemical constituents into the deep sea. Many of these

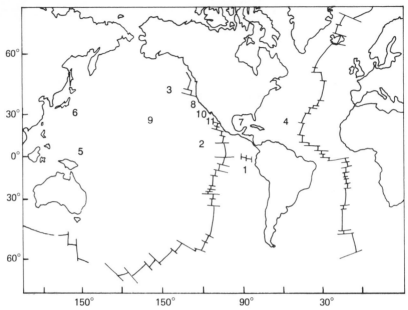

Fig. 1. Map showing the location of a variety of hydrothermal vents and cold-water seeps. (1) Galapagos Rift zone (00°48′N, 86°13′W; 2480 m). (2) East Pacific Rise with vents located at: 10°57′N (2400 m); 12°49′N (2600 m); 20°50′N (2615 m); 18°31′S. (3) Juan de Fuca Ridge including: southern Juan de Fuca Ridge (46°0′N, 130°01′W; 1544 m), northern Juan de Fuca Ridge (46°53′N, 129°17′W; 2370 m), axial seamount – central Juan de Fuca (45°57′N, 130°01′W; 1545 m), Endeavour Ridge (47°57′W, 129°04′W; 2200 m), Explorer Ridge (49°44′W, 130°18′W, 2000 m). (4) Mid-Atlantic Ridge TAG site (26°N; 3200 m) and snake pit site (23°22′N). (5) Manus back-arc basin (3°10′S, 150°17′E; 2500 m). (6) Eurasian–Philippine plate boundary (Tenryu Canyon, Sagami Bay, Japan Trench). (7) Gulf of Mexico including: Orca Basin (26°55′N, 91°20′W; 2400 m), west Florida Escarpment (26°02′N, 84°55′W; 3270 m), Louisiana slope (27°40′N, 91°32′W; 650 m). (8) Oregon subduction zone (44°41′N, 125°18′W; 2036 m). (9) Loihi Seamount (18°55′N, 155°16′W; 1000 m). (10) San Clemente Basin (32°14′N, 117°44′W; 1800 m). (11) Guaymas Basin (27°02′N, 111°22′W; 2020 m).

dissolved ions and solutes (HS^-, NH_4^+, Fe^{2+}, Mn^{2+}, dissolved organic matter (DOM)) and volcanic gases (H_2, CO, CH_4) have the potential to support bacterial chemosynthetic primary production (Table 1).

Initial geochemical investigations of the Galapagos Rift hydrothermal vents indicated that the temperature and bulk fluid chemistry of the water samples were the result of a two end-member mixing process involving: (1) the superheated geothermal fluids produced at depth through the interaction of descending seawater and hot basalt, and (2) ambient bottom seawater (Corliss et al., 1979). If all hydrothermal vents in a given region have end-members of uniform chemical properties, then the concentration of conservative

Table 1. *Potential bacterial metabolic processes at deep-sea hydrothermal vents and cold seeps*

Conditions	Electron (energy) donor	Electron acceptor	Carbon source	Metabolic process
Aerobic	H_2	O_2	CO_2	Hydrogen oxidation
	HS^-, S^0, $S_2O_3^{2-}$, $S_4O_6^{2-}$	O_2	CO_2	Sulfur oxidation
	Fe^{2+}	O_2	CO_2	Iron oxidation
	Mn^{2+}	O_2	?	Manganese oxidation
	NH_4^+, NO_2^-	O_2	CO_2	Nitrification
	CH_4 (and other C-1 compounds)	O_2	CH_4, CO_2, CO	Methane (C-1) oxidation
	Organic compounds	O_2	Organic compounds	Heterotrophic metabolism
Anaerobic	H_2	NO_3^-	CO_2	Hydrogen oxidation
	H_2	S^0, SO_4^{2-}	CO_2	Sulfur and sulfate reduction
	H_2	CO_2	CO_2	Methanogenesis
	CH_4	SO_4	?	Methane oxidation
	Organic compounds	NO_3^-	Organic compounds	Denitrification
	Organic compounds	S^0, SO_4^{2-}	Organic compounds	Sulfur and sulfate reduction
	Organic compounds	Organic compounds	Organic compounds	Fermentation

components that do not react during ascent and mixing (e.g. silica) should conform to a single mixing curve when plotted against temperature (Fig. 2a). The chemical variability observed among the individual hydrothermal vents can then be interpreted as evidence for variable mixing between the two end-member fluids. This process is ultimately controlled by subseabed plumbing and by the mechanisms of hydrothermal circulation, and can be expected to vary both spatially and temporally.

Yet the chemical analysis of selected vent water samples indicates that differences may exist in the high temperature end-member fluid composition. For example, the barium concentrations in hydrothermal fluids from the Clambake vent are twice as high (at a given temperature) as the concentrations measured in three other vent fields within a 1.5 km radius (Corliss *et al.*, 1979; Fig. 2b). Consequently, while there is a certain amount of chemical uniformity and, therefore, predictability in the Galapagos Rift hydrothermal vents,

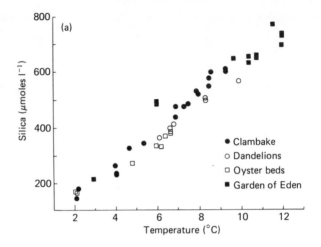

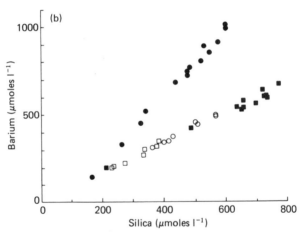

Fig. 2. Compositional trends for selected chemical species at the Galapagos Rift hydrothermal vents. (*a*) Plot of silica concentration against temperature for the four vent fields indicated. (*b*) Plot of barium concentration against silica concentration. (Symbols as in *a*) (From Corliss *et al.*, 1979.)

there also exists considerable variation in both absolute and relative composition of the vent fluids.

Of considerable importance for a critical evaluation of the hypothesis of chemosynthetic bacterial primary production is information on the diversity and concentration of potential electron donors and electron acceptors which might be utilized in these specialized habitats. Although hydrogen sulfide was detected in three of the four vents initially discovered at the Galapagos Rift (Corliss *et al.*, 1979), its concentration did not vary systematically with vent temperature. Dissolved oxygen (O_2), which is generally considered to be the most efficient electron acceptor for bacterial chemolithoautotrophy, extrapolates, at these sites, to a concentration of zero at a temperature of 8–9 °C (Fig. 3). Nitrate is absent at temperatures above 11–12 °C (Corliss *et al.*, 1979). If these two member mixing models can be considered to be accurate chemical descriptions of hard lava hydrothermal systems in general, then the zone of active aerobic bacterial production must be considered to be a small portion of the total hydrothermal vent habitat (Karl, 1985).

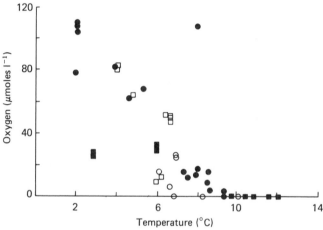

Fig. 3. Composite plot of dissolved oxygen (O_2) against temperature at four Galapagos Rift hydrothermal vent fields. (Symbols as in Fig. 2*a*.) (From Corliss *et al.*, 1979.)

Although the warmest hydrothermal vent waters measured during the initial Galapagos Rift expedition were only ~19 °C, an extrapolation of the silica *versus* temperature data (see Fig. 2*a*) to the quartz solubility curve suggested an end-member fluid temperature of ~350 °C. The discovery of massive sulfide-mineral deposits on the seafloor at the ridge axis of the East Pacific Rise (EPR) spreading center (21 °N latitude), during a 1978 expedition with the French manned-submersible *CYANA*, was strong evidence for the presence of high-temperature solutions (Francheteau *et al.*, 1979; Hekinian *et al.*, 1980). A subsequent cruise to the EPR 21 °N site using the manned submersible *ALVIN* discovered hydrothermal vents with temperatures of 380 ± 30 °C (Spiess *et al.*, 1980) and with the precise end-member fluid composition predicted by Edmond *et al.* (1979)

on the basis of data collected from the low-temperature Galapagos Rift vent water samples. Here, the high-temperature hydrothermal fluids flow from discrete mineralized chimneys at rates of 1–$2\,m\,s^{-1}$ (Fig. 4). The end-member fluids are clear until they contact cold ambient bottom seawater at which time they form fine-grain metal sulfide precipitates, hence the term 'black smoker'. The chemical composition of these high temperature end-member fluids has been summarized by Von Damm *et al.* (1985).

The sulfur chemistry in vent fluids is complicated by at least two separate sources of HS^- (seawater sulfate and basaltic sulfur) and by at least two sinks (deposition of sulfate as anhydrite ($CaSO_4$) and the formation of metal sulfide minerals). Sulfide concentrations in the black smoker fluids ranged from 6.7 to $8.7\,mmoles\,kg^{-1}$ and pH from 3.3 to 3.8 (McDuff & Edmond, 1982; Von Damm *et al.*, 1985). Furthermore, dissolved methane (CH_4) and hydrogen (H_2) measurements indicated concentrations of 50–65 μmol (STP) kg^{-1} and 355–1700 μmol (STP) kg^{-1}, respectively, with considerable variation among the individual vent samples (Welhan & Craig, 1983). The conspicuous lack of significant amounts of C-2 hydrocarbons at 21 °N and the ^{13}C-enriched character of the methane ($\delta^{13}C = -15$ to $-17.5\,‰\,vs$ PDB; Welhan & Craig, 1983) argue against a thermocatalytic or biogenic origin for methane. The CH_4 flux calculated from the high temperature vents is sufficient to replace the deep-sea CH_4 in $\sim 30\,yr$, implying a very rapid bacterial consumption in hydrothermal plumes (Welhan & Craig 1979).

The heat flux for a single black smoker is on the order of $4 \times 10^7\,J\,s^{-1}$, a value which is 3–6 times the total theoretical heat flow for a $1\,km$ segment of mid-ocean ridge that extends $30\,km$ on each side (a distance of $30\,km$ is equivalent to a crustal age of $10^6\,yr$; Macdonald *et al.*, 1980). Clearly, black smokers must be highly ephemeral and probably have an average lifetime of approximately $10\,yr$ (Macdonald, 1982). In 1979, two active black smokers (350 °C) were observed and sampled at the 21°N site designated NGS. However, in 1981 that same vent field was inactive and devoid of macrofauna (Von Damm *et al.*, 1985).

In addition to the black smokers, Galapagos-type warm water vents ($\leqslant 25$ °C), sulfide mound thermal springs and white smoker edifices (< 300 °C) were also observed at the 21 °N EPR site (see Fig. 4). Furthermore, much of the hydrothermal fluid effluent appeared to be discharged as a 'line source' along cracks and fissures in the pillow basalt landscape rather than from Galapagos-type 'point

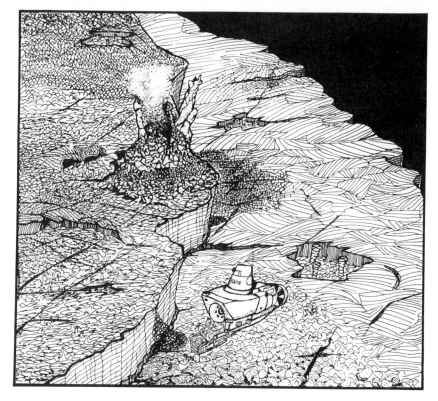

Fig. 4. Artists' conceptions of the large-scale geologic setting of a hydrothermal vent field (11 °N, EPR) and the smaller-scale biological communities in the vicinity of a black smoker chimney. Upper drawing from McConachy *et al.* (1986); lower drawing from Haymon & MacDonald (1985).

source' discharges. The similarity of the 21 °N EPR vent fauna to those previously sampled at the Galapagos Rift was the first suggestion that hydrothermal vent communities might be widespread.

Since 1980, direct evidence for the existence of hydrothermal vents has been obtained for numerous hard lava systems (Fig. 1). Additional discoveries and descriptions are likely to follow. For example, Lupton & Craig (1981) have demontrated that the ^{3}He derived from hydrothermal injections can be detected more than 2000 km west of the EPR spreading center. The extent and magnitude of the ^{3}He plume between 15 °S and 20 °S are interpreted as strong indications that the largest and most active hydrothermal fields have yet to be discovered (Lupton & Craig, 1981).

Although the hydrothermal vents of the Juan de Fuca–Explorer–Endeavor Ridge system in the northeast Pacific Ocean share certain characteristics with the EPR spreading centers, there are also significant differences in the water chemistry and faunal communities. This was especially evident at the Explorer Ridge where at least one of the high-temperature vents (291 °C) contained no detectable HS$^-$ (Tunnicliffe et al., 1986). This is believed to be the result of HS$^-$ removal by zinc. No animals were present at these 'abiotic' vents (Tunnicliffe et al., 1986). Differences in the faunal community assemblages between the Explorer Ridge and the Juan de Fuca Ridge were attributed, in part, to vent age; the latter displayed geological features of a relatively old system while the former may be newly formed.

Sediment-covered submarine hot springs

In contrast to the hydrothermal vents which have been investigated at sediment-starved mid-ocean ridge spreading centers, hot springs have been discovered or are expected to occur at several sediment-covered locations, including a buried ridge axis, a back arc spreading center and an active hotspot (Fig. 1). In these specialized habitats, hot geothermal fluids interact with the sediments before emerging from the seafloor. This interaction results in a chemical composition which may be quite different from the hydrothermal vents produced by seawater–basalt interactions alone (Edmond & Von Damm, 1985). Thermocatalytic breakdown of sedimentary organic matter results in petroleum, NH_4^+ and other potential substrates for bacterial metabolism. Because these carbon and energy sources ultimately depend upon surface derived organic matter, at least some of the microbial growth in these regions clearly cannot be considered chemosynthetic primary production.

Cold seeps

Recently, dense populations of hydrothermal vent-type organisms have been discovered in several distinct ambient temperature regions of the deep sea (Fig. 1). Although the geologic setting and geochemical characteristics of these communities are quite different from the hydrothermal vent habitats, the striking similarity of the large epibenthic animals has raised questions regarding the role and nature of bacterial production in these 'cold seep' regions.

At least four separate cold-seep habitats have been described. The first site is an anoxic, hypersaline basin in the northern Gulf of Mexico (Orca Basin; Shokes *et al.*, 1977) with a chemical composition similar to the Red Sea brines and with elevated concentrations of methane and dissolved/particulate organic matter but a low concentration of HS^- (<1 μm; Sackett *et al.*, 1979). A separate hypersaline, HS^- enriched, NH_4^+ enriched (3–4 mM) sedimentary habitat has also been discovered on the passive continental margin at the base of the Florida Escarpment (Paull *et al.*, 1984; Hecker, 1985). Although active fluid flow was not observed, it was inferred from the steep chemical gradients that were measured in sediment pore water profiles. The elevated salinities observed in this region suggest that the fluids are derived from the adjacent Florida carbonate platform rather than from the abyssal plain sediments (Paull *et al.*, 1984). Since the HS^- and NH_4^+ are probably derived from organic matter diagenesis, these substrates would not be expected to support chemosynthetic primary production. Nevertheless, they might support a novel production cycle based on recycled organic matter. A third cold-seep habitat has recently been found at tectonically active plate subduction zones off Oregon, where ancient seawater (i.e. connate water) is discharged as pelagic sediments and then compacted and accreted to the leading edge of the plate (Suess *et al.*, 1985; Kulm *et al.*, 1986). These cold-water seeps are not enriched in HS^- but they do contain substantial concentrations of methane believed to be of biogenic origin (Suess *et al.*, 1985). It has been hypothesized that free-living and symbiotic methane-oxidizing bacteria serve as a localized source of food for this unusual benthic community. Preliminary reports of the existence of similar benthic communities in the region of active accretion between the Eurasian and Philippine plates (Tenryu Canyon, Sagami Bay and Japan Trench; Swinbanks, 1985a, b) suggest that this may be a general feature of convergent plate boundaries. Finally, several regions of diffuse oil and gas seepage on the Louisiana continental slope also appear to support the growth of dense populations of macrobenthic animals (Kennicutt et al., 1985).

Perhaps the only general conclusion which can be drawn regarding bacterial production at hard lava oceanic ridges, sediment-covered hot springs and the various cold-seep habitats is that it is unlikely that a single common metabolic process controls the flow of carbon and energy in these diverse deep-sea habitats. Consequently, it is incumbent upon the individual investigators to conduct a careful, region-specific evaluation of the chemosynthetic primary production hypothesis rather than to accept it as fact. Even if the process of chemolithoautotrophy at a particular hot vent or cold seep is established as fact, it may not be appropriate to extrapolate these results to other sites given the above-mentioned diversity in the geophysical properties and geochemical characteristics.

DIVERSITY OF MICROBIAL HABITATS

Deep-sea hydrothermal vents comprise numerous potential habitats for the growth of chemolithoautotrophic microorganisms. While there is a certain degree of chemical similarity among these diverse geomorphological regions, the precise chemical, physical and resultant microbiological signatures of the individual sites are likely to be determined by a complex balance between magma chamber plumbing, hydrothermal circulation and the presence or absence of pelagic sediments. Consequently, it may be inappropriate, if not misleading, to assume that field results from different locations or even between two individual vents from a given site reflect similar microbial processes. Furthermore, since the temporal patterns of fluid discharge have not been studied systematically, it may even be inappropriate to assume constancy for an individual hydrothermal vent over time scales of hours, days or years. While it is now believed that individual hydrothermal vents are ephemeral, with 'life spans' of the order of approximately 10–100 years, it is unknown whether there is a systematic succession of microorganisms which corresponds to vent age. As a result, individual vents must be carefully sampled and properly characterized before a thorough understanding of the complex ecological relationships which are likely to exist can emerge.

To date, this careful sampling and thorough characterization has not been possible.

At least four distinct hydrothermal vent habitats and associated microbial communities have been described: (1) free-living bacterial populations associated with the discharged vent fluids and presumably growing and reproducing within the subseabed system (Karl, Wirsen & Jannasch, 1980; Ruby, Wirsen & Jannasch, 1981; Harwood, Jannasch & Canale-Parola, 1982; Ruby and Jannasch, 1982; Tuttle, Wirsen & Jannasch, 1983); (2) microbial mats growing on rock, chimney, sediment or animal surfaces which are exposed to vent water (Jannasch & Wirsen, 1981; Grassle, 1982); (3) symbiotic associations of microorganisms and vent fauna (Cavanaugh et al., 1981; Cavanaugh, 1985; Desbruyers et al., 1983) such as the vestimentiferan tube worms (*Riftia pachyptila*), giant clams (*Calyptogena magnifica*) and polychaete worms (*Alvinella pompejana*); and (4) microorganisms within deep-sea hydrothermal vent plumes (Winn, Karl & Massoth, 1986; Cowen, Massoth & Baker, 1986).

Due to the absence of rapidly flowing pore waters, cold-seep-region habitats are dominated by sedimentary populations of microorganisms, bacterial mats and possible symbiotic associations with epibenthic and infaunal communities. Although it is likely that sediment-derived nutrients are injected into the overlying water column, the absence of a warm buoyant fluid discharge prevents the formation of a distinct low density plume analogous to those observed in volcanic settings. However, it is conceivable that saline water discharge may result in the formation of a coherent high density plume which would be expected to sink rather than rise. A similar benthic plume might also be expected to occur at high temperature vents that have experienced boiling thereby leaving a residual hypersaline solution.

As most vent water samples are collected at points where the fluid discharge mixes with ambient bottom seawater, we have no quantitative information on the precise chemical and physical properties of the environment (or environments) where microorganisms are likely to proliferate. It is probable that dense microbial mats or lenses grow in response to the gradients created and maintained by subseabed mixing of vent fluids and bottom seawater; however, these environments have been impossible to sample. In order to test adequately hypotheses regarding potential pathways for the flux of carbon and energy, it is imperative to obtain corroborative information on the chemical composition of the suspected source habitats (Karl, 1985).

METHODS OF SAMPLING

The initial discovery and much of the early exploration of hydrothermal vents relied to a large extent upon towed instrument sensors, fine-scale bathymetric mapping, ocean bottom instruments and the collection of water samples from conventional surface vessels. Unfortunately, the zone of neovolcanic activity and, hence, active hydrothermal fluid discharge, is generally restricted to a very narrow portion of the seafloor (1–2 km; Macdonald, 1982) so that sophisticated, manned submersibles (e.g. *ALVIN, SEACLIFFE, TURTLE, CYANA, NAUTILE, PISCES IV, JOHNSON SEALINK, MAKALI I, PISCES V, SHINKAI 2000*) are essential for detailed sampling and instrument deployment. Two major disadvantages of using manned submersibles for hydrothermal vent and cold seep research are their high cost and limited accessibility. The US Navy nuclear powered research and engineering vehicle, *NR-1*, has recently been commissioned for academic research efforts. The unique advantage of *NR-1* is her ability to remain submerged for up to 30 days compared to <12 h for conventional research submarines.

Future research progress is clearly limited by the availability of sample materials for direct *in situ* experimentation and not by the availability of experimental methods, motivated scientists or testable hypotheses. Despite recent and significant improvements in sampling capabilities (Von Damm *et al.*, 1985; Lane *et al.*, 1986; Wirsen, Tuttle & Jannasch, in press), a major technological limitation has been the inability to obtain large-volume samples of undiluted vent water which are suitable for simultaneous chemical and microbiological analyses. Furthermore, the walls and internal surfaces of the hydrothermal vent system, rather than the flowing waters, *per se*, are most likely to be the principal habitats for microbial growth. These are, at best, difficult to sample. Consequently, the results available to date on biomass and metabolic activity of discharged (or dislodged) microbial cells must be taken to represent minimum estimates for the vent ecosystem as a whole.

EVIDENCE FOR BACTERIAL CHEMOLITHOAUTOTROPHY

By analogy to subaerial volcanic systems, it was first suggested that a significant amount of organic carbon utilized by the localized vent habitats could be produced by the process of chemosynthetic primary production (Lonsdale, 1977). It is also possible, however, that bacterial primary production at deep-sea vents may occur via chemoorganotrophy at the expense of heterotrophic utilization of organic substrates (including methane) produced abiotically during high-temperature rock–water interactions. Nowhere else on earth is there an ecosystem that is not, at least partially, dependent upon photosynthesis. The scientific question of fundamental significance with respect to life at deep-sea vents and cold seeps, then, is whether the energy utilized for growth is derived from geothermal sources or whether microbial production requires allochthonous substrates ultimately derived from photosynthesis. However, irrespective of thermodynamic considerations, the ecological consequences of localized bacterial production of organic matter are indeed profound.

Microscopic analyses of the first deep-sea hydrothermal vent water samples collected during the 1977 Galapagos Rift expedition (analyzed by J. Baross cited in Corliss et al., 1979) revealed the presence of 'high concentrations of sulfur-oxidizing and heterotrophic bacteria (from 10^8 to 10^9 bacteria ml^{-1})' as determined by epifluorescence microscopy of glutaraldehyde-preserved water samples (Corliss et al., 1979). These data supported, although by no means rigorously confirmed, the hypothesis of bacterial chemosynthetic primary production. Since 1977, additional water samples have been analyzed for the abundance and in situ metabolic activities of bacterial populations contained within hydrothermal fluids, attached to rock and animal surfaces and associated with hydrothermal vent animals. A summary of the experimental data collected from deep-sea vents is presented below. To date, there have been no systematic microbiological experiments performed at off-axis vent sites or cold water seep habitats.

Distribution and abundance of bacteria in low-temperature ($<50\,°C$) vent waters

When the slightly turbid hydrothermal vent waters are sampled, filtered and directly observed using either bright-field, epifluorescence or electron microscopy, much of the turbidity of the solution can be ascribed to the presence of bacterial cells (Jannasch & Wirsen, 1979; Grassle et al., 1979; Karl, Wirsen & Jannasch, 1980; Fig. 5). An important qualitative observation is the occasional presence of large clumps of bacterial cells and the apparent heterogeneity in cell size and morphology. Cell counts of suspended bacteria in vent waters up to 1 m above the seafloor were 5×10^5 to $10^6\,ml^{-1}$ (Jannasch & Wirsen, 1979; Karl, Wirsen & Jannasch, 1980), approximately one to two orders of magnitude greater than in control non-vent bottom water. Microscopic examination of bacterial mats, rock/ animal surfaces and artificial surfaces deployed *in situ* for a period of 10.5 months has also revealed the presence of dense populations of morphologically distinct bacteria (Jannasch & Wirsen, 1981; Jannasch, 1984).

Quantitative data on adenosine 5'-triphosphate (ATP) concentrations present in the vent waters indicated that the total biomass of microorganisms in the flowing hydrothermal fluids from Garden-of-Eden vent (Galapagos Rift) exceeded that in the control deep water by a factor of >300, and was approximately 4 times greater than the ATP concentration measured for the photosynthetically active surface waters (Karl, Wirsen & Jannasch, 1980). If the ATP data are extrapolated to total living microbial biomass (using a C:ATP ratio of 250 (Karl, 1980), the evidence for a localized bacterial-based food supply is quite overwhelming (Fig. 6). Subsequent microbial biomass investigations of various geographically distinct hydrothermal vent ecosystems at 11° and 21°N on the EPR, the Guaymas Basin in the Gulf of California and the Endeavour Ridge, consistently revealed an elevated presence of microorganisms in discharged vent fluids, relative to ambient bottom seawater (Karl, Burns & Orrett, 1983; Karl *et al.*, 1984; Karl, 1985; Winn & Karl, 1985; Winn, Karl & Massoth, 1986; Table 2). Biomass enrichment factors (ATP in vent water: ATP in ambient deep-sea water) ranged from <3 to >500 for 23 individual vent samples analyzed to date. The maximum ATP concentration of 0.5–1 μg ATP l^{-1} (equivalent to 125–250 μg bacterial C l^{-1}) measured at geographically distinct hydrothermal vents suggests that the potential for biomass production may be similar for a wide variety of hydrothermal vent fields.

These biomass and bacterial cell number data confirm the presence of bacteria in hydrothermal vent fluids and, hence, support the model of localized microorganism production in the deep sea. Alone, however, biomass data can neither confirm the presence nor predict *in situ* metabolic activities of chemolithoautotrophic bacteria.

Enrichment culture isolation of 'target' bacterial species

Microbiologists have traditionally relied upon enrichment culture techniques to obtain pure culture isolates of a given species or physiological group (Schlegel & Jannasch, 1967). Once isolated, these specific bacterial strains can then be used in laboratory experiments

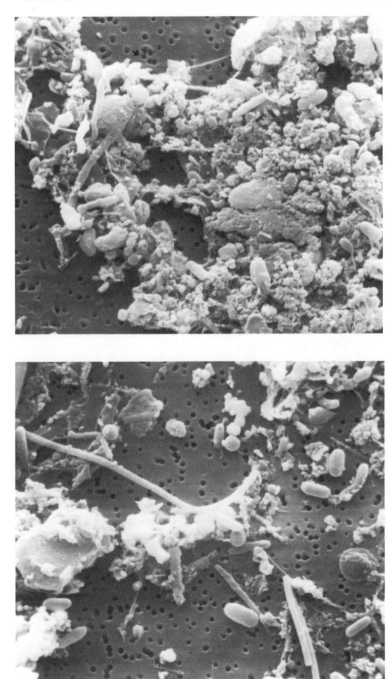

Fig. 5. Scanning electron micrographs of particulate materials from the turbid Galapagos Rift zone vent waters collected on Nuclepore filters. Amorphous particulate material in the upper photo is probably sulfur. Average pore diameter is 0.2 μm.

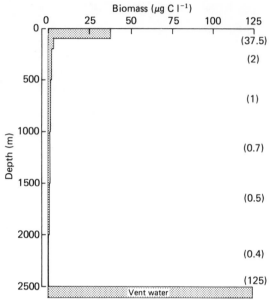

Fig. 6. Distributions of total microbial biomass (ATP × 250) of the water column overlying the Galapagos Rift zone and of a sample collected from the Garden-of-Eden vent (vent water). The samples were collected in Niskin bottles and extracted with boiling phosphate buffer. ATP was measured using the firefly bioluminescence assay system. Values in parentheses correspond to the average value of biomass carbon ($\mu g\ C\ 1^{-1}$) of the depth intervals shown.

designed to study specific physiological, biochemical or genetic characteristics of the isolate.

Successful elective culture of a specific bacterial group is conclusive evidence for the existence of that species in the natural habitat from which the sample was collected. However, a negative result does not necessarily imply that a specific physiological group is absent. In theory, positive enrichment of a particular species or physiological group could result from the presence of a single viable cell or spore. Consequently, unless performed in a semi-quantitative manner (using a vanishing dilution or most probable number (MPN) technique), the ecological significance of elective culture results are of limited value. Furthermore, the mere presence of a given strain does not confirm its *in situ* metabolic activity or capacity for growth and reproduction. These must be evaluated directly under *in situ* conditions.

Elective culture isolation experiments performed using samples collected from deep-sea hydrothermal vents (sometimes, by necessity, under non-sterile conditions) have revealed the presence of

Table 2. *Summary of data on the ATP concentrations measured for water samples collected from geographically distinct hydrothermal vent fields (From D. Karl & C. Winn, unpublished data)*

Vent location	ALVIN dive no.	Date	Number of water samples	Range of ATP concentrations (ng 1^{-1})
Galapagos Rift				
Clambake	878–9	1–79	11	9.0–52.7
Mussel bed	800	1–79	4	9.0–24.4
Garden-of-Eden	883–4	1–79	20	26–906
Rose garden	983	11–79	3	6.7–30.4
Control, deep water		1–79	3	1.55–1.95
21°N, EPR				
Clam acres	1212	4–82	3	4.4–13.4
Clam acres	1214	4–82	4	6.5–18.8
Clam acres	1216	4–82	1	12.5–15.5
Clam acres	1221	4–82	3	7.0–10.5
Clam acres	1225	4–82	6	20.7–68.9
Clam acres	1228	4–82	1	10.4–17.8
Control, deep water		4–82	4	1.8–2.2
11°N, EPR				
	1381	5–84	8	8.6–26.2
	1385	5–84	1	6.8–11.8
Endeavor Ridge				
	1418–9	7–84	2	24.2–500
Guaymas Basin				
	1603	7–85	2	4.3–25.8
	1610	8–85	1	17.3–46.0
	1615	8–85	1	114.3–208.3

a physiologically diverse microbial community (Table 3). Several target species known to be capable of chemolithoautotrophic growth have been recovered from hydrothermal fluids. The majority of the sulfur-oxidizing bacteria are mixotrophic (i.e. heterotrophic bacteria which exhibit growth stimulation upon the addition of reduced sulfur compounds) rather than the putative obligate chemolithoautotrophs (Ruby, Wirsen & Jannasch, 1981; Jannasch & Wirsen, 1985; Wirsen, Tuttle & Jannasch, 1986). Extinction dilution experiments using a thiosulfate enrichment medium indicated the presence of a viable sulfur-oxidizing population that was one to three orders of magnitude less than that determined by direct microscopy (Wirsen, Tuttle & Jannasch, 1986). The isolation of additional groups, suspected to

Table 3. *Enrichment culture isolation of specific groups of microorganisms in samples collected from deep-sea hydrothermal vents*

Vent location	Sample source	Isolate	Comment	Reference
Galapagos Rift	Mussel bed vent	Free-living spirochete	Obligately anaerobic, heterotrophic, low growth rate and cell yield; can grow in presence of HS^-	Harwood, Jannach & Canale-Parola, 1982
Galapagos Rift	Mussel bed vent water	*Thiobacillus* sp. (strains TB49r, TB49s TB49c)	Chemoautolithotrophic $S_2O_3^{2-}$ oxidizer	Jump & Tuttle, 1981
Galapagos Rift	Rose-garden vent from mussel periostracum	*Thiomicrospira* sp. (strain L-12)	Obligately chemolithotrophic sulfur oxidizer, microaerophilic, tolerant of HS^- up to 300 μM; growth optimum at 25 °C and and 1 atm, barotolerant to 250 atm	Ruby & Jannasch, 1982
Galapagos Rift	Various	95 strains of sulfur-oxidizing bacteria (including 12 obligate chemoautolithotrophs and 83 heterotrophs)	Obligate chemoautolithotrophs were microaerophilic and used HS^-, S^0, $S_4O_6^{2-}$ or $S_2O_3^{2-}$ as an energy source; some heterotrophs produced acid and others produced base when grown on $S_2O_3^{2-}$ + yeast extract; CO_2 incorporation rate for heterotrophs was <5% that measured for autotrophic growth	Ruby, Wirsen & Jannasch, 1981

Location	Sample	Organism	Description	Reference
Galapagos Rift	From glass contact slides left *in situ* for 10 mo at mussel bed vent and from periostracum of a mussel	Manganese oxidizer	Utilize Mn^{2+} with an inducible enzyme system; oxidation requires O_2; showed ATP synthesis at expense of Mn^{2+} oxidation; no evidence for autotrophic growth; they appear to be mixotrophs	Ehrlich, 1983
Galapagos Rift	Scrapings from mussel bed organisms	*Hyphomonas hirschiana, Hyphomonas jannaschiana*	Aerobic, heterotrophic prosthecate bacteria; reproduce by budding	Weiner *et al.*, 1985
21 °N, East Pacific Rise	Base of white smoker chimney	*Methanococcus jannaschii* (DSM #2661)	Extremely thermophilic methanogen; optimum growth at 85 °C; obligately anaerobic; grow autotrophically on 80% H_2 + 20% CO_2 with CH_4 formation, doubling time of 26 min at 85 °C	Jones *et al.*, 1983
21 °N, East Pacific Rise	Scrapings from vestimentiferan tube worm	*Thiomicrospira crunogena* (ATCC #35932)	Obligate chemolithoautotrophic sulfur oxidizer, oxidizes $S_2O_3^{2-}$, HS^- and S^0; CO_2 is primary source of carbon; oxidation requires O_2; potentially high growth rate (0.8 h^{-1}); will grow from 4 °–38.5 °C, 1–250 atm	Jannasch *et al.*, 1985
Juan de Fuca	Hydrothermal fluids from sulfide mounds and chimneys	Thermophilic, heterotrophic bacteria	Grown on acetate–GELRITE medium at temperatures up to 120 °; strictly anaerobic	Deming & Baross, 1986

be present and active under the defined environmental conditions, has been unsuccessful. These groups include the large, filamentous mat-forming sulfur bacteria (e.g. *Beggiatoa, Thiothrix, Thiovulum*) which have been collected but not isolated, the putative bacterial symbionts of the vestimentiferan tube worms and giant clams (see section on symbiosis) and other more exotic or novel microbes (e.g. facultative sulfur-respiring heterotrophs, anaerobic methane-oxidizers, sulfur-respiring methanogens). Successful isolation of members of the latter groups will probably require more creative approaches and a large element of chance. Relatively few microbiologists would regard the isolation of new organisms as a primary goal of their research (Williams, Goodfellow & Vickers, 1984). This is especially true for ecological studies at deep-sea vents where the most fundamental hypotheses can be tested only by direct *in situ* experimentation.

At the present time it is neither known, nor easily determined, whether the species which have been isolated from hydrothermal vents are important components of the natural ecosystem. Future experiments using fluorescent antibodies produced against the pure culture isolates or the analysis of 16S ribosomal RNA nucleotide sequences might help to characterize the natural microbial assemblages. Furthermore, the vent water isolates could also be used in 'caged' culture experiments to evaluate their ability to grow and to reproduce in the suspected natural habitats.

Microbial production and assimilation of carbon dioxide

Critical 'proof' for the hypothesized productivity of chemolithoautotrophic bacteria might be provided by direct measurements of the uptake and assimilation of radiolabeled $^{14}CO_2$ (Jannasch & Wirsen, 1979; Karl, Wirsen & Jannasch, 1980). These technically simple and straightforward experiments are often difficult to interpret. This is primarily because the dark assimilation of $^{14}CO_2$ cannot be equated readily with chemolithoautotrophic metabolism because all microorganisms assimilate $^{14}CO_2$ during normal growth and metabolism (Tuttle & Jannasch, 1973). Consequently an analytical method for uniquely determining $^{14}CO_2$ uptake by chemolithoautotrophic microorganisms must be devised before the measurement of $^{14}CO_2$ uptake can be related quantitatively to chemosynthetic primary production. It is not unexpected that $^{14}CO_2$ is assimilated by hydrothermal vent microbial populations (Table 4).

Table 4. $^{14}CO_2$ assimilation capacity of deep-sea hydrothermal vent microbial populations (GRZ = Galapagos Rift zone; 21 °N = East Pacific Rise at 21 °N)

Location	Incubation conditions	Substrate addition	Sample no.	CO_2 incorporated (nmol l^{-1} d^{-1})	Reference
Mussel bed (GRZ)	Dark, in situ syringe sample incubation, 23 days	None	1	76.1	Tuttle et al., 1983
		None	2	299.1	
		0.1 mM $S_2O_3^{2-}$	1	271.1	
		0.1 mM $S_2O_3^{2-}$	2	384.2	
		1 mM $S_2O_3^{2-}$	1	233.2	
		1 mM $S_2O_3^{2-}$	2	284.4	
Rose Garden (GRZ)	Dark, in situ syringe sample incubation, 6.2 d	None	1	3.4, 33.2[a]	Tuttle, Wirsen & Jannasch, 1983
		None	2	398.1	
		0.1 mM $S_2O_3^{2-}$	1	12.2, 9.2	
		0.1 mM $S_2O_3^{2-}$	2	10.9, 2.5	
		1 mM $S_2O_3^{2-}$	1	79.6, 134.2	
		1 mM $S_2O_3^{2-}$	2	70.2, 97.2	
Rose Garden (GRZ)	Dark, in situ serum bottle incubation, 6.2 d	None	1	146.1	Tuttle, Wirsen & Jannasch, 1983
		None	2	2.7	
		None	3	26.7	
		0.5 mM Na_2S	1	140.8	
		10 mM $S_2O_3^{2-}$	1	154.0	
Rose Garden (GRZ)	Dark, in situ (3.6°, 245 atm) or on ship (3 °C, 1 atm); all samples 6.2 d	None	1 (in situ)	21.0	Tuttle, Wirsen & Jannasch, 1983
		None	2 (in situ)	21.3	
		None	3 (ship)	35.7	
		1 mM $S_2O_3^{2-}$	1 (in situ)	21.7	
		1 mM $S_2O_3^{2-}$	2 (in situ)	37.3	
		1 mM $S_2O_3^{2-}$	3 (ship)	41.0	
		1 mM $S_4O_6^{2-}$	1 (in situ)	7.7	
		1 mM $S_4O_6^{2-}$	2 (in situ)	8.3	
		1 mM $S_4O_6^{2-}$	3 (ship)	7.0	
Holger's hole (21°N)	Dark, 1 atm, 23 °C, 100 h	None	1	40	Tuttle, 1985
		1 mM $S_2O_3^{2-}$	1	140	
		1 mM $S_2O_3^{2-}$	2	140	
White smoker (21°N)	Dark, 1 atm, 23 °C, 24 h	None	1	210	Tuttle, 1985
		1 mM $S_2O_3^{2-}$	1	190	
		1 mM $S_4O_6^{2-}$	1	430	
White smoker (21°N)	Dark, 1 atm, 3 °C or 23 °C; all samples 24 h	None	1 (3 °C)	10	Tuttle, 1985
		None	2 (23 °C)	230	
		1 mM $S_2O_3^{2-}$	1 (3 °C)	10	
		1 mM $S_2O_3^{2-}$	2 (23 °C)	340	
		1 mM $S_4O_6^{2-}$	1 (3 °C)	20	
		1 mM $S_4O_6^{2-}$	2 (23 °C)	380	

[a]Replication between subsamples from the same syringe

One approach which has been used to evaluate the presence and relative significance of chemolithoautotrophic bacterial production is to measure rates of $^{14}CO_2$ uptake in the presence of added amounts of known chemolithoautotrophic substrates (Tuttle *et al.*, 1983; Tuttle, 1985; Jannasch & Mottl, 1985; Wirsen, Tuttle & Jannasch, 1986; Table 4). Experiments performed to date have employed, exclusively, the addition of reduced sulfur compounds even though the experimental design is amenable to evaluate any organic or inorganic substrate. In general these experiments have yielded inconsistent and variable results which are difficult to interpret within the framework of the chemolithoautotrophic hypothesis. Lack of stimulation of $^{14}CO_2$ uptake following the addition of appropriate growth substrates could result from: (1) the initial presence of an excess concentration of substrate in the natural habitat (i.e. the population is not substrate limited); (2) the selection of an inactive or less preferred substrate in the experimental treatment; or (3) the absence of a significant population of target bacteria. Stimulation of $^{14}CO_2$ uptake, on the other hand, is not rigorous proof for *in situ* chemosynthetic primary production, because the addition of a specific exogenous substrate to a natural assemblage of microorganisms is equivalent to performing an enrichment culture experiment. Long-term incubation with $S_2O_3^{2-}$ has yielded, not unexpectedly, an increase in the number (and presumably activity) of thiosulfate-oxidizing bacteria (Tuttle & Jannasch, 1977). Consequently, it is important to keep the incubation period short (<0.1 of the potential doubling time) and also to monitor the kinetics of $^{14}CO_2$ assimilation. Furthermore, it is essential that the added substrates are highly purified since even a 1% contamination by reduced organic carbon (not unusual in many analytical grade reagents) could represent an addition of potential heterotrophic substrates that would obviously influence the interpretation of the experimental results.

The absence of a systematic stimulation of CO_2 uptake upon the addition of selected chemolithoautotrophic substrates (Table 4) may be interpreted as direct evidence against the role of reduced sulfur compounds in the metabolism of hydrothermal vent microorganisms. Karl (1985) has previously concluded, based on a totally independent line of reasoning, that the role of sulfur compounds may be less important than currently perceived.

Tuttle & Jannasch (1979) have emphasized that the measurement of net assimilation of carbon dioxide can be assumed to represent chemolithoautotrophic growth, but also that ^{14}C tracer experiments represent non-isotopic equilibrium systems which cannot adequately resolve gross uptake (the combined activities of both heterotrophs and autotrophs) from strictly chemolithoautotrophic metabolism. However, if simultaneous and independent measurements were available for the rate of net microbial carbon production and for the rate of total carbon dioxide assimilation (both expressed in units of carbon produced per unit volume per unit time) one could then determine whether carbon dioxide assimilation represented a significant percentage of the total productivity. In heterotroph-dominated systems the ratio of carbon dioxide assimilated to total carbon produced should be <0.10 (Romanenko, 1964; Overbeck, 1979) whereas in autotroph-dominated systems the ratio would be expected to approach 1.0. The use of ATP (as a measure of total microbial biomass; Karl, 1980) and [^{3}H]-adenine (as a measure of total microbial production; Karl & Winn, 1984) are uniquely suited for such investigations because these experimental methods do not discriminate among physiologically diverse groups of microorganisms.

To date, these two methods of estimating bacterial production have not been systematically coordinated. However, preliminary results obtained for water samples collected from the 21 °N hydrothermal vent site indicate that carbon dioxide assimilation ($\sim$100–200 nmoles C l^{-1} d^{-1}; Table 4) represents only a small percentage of the total carbon production estimate based on DNA synthesis ($\sim$6000 nmoles C l^{-1} d^{-1}; Karl, 1985). These data suggest a potential for the assimilation of carbon sources other than carbon dioxide and are consistent with the substantial chemoorganotrophic metabolic activity which has been measured at the Galapagos Rift zone (Tuttle, Wirsen & Jannasch, 1983). In the future, simultaneous measurements of carbon dioxide assimilation (using $^{14}CO_2$) and total microbial production (using [^{3}H]-adenine) could be achieved in a single sample using dual-labeled procedures.

In vitro enzyme activities

Ribulose-1,5-bisphosphate carboxylase (RuBPCase) and phosphoenol pyruvate carboxylase (PePCase) are key enzymes that catalyze carboxylation reactions in microbial cells. Many (but not all) chemolithoautotrophic bacteria contain RuBPCase which is the primary carboxylating enzyme in the Calvin–Benson cycle (e.g. reductive pentose phosphate cycle). By comparison, PePCase is present in all autotrophic and heterotrophic organisms. It is suggested that the presence of RuBPCase might, therefore, indicate the presence of selected groups of chemolithoautotrophic bacteria (e.g. sulfur oxidizers, ammonium oxidizers) while the ratio of RuBPCase to PeP-Case might indicate the relative importance of such autotrophic bacteria in a mixed microbial assemblage (Morris *et al.*, 1985). The results of laboratory experiments using marine nitrifying bacteria revealed a positive correlation between *in vitro* RuBPCase activity and the rate of *in vivo* chemoautotrophic carbon dioxide fixation (Glover, 1983); however, the amount of carbon dioxide fixed per unit of enzyme activity was species-specific. The *in vitro* activity of PePCase, on the other hand, was independent of the rate of carbon dioxide fixation. In applying this method to natural ecosystems it is essential to bear in mind that the enzyme activity measured represents the contribution from growing cells, metabolically active but non-proliferating cells, intact dead cells, cell fragments and free enzymes adsorbed onto particulate matter.

Tuttle (1985) has adapted this approach for application to deep-sea hydrothermal vent microbial populations. At the onset of these experiments it was first established that sulfur-oxidizing bacteria isolated from hydrothermal vent waters did, in fact, contain RuBPCase activity when grown under autotrophic conditions in the laboratory. On the basis of these experiments, Tuttle (1985) proposed the following reference data: (1) 0.06 units of RuBPCase activity is equivalent to $\sim 10^9$ cells of *Thiomicrospira* sp. or *Thiobacillus* sp. in late exponential phase; (2) RuBPCase in autotrophically grown cells exceeded PePCase by a factor of ~ 40; and (3) heterotrophs, lacking RuBP-Case, exhibited high but variable amounts of PePCase activity which often exceeded the RuBPCase activity of an equivalent number of autotrophic sulfur oxidizers. The field results revealed the presence of both RuBPCase and PePCase in discharged hydrothermal vent fluids. The variability observed in these experiments makes it difficult to formulate any general conclusions. However, based on the results available to date there does not appear to be overwhelming evidence for the existence of a free-living microbial community dominated by chemolithoautotrophs (Tuttle, 1985). In fact, the ratios of RuBP-Case to PePCase suggest the presence of a substantial population of heterotrophic bacteria in the hydrothermal vent waters. These data are consistent with direct measurements of the heterotrophic uptake of acetate and glucose in samples collected from the Galapagos Rift vents (Tuttle, Wirsen & Jannasch, 1983; Wirsen, Tuttle & Jannasch, 1986) and with the previously mentioned disparity between total microbial carbon production and carbon dioxide assimilation. The observed stimulation of chemoorganotrophic activity upon the addition of $S_2O_3^{2-}$ is indicative of the presence of mixotrophic sulfur bacteria and this result is consistent with the results of enrichment culture isolation experiments (Ruby, Wirsen & Jannasch, 1981).

In contrast to the freely flowing vent waters, sample materials collected from various tissues of *Calyptogena* and *Riftia* revealed a preponderance of RuBPCase activity (Tuttle, 1985). The RuBP-Case:PePCase ratios ranged from 1.47 for *Calyptogena* stomach tissue to 50.3 for a sample collected from the anterior portion of the trophosome. The potential role of symbiotic chemolithoautotrophic bacteria will be discussed in a subsequent section of this review.

Stable isotope data

The measurement of the naturally occurring stable isotope abundances in samples collected from deep-sea vents and cold seeps provides a potential mechanism for elucidating the sources and utilization pathways for dietary carbon (Rau & Hedges, 1979; Rau, 1981a), nitrogen (Rau 1981b) and sulfur (Fry, Gest & Hayes, 1983). Since thermodynamic properties of substances depend upon their mass (Craig, 1953), isotopic fractionation is expected to occur during chemical, physical and biological processes. For example, carbon isotopes are known to be fractionated during the process of photosynthesis (Park & Epstein, 1960). Although the data are much more limited, chemolithoautotrophic bacteria appear to use $^{12}CO_2$ preferentially to an even greater extent than do photoautotrophs, thereby synthesizing organic matter with a relative ^{13}C content (referred to as $\delta^{13}C$, measured relative to a PDB (Pedee belemnite) marine carbonate standard) which can be 30‰ lower than that of available inorganic carbon pool (Rau, 1985).

Similarly, the ^{15}N content of animal tissues is known to increase (i.e. $\delta^{15}N$ becomes more positive) as materials flow from primary producers through primary consumers and to higher trophic levels (Minagawa & Wada, 1984). Consequently, it may be possible to use $\delta^{13}C$ and $\delta^{15}N$ measurements to distinguish localized chemolithoautotrophic bacterial production from a food web based on secondary heterotrophic metabolism.

Rau & Hedges (1979) published the first $\delta^{13}C$ results from hydrothermal vent animal tissue which demonstrated a striking depletion

Table 5. *Relative ^{13}C abundances of inorganic and organic carbon pools from various source materials and hydrothermal vent animal tissues*

Sample source	^{13}C (‰)	Reference
Dissolved inorganic carbon	−1.7 to −3.5	Williams et al., 1981
Particulate organic detritus	−20 to −25 (mean −22)	Williams et al., 1981
Non-vent marine organisms	−8 to −25	Rau & Hedges, 1979
Non-vent molluscs	−16 to −18	Rau & Hedges, 1979
Dissolved inorganic carbon, Galapagos Rift (end-member)	−5.1 to −5.9	Craig et al., 1980
Dissolved inorganic carbon, 21 °N site (end-member)	−7	Craig et al., 1980
Thermogenic methane	>0 to −55	Frank, Gormly & Sackett, 1974
Biogenic methane	−60 to −110	Whiticar, Faber & Schoell, 1986
21 °N Methane (end-member)	−15.0 to −17.6	Welhan & Craig, 1983
Guaymas Basin methane	−40 to −50	Welhan & Craig, 1983
Guaymas Basin methane	−69 to −73	Simoneit et al., 1979
Galapagos Rift mussel tissue	−32.7 to −33.9	Rau & Hedges, 1979
Galapagos Rift mussel tissue, 1 m from vent	−33.8 to −33.9	Williams et al., 1981
Galapagos Rift mussel tissue, 8 m from vent	−32.8	Williams et al., 1981
Galapagos Rift clam tissue	−32.0 to −32.1	Rau, 1981a
Galapagos Rift vestimentiferan tube worm tissue	−10.8 to −11.0 (muscle) −10.9 to −11.1 (trophosome)	Rau, 1981a
Galapagos Rift vestimentiferan tube worm tissue	−10.8 to −11.0	Williams et al., 1981
Galapagos Rift crab tissue	−13.7 to −17.6	Rau, 1985
21 °N, EPR clam tissue	−32.6 to −32.7	Rau, 1981a
21 °N, EPR pompeii worm tissue	−11.2	Desbruyeres et al., 1983

of ^{13}C when compared to other known sources of oceanic organic matter. These data were consistent with the hypothesis that a localized source of chemolithoautotrophically produced bacterial carbon was the basis for the Galapagos Rift hydrothermal vent food web. Subsequent measurements of clam tissues collected from the Galapagos Rift and from the 21 °N hydrothermal vent field confirmed these depleted ^{13}C values (Rau, 1981a; Table 5). In contrast to the bivalves, samples of several tissues from the vestimentiferan tube worm had isotope ratios that were significantly higher than those previously measured for open ocean biota. This large carbon isotopic dissimilarity between the tube worms and the large clams, which inhabit the same vent field, was interpreted by Rau (1981a) as evidence for at least two independent food source materials.

Subsequent measurements by Williams *et al.* (1981) confirmed these $\delta^{13}C$ values for hydrothermal vent animal tissues. A novel aspect of their study, however, was the measurement of ^{14}C (a radioactive isotope of carbon) in addition to the concentrations of ^{12}C and ^{13}C. Since the $\Delta^{14}C$ content (expressed as the deviation [‰] from the ^{14}C activity of nineteenth-century wood) of the dissolved inorganic carbon (DIOC) derived from magmatic sources is much different from either DIOC of ambient bottom seawater ($\Delta^{14}C = -1000$ and -233‰, respectively) or particulate matter derived from surface waters ($\Delta^{14}C = +20$‰), it is possible to calculate the relative contribution of each individual carbon source to the production of animal tissue. Their results indicated that the principal source of dietary carbon for both mussels and tube worms was derived from the DIOC pool in the vent effluent waters. Since the tissue ^{14}C values were lower than the ambient deep-water DIOC value, at least some of the carbon incorporated into animal tissue must have been derived from magmatic sources (Williams *et al.*, 1981). The nearly identical ^{14}C activities of a sample of vestimentiferan tissue (-270 ± 20‰) and two separate samples of mussel tissue (-263 ± 8‰ and -270 ± 6‰) suggest that the tube worms are utilizing the same DIOC sources even though the $\delta^{13}C$ value of the tube worm tissue was 23‰ greater than the mussel tissue (Williams *et al.*, 1981; Table 5).

Similar carbon isotope measurements have recently been conducted at three cold-water seep sites (Table 6). Kennicutt *et al.* (1985) have reported $\delta^{13}C$ values for animal tissues sampled from the Louisiana seep that are nearly identical to the values previously measured at the Galapagos Rift and 21 °N hydrothermal vents. Although the authors use their data to suggest the presence of a local chemolithoautotrophic food source, other interpretations (e.g. heterotrophic oxidation of isotopically light hydrocarbons or methane) seem equally plausible. Measurements of samples collected from the Florida Escarpment cold-seep site revealed animal tissues with an extremely negative ^{13}C value (Paull *et al.*, 1985; Table 6). The authors attribute these observations to a localized chemosynthetic food source. It is hypothesized that the assimilation of isotopically depleted biogenic methane is partially responsible for these isotope-depleted animal tissues. Measurement of ^{14}C activity, however, indicates that a majority ($\sim$60%) of the tissue carbon is 'modern' which

Table 6. *Relative ^{13}C abundances in animal tissues collected from cold-water seep habitats*

Sample source	^{13}C (‰)	Reference
Florida Escarpment		
Mussel tissue	−74.3 (±2.0; $n = 10$)	Paull *et al.*, 1985
Gastropod tissue	−59.9 (±0.7; $n = 2$)	Paull *et al.*, 1985
Vestimentiferan tissue	−42.7 (±0.7; $n = 3$)	Paull *et al.*, 1985
Sedimentary organic carbon	−80.4 (±0.4; $n = 3$)	Paull *et al.*, 1985
Louisiana Slope		
Clam tissue	−31.2 to −35.4	Kennicutt *et al.*, 1985
Tube worm (tissue)	−27.0	Kennicutt *et al.*, 1985
(tube)	−28.1	
Various fish species	−17.2 to −17.9	Kennicutt *et al.*, 1985
Mussel gill tissue	−51 to −57	Childress *et al.*, 1986
Methane	−45	Childress *et al.*, 1986
Orca Basin		
Dissolved inorganic carbon	−15 to −19	Sackett *et al.*, 1979
Dissolved organic carbon	−23 to −27	Sackett *et al.*, 1979
Particulate organic carbon	−17 to −22.6	Sackett *et al.*, 1979
Methane	−72 to −74.5	Sackett *et al.*, 1979
Oregon Subduction Zone		
Clam gill tissue	−51.6	Kulm *et al.*, 1986
Clam periostracum	−35.7	Kulm *et al.*, 1986
Tube worm (tissue)	−31.9	Kulm *et al.*, 1986
(tube)	−26.7	

restricts the potential contribution of fossil methane sources. This source limitation is inconsistent with the $\delta^{13}C$ data as currently interpreted. Consequently, a fairly complex carbon flux scenario must be devised to accommodate the two independent data sets (Paull *et al.*, 1985).

The results of these carbon isotopic data, as well as the data on $\delta^{15}N$ and $\delta^{34}S$ abundances not explicitly discussed here, clearly indicate a localized source of food for the hydrothermal vent and cold-seep habitats that appears to be separate from organic matter derived from photosynthesis. A major limitation of these investigations concerns the uncertainty in the isotopic composition of all the potential sources of carbon (e.g. vent water DOC, carbon monoxide, methane) supporting microbial production. Furthermore, it seems obvious that one needs to measure directly the $\delta^{13}C$ (and $\delta^{15}N$, $\delta^{34}S$) of the presumed food (i.e. the discharged bacterial cells) but, to date, this has not been possible. Furthermore, it is conceivable that several independent microbial metabolic processes occur simultaneously which may result in an isotopic signature which is not characteristic of any simple food web. An example of such antagonistic metabolic reactions might be the simultaneous production and oxidation of methane. Both processes fractionate carbon isotopes but in opposite directions. Finally, it must be kept in mind that we have not yet measured, directly or indirectly, the extent to which vent- or seep-associated bacteria fractionate either carbon, nitrogen or sulfur isotopes during metabolism and growth. These experiments could be performed by allowing well characterized vent isolates to grow *in situ* (cage or dialysis-type culture) and by measuring directly the relative isotopic content of the biomass that is produced.

Chemosynthetic symbiosis

During the initial anatomical and histological investigations of the giant vestimentiferan tube worm *Riftia pachyptila* elemental sulfur crystals were discovered in the trophosome. It was suggested that the sulfur deposits might be the result of the metabolism of symbiotic sulfur-oxidizing bacteria (Jones, 1981). Subsequent microscopic, biochemical and enzymatic evidence has been presented confirming the presence of chemolithoautotrophic (RuBPCase-containing) bacteria (Cavanaugh *et al.*, 1981; Felbeck 1981; Cavanaugh, 1985). Since this animal lacks a mouth and a digestive system it was hypothesized that the endosymbiotic bacteria provided organic compounds at the expense of inorganic substrates (HS^-, oxygen and carbon dioxide) supplied by the host, thus providing the first example of a chemolithoautotrophic bacterial–animal symbiosis. Since that initial discovery, many diverse examples of 'chemosynthetic symbiosis' have been reported from reducing sediments and hydrothermal vent habitats (Cavanaugh, 1985). More recently, Childress *et al.* (1986) have demonstrated the existence of a methane-based bacterial symbiotic association with a Louisiana slope mussel. It is likely that other substrate-specific symbioses will also be discovered.

In spite of the apparent widespread occurrence of this phenomenon we know relatively little about the nature of the bacteria associated with the host tissues, their ability to fractionate carbon, nitrogen or sulfur isotopes or the detailed mechanism of the presumed transfer of carbon and energy between host and symbiont.

While these bacterial–animal associations are clearly an example of symbiosis ('living together') the evidence for mutualism is less convincing. Relevant data, previously summarized (Jannasch & Nelson, 1984; Jannasch 1984b, 1985; Jannasch & Taylor 1984; Cavanaugh, 1985; Jannasch & Mottl, 1985; Jannasch & Wirsen, 1985), will not be reviewed here. Until the primary symbiont is available for pure culture laboratory study it may be inappropriate to speculate on the quantitative role of this presumed bacterial chemolithoautotrophic symbiosis to the total metabolism of the specific animal host. A recent report of the ability of *Solemya reidi* (a clam) mitochondria to assimilate HS^- and generate ATP (Powell & Somero, 1986) may require that we re-evaluate the nutritional role of these bacterial endosymbiotic associations.

Black smokers and hydrothermal plumes

Ever since the high-temperature vents (>100 °C) were first discovered in 1979 at the 21 °N site, important questions have been raised regarding the possible presence of bacteria in these superheated seawaters. Although the data available thus far are, to a certain extent, contradictory, there does appear to be strong evidence for the existence of thermophilic bacteria in and around the black smoker habitat (Baross, Lilley & Gordon, 1982; Baross & Deming, 1983; Jones *et al.*, 1983; Lilley, Baross & Gordon, 1983; Karl *et al.*, 1984; Deming & Baross, 1986). Perhaps the most important question, however, regards the exact temperature at which thermophilic bacteria from these environments are capable of growing.

Both the *in situ* field experiments and laboratory cultivation at temperatures in excess of 100 °C are technically very difficult to perform. Interpretation of measurements on *in situ* collections of high-temperature vent water samples is further complicated by the difficulty of obtaining pure vent water samples which are undiluted by ambient seawater. The latter contains a diverse assemblage of microorganisms and clearly represents a potential point source of 'contamination'. Likewise, the reported cultivation of vent-derived organisms at 250 °C (Baross & Deming, 1983) has generated much controversy (Trent, Chastain & Yayanos, 1984; White 1984) which, in my opinion, remains unresolved at the present time.

Lane *et al.* (1986) have recently conducted an experiment at the Guaymas Basin hydrothermal vent field wherein a titanium growth chamber (referred to as 'vent cap') was placed over two separate high-temperature hydrothermal vents (156 and 300 °C). The results of this *in situ* growth experiment indicated that bacteria were able to grow in close proximity to, but not directly in, the high-temperature waters. Similar conclusions were obtained by Karl, Wirsen & Jannasch (1986) based upon field experiments performed during the same cruise. In addition, the data presented by Karl, Wirsen & Jannasch (1986) indicated that undiluted water samples collected from eight separate high-temperature (150–350 °C) vents contained no recognizable bacterial cells and no measurable ATP. Paired water samples collected at a distance of 20–30 cm above the point of hot fluid discharge, however, contained a large number of bacteria, had ATP concentrations in excess of 300 ng l^{-1} and were metabolically active over a wide range of incubation temperatures (25–80 °C; Karl, Wirsen & Jannasch, 1986). These results indicate that the origin of the bacteria associated with high-temperature vent plumes is probably via entrainment of cells from peripheral habitats rather than from the hot geothermal fluids. The entrained microorganisms associated with these hydrothermal fluids may be responsible for the removal of selected metals, reduced gases and dissolved organic matter as the warm buoyant plumes rise and move off-axis (Winn, Karl & Massoth, 1986; Cowen, Massoth & Baker, 1986). Measurement of the growth and metabolic activities of these hydrothermal plume-associated microorganisms represents an important area for future research.

SUMMARY AND FUTURE RESEARCH PROSPECTS

In preparing this review I have been encouraged by the enormous progress which has been made toward elucidating the basis for life at deep-sea hydrothermal vents and cold-seep habitats. Given the fact that we have only recently discovered these novel ecosystems and considering the difficulties in obtaining samples for *in situ* experimentation, I consider the published record of research achievement to be quite remarkable. However, at the same time I have also been impressed by the many important, fundamental scientific questions which remain unresolved or at best only partially answered. First and foremost among the list of future research objectives is the need for a comprehensive and systematic evaluation of the chemolithoautotrophic production hypothesis and for a detailed chemical description of the suspected microbial habitats.

One might arbitrarily conclude that geothermal energy-based bacterial processes are the most logical explanation to account for the observed high bacterial productivity. However, at the present time we must conclude, based on the experimental data available, that the hypothesis of chemosynthetic primary production has not been rigorously tested and, therefore, cannot be accepted as the principle mode of bacterial nutrition at the deep-sea vents and cold seeps. Although chemolithoautotrophic bacteria are clearly present in both habitats, their quantitative contribution to total microbial production is, in my opinion, unknown at the present time. Considering the preponderance of heterotrophic sulfur-oxidizing bacteria (mixotrophs) and the evidence for the importance of anaerobic metabolism, it is probable that the role of free-living chemolithoautotrophic sulfur-oxidizing bacteria has been overestimated. Future experiments should also evaluate the importance of additional potential substrates including hydrogen, methane, reduced organic compounds and, perhaps, pyrophosphate (Liu, Hart & Peck, 1982).

The discussion of bacterial productivity has thus far revolved around carbon assimilation when, in fact, carbon rarely, if ever, limits the productivity of natural microbial assemblages. Future field experiments should focus on the utilization of nitrogen (especially dinitrogen fixation), phosphorus and other essential nutrients. Irrespective of the amount of geothermal energy available to the microbial community, the production of new cellular materials will ultimately be controlled by the availability of these essential nutrients. Under certain conditions of energy (electron) availability accompanied by growth nutrient limitation it is conceivable that the microorganisms present in that particular environment will produce excessive amounts of carbon-rich storage products or excrete nitrogen- and phosphorus-poor soluble organic compounds. Excretion of DOM is known to occur during the normal, energy-sufficient growth of sulfur-oxidizing bacteria. It is interesting to speculate that bacterial–animal symbioses evolved at the vents in response to a condition of inorganic nutrient limitation. In any case, it appears likely that the endosymbiotic bacteria neither grow rapidly nor produce large amounts of biomass but, rather, convert carbon dioxide to soluble reduced carbon compounds for the nutrition of the host animal. The specialized role of endosymbiotic bacteria may help explain our lack of success at culturing these bacteria in spite of their ubiquitous occurrence.

From a thermodynamic perspective it is essential to identify the source of both electron acceptors and electron donors before commenting on the relative importance of chemolithoautotrophic bacterial production. If reduced inorganic substrates are not important as electron donors at deep-sea vents and cold seeps, then what are the alternatives? Lonsdale (1977) discussed the potential importance of thermal advection as a means of focussing photosynthetically derived suspended particulate matter near hydrothermal vent sites. While this hypothesis has enjoyed modest support (Enright et al., 1981), it probably does not represent a process of quantitative significance (Williams et al., 1981). Furthermore this advection hypothesis does not explain the presence of animal communities at numerous cold-seep habitats which exist in the absence of thermal plumes. The potential role of chemoorganotrophic bacterial metabolism, both primary chemoorganotrophy at the expense of abiotically produced organic compounds and the more 'conventional' utilization of recycled photosynthetically derived organic substrates, has not been evaluated carefully or systematically. The possibility of chemical evolution (abiogenic synthesis of organic substrates) at deep-sea vents has been hypothesized (Corliss, Baross & Hoffman, 1981) but has not been tested. However, water samples collected from the geothermally active Red Sea brines have revealed high concentrations of thiocyanate (2.4×10^{-5} M) and glycine (1×10^{-6} M) suggesting an abiotic origin (Ingmanson & Dowler, 1980). A second, less exotic but at the present time equally plausible, source of chemoorganotrophic substrates is the large pool of DOM in the deep ocean (~ 0.3 to 0.5 mg C 1^{-1}). It is conceivable that DOM transported into the vent habitat by percolation of seawater through the fractured landscape represents a major supply of carbon and energy for vent microorganisms. Clearly this ecosystem model, if confirmed, would represent a significant departure from the current views of the vent ecosystem as being independent from photosynthesis. While I do not necessarily claim this to be fact, it is interesting to point out that the productivity and biomass of bacterial cells which have been measured at deep-sea vents (Fig. 6 and Table 2) are not inconsistent with this production mechanism (i.e. maximum biomass of bacteria in vent waters is $\sim 100–200$ μg C 1^{-1} compared to $300–500$ μg C 1^{-1} total DOM). In this regard, the measurement of δ^{13}C or Δ^{14}C may be essential, provided all potential source pools are also measured. It is also conceivable, and highly probable, that mixotrophic organisms simultaneously utilizing both organic and inorganic electron donors have evolved at deep-sea vents and cold-seep habitats. The possibility of additional novel pathways for energy metabolism and biosynthesis should not be ignored in the design of future *in situ* experiments. Once the question of bacterial production is resolved, it will be necessary to focus further on the remainder of the food web in order to evaluate the relationship (if any) between microbial biomass and the higher trophic levels.

A major limitation, which is likely to continue to plague *in situ* microbiological studies of deep-sea vents and cold seeps, is the difficulty in obtaining samples representative of the source habitats. This general concern for reliable collection of samples can also be viewed in a much broader context when one considers the effort and expense invested in the exploration of these novel ecosystems. Future microbiological studies should also include the simultaneous investigation of more accessible shallow-water marine geothermal springs and hydrocarbon seeps, at least to the extent that these latter ecosystems can be considered true 'analogs'. The development of additional, unmanned autonomous submersibles with improved sampling and remote manipulation capabilities should also be encouraged. Finally, the construction of a long-term ocean bottom observatory (J. Delaney, personal communication) at an active spreading center, seamount or cold-seep habitat would provide an opportunity to conduct a comprehensive investigation of the spatial and temporal variability of the microbial habitat and an in-depth study of microbial colonization, production and succession.

ACKNOWLEDGEMENTS

I thank C. D. Winn, G. McMurtry & G. Tien for helpful comments and criticisms, C. D. Winn, K. Orrett & D. Burns for participating in the research described herein, and L. Wong for assistance in the preparation of this manuscript. This research effort was supported, in part, by National Science Foundation grants OCE 78-20721, OCE 80-24255, OCE 83-11219 and OCE 83-51751 awarded to the author.

REFERENCES

BAROSS, J. A. & DEMING, J. W. (1983). Growth of 'black smoker' bacteria at temperatures of at least 250 °C. *Nature*, **303**, 423–6.

BAROSS, J. A., LILLEY, M. D. & GORDON, L. I. (1982). Is the CH_4, H_2 and CO venting from submarine hydrothermal systems produced by thermophilic bacteria? *Nature*, **298**, 366–8.

BOLGER, G. W., BETZER, P. R. & GORDON, V. V. (1978). Hydrothermally-derived manganese suspended over the Galapagos spreading center. *Deep-Sea Research*, **25**, 721–33.

CAVANAUGH, C. M. (1985). Symbioses of chemoautotrophic bacteria and marine invertebrates from hydrothermal vents and reducing sediments. *Biological Society of Washington Bulletin*, No. 6, 373–88.

CAVANAUGH, C. M., GARDINER, S. L., JONES, M. L., JANNASCH, H. W. & WATERBURY, J. B. (1981). Prokaryotic cells in the hydrothermal vent tube worm *Riftia pachyptila* Jones: Possible chemoautotrophic symbionts. *Science*, **213**, 340–2.

CHILDRESS, J. J., FISHER, C. R., BROOKS, J. M., KENNICUTT, M. C., BIDIGARE, R. & ANDERSON, A. (1986). A methanotrophic marine molluscan (Bivalvia, Mytilidae) symbiosis: Mussels fueled by gas. *Science*.

CORLISS, J. B., BAROSS, J. A. & HOFFMAN, S. E. (1981). An hypothesis concerning the relationship between submarine hot springs and the origin of life on Earth. *Oceanologica Acta*, No. SP, 59–69.

CORLISS, J. B., DYMOND, J., GORDON, L. I., EDMOND, J. M., VON HERZEN, R. P., BALLARD, R. D., GREEN, K., WILLIAMS, D., BAINBRIDGE, A., CRANE, K. & VAN ANDEL, T. H. (1979). Submarine thermal springs on the Galapagos Rift. *Science*, **203**, 1073–83.

COWEN, J. P., MASSOTH, G. J. & BAKER, E. T. (1986). Bacterial scavenging of Mn and Fe in a mid- to far-field hydrothermal particle plume. *Nature*.

CRAIG, H. (1953). The geochemistry of the stable carbon isotopes. *Geochimica et Cosmochimica Acta*, **3**, 53–92.

CRAIG, H., WELHAN, J. A., KIM, K., POREDA, R. & LUPTON, J. E. (1980). Geochemical studies of the 21 °N EPR hydrothermal fluids. *EOS, Transactions American Geophysical Union*, **61**, 992.

DEMING, J. W. & BAROSS, J. A. (1986). Solid medium for culturing black smoker bacteria at temperatures to 120 °C. *Applied and Environmental Microbiology*, **51**, 238–43.

DESBRUYERES, D., GAILL, F., LAUBIER, L., PRIEUR, D. & RAU, G. H. (1983). Unusual nutrition of the 'Pompeii worm' *Alvinella pompejana* (polychaetous annelid) from a hydrothermal vent environment: SEM, TEM, ^{13}C and ^{15}N evidence. *Marine Biology*, **75**, 201–5.

EDMOND, J. M., MEASURES, C., McDUFF, R. E., CHAN, L. H., COLLIER, R., GRANT, B., GORDON, C. I. & CORLISS, J. B. (1979). Ridge crest hydrothermal activity and the balances of the major and minor elements in the ocean: The Galapagos data. *Earth and Planetary Science Letters*, **46**, 1–18.

EDMOND, J. M., & VON DAMM, K. L. (1985). Chemistry of ridge crest hot springs. In *The Hydrothermal Vents of the Eastern Pacific: An overview – Bulletin of the Biological Society of Washington*, No. 6, 43–7.

EHRLICH, H. L. (1983). Manganese-oxidizing bacteria from a hydrothermally active area on the Galapagos Rift. *Environmental Biogeochemistry and Ecology Bulletin (Stockholm)*, **35**, 357–66.

ENRIGHT, J. T., NEWMAN, W. A., HESSLER, R. R. & McGOWAN, J. A. (1981). Deep-ocean hydrothermal vent communities. *Nature*, **289**, 219–20.

FELBECK, H. (1981). Chemoautotrophic potential of the hydrothermal vent tube worm, *Riftia pachyptila* Jones (Vestimentifera). *Science*, **213**, 336–8.

FRANCHETEAU, J., NEEDHAM, H. D., CHOUKROUNE, P., JUTEAU, T., SEQURET, M., BALLARD, R. D., FOX, P. J., NORMARK, W., CARRANZA, A., CORDOBA, D., GUERRERO, J., RANGIN, C., BOUGAULT, H., CAMBON, P. & HEKINIAN, R. (1979). Massive deep-sea sulphide ore deposits discovered on the East Pacific Rise. *Nature*, **277**, 523–8.

FRANK, D. J., GORMLY, J. R. & SACKETT, W. M. (1974). Reevaluation of carbon-isotope compositions of natural methanes. *The American Association of Petroleum Geologists Bulletin*, **58**, 2319–25.

FRY, B., GEST, H. & HAYES, J. M. (1983). Sulphur isotopic compositions of deep-sea hydrothermal vent animals. *Nature*, **306**, 51–2.

GLOVER, H. E. (1983). Measurement of chemoautotrophic CO_2 assimilation in marine nitrifying bacteria: an enzymatic approach. *Marine Biology*, **74**, 295–300.

GRASSLE, J. F. (1982). The biology of hydrothermal vents: A short summary of recent findings. *Marine Technology Society Journal*, **16**, 33–8.

GRASSLE, J. F., BERG, C. J., CHILDRESS, J. J., GRASSLE, J. P., HESSLER, R. R., JANNASCH, H. W., KARL, D. M., LUTZ, R. A., MICKEL, T. J., RHOADS, D. C., SANDERS, H. L., SMITH, K. L., SOMERO, G. N., TURNER, R. D., TUTTLE, J. H., WALSH, P. J. & WILLIAMS, A. J. (1979). Galapagos '79: Initial findings of a deep-sea biological quest. *Oceanus*, **22**, 2–10.

HARWOOD, C. S., JANNASCH, H. W. & CANALE-PAROLA, E. (1982). Anaerobic spirochete from a deep-sea hydrothermal vent. *Applied and Environmental Microbiology*, **44**, 234–7.

HAYMON, R. M. & MACDONALD, K. C. (1985). The geology of deep-sea hot springs. *American Scientist*, **73**, 441–9.

HECKER, B. (1985). Fauna from a cold sulfur-seep in the Gulf of Mexico: comparison with hydrothermal vent communities and evolutionary implications. In *The Hydrothermal Vents of the Eastern Pacific: A Overview – Bulletin of the Biological Society of Washington*, No. 6, 465–73.

HEKINIAN, R., FEVRIER, M., BISCHOFF, J. L., PICOT, P. & SHANKS, W. C. (1980). Sulfide deposits from the East Pacific Rise near 21 °N. *Science*, **207**, 1433–44.

HESS, H. H. (1962). History of the ocean basins. In *Petrological Studies: A Volume to Honor A. F. Buddington*, ed. A. E. Engel, H. L. James & B. F. Leonard, pp. 599–620. New York: Geological Society of America.

INGMANSON, D. E. & DOWLER, M. J. (1980). Unique amino acid composition of Red Sea brine. *Nature*, **286**, 51–2.

JANNASCH, H. W. (1984a). Chemosynthetic microbial mats of deep-sea hydrothermal vents. In *Microbial Mats: Stromatolites*, ed. Y. Cohen, R. W. Castenholz & H. O. Halvorson, pp. 121–31. New York: Alan R. Liss.

JANNASCH, H. W. (1984b). Microbes in the oceanic environment. In *The Microbe 1984. Part II – Prokaryotes and Eukaryotes*, ed. D. P. Kelly & N. G. Carr, pp. 97–122. Cambridge University Press.

JANNASCH, H. W. (1985). The chemosynthetic support of life and the microbial diversity at deep-sea hydrothermal vents. *Proceedings of the Royal Society of London, Bulletin*, **225**, 277–97.

JANNASCH, H. W. & MOTTL, M. J. (1985). Geomicrobiology of deep-sea hydrothermal vents. *Sciences*, **229**, 717–25.

JANNASCH, H. W. & NELSON, D. C. (1984). Recent progress in the microbiology of hydrothermal vents. In *Current Perpectives in Microbial Ecology*, ed. M. J. Klug & C. A. Reddy, pp. 170–6. Washington D.C.: American Society for Microbiology.

JANNASCH, H. W. & TAYLOR, C. D. (1984). Deep-sea microbiology. *Annual Review of Microbiology*, **38**, 487–514.

JANNASCH, H. W. & WIRSEN, C. O. (1979). Chemosynthetic primary production at East Pacific sea floor spreading centers. *BioScience*, **29**, 592–8.

JANNASCH, H. W. & WIRSEN, C. O. (1981). Morphological survey of microbial mats near deep-sea thermal vents. *Applied and Environmental Microbiology*, **41**, 528–38.

JANNASCH, H. W. & WIRSEN, C. O. (1985). The biochemical versatility of chemosynthetic bacteria at deep-sea hydrothermal vents. In *The Hydrothermal Vents of the Eastern Pacific: An Overview – Bulletin of the Biological Society of Washington*, No. 6, 325–34.

JANNASCH, H. W., WIRSEN, C. O., NELSON, D. C. & ROBERTSON, L. A. (1985). *Thiomicrospira crunogena* sp. nov., a colorless, sulfur-oxidizing bacterium from a deep-sea hydrothermal vent. *International Journal of Systematic Bacteriology*, **35**, 422–4.

JENKINS, W. J., EDMOND, J. M. & CORLISS, J. B. (1978). Excess ^{3}He and ^{4}He in Galapagos submarine hydrothermal waters. *Nature*, **272**, 156–8.

JONES, M. L. (1981). *Riftia pachyptila* Jones: Observations on the vestimentiferan worm from the Galapagos Rift. *Science*, **213**, 333–6.

JONES, W. J., LEIGH, J. A., MAYER, F., WOESE, C. R. & WOLFE, R. S. (1983). *Methanococcus jannaschii* sp. nov., an extremely thermophilic methanogen from a submarine hydrothermal vent. *Archives for Microbiology*, **136**, 254–61.

JUMP, A. & TUTTLE, J. H. (1981). Characterization of the thiosulfate-oxidizing bacteria isolated from marine environments. In *American Society for Microbiology, Abstracts of the Annual Meeting*, p. 187.

KARL, D. M. (1980). Cellular nucleotide measurements and applications in microbial ecology. *Microbiological Reviews*, **44**, 739–96.

KARL, D. M. (1985). Effects of temperature on the growth and viability of hydrothermal vent microbial communities. In *The Hydrothermal Vents of the Eastern Pacific: An Overview – Bulletin of the Biological Society of Washington*, No. 6, 345–53.

KARL, D. M., BURNS, D. J. & ORRETT, K. (1983). Biomass and *in situ* growth characteristics of deep sea hydrothermal vent microbial communities. In *American Society for Microbiology, Abstracts of the Annual Meeting*, p. 235.

KARL, D. M., BURNS, D. J., ORRETT, K. & JANNASCH, H. W. (1984). Thermophilic microbial activity in samples from deep-sea hydrothermal vents. *Marine Biology Letters*, **5**, 227–31.

KARL, D. M. & WINN, C. D. (1984). Adenine metabolism and nucleic acid synthesis: applications to microbiological oceanography. In *Heterotrophic Activity in the Sea*, ed. J. E. Hobbie & P. J. LeB. Williams, pp. 197–215. New York: Plenum Press.

KARL, D. M., WIRSEN, C. O. & JANNASCH, H. W. (1980). Deep-sea primary production at the Galapagos hydrothermal vents. *Science*, **207**, 1345–7.

KARL, D. M., WIRSEN, C. O. & JANNASCH, H. W. (1986). Microbiology of high temperature hydrothermal vents. *American Society for Microbiology, Abstracts of the Annual Meeting*, p. 253.

KENNICUTT, M. C., BROOKS, J. M., BIDIGARD, R. R., FAY, R. R., WADE, T. L. & McDONALD, T. J. (1985). Vent-type taxa in a hydrocarbon seep region on the Louisiana slope. *Nature*, **317**, 351–3.

KLINKHAMMER, G., BENDER, M. & WEISS, R. F. (1977). Hydrothermal manganese in the Galapagos Rift. *Nature*, **269**, 319–20.

KULM, L. D., SUESS, E., MOORE, J. C., CARSON, B., LEWIS, B. T., RITGER, S. D., KADKO, D. C., THORNBURG, T. M., EMBLEY, R. W., RUGH, W. D., MASSOTH, G. J., LANGSETH, M. G., COCHRANE, G. R. & SCAMMAN, R. L. (1986). Oregon subduction zone: venting, fauna, and carbonates. *Science*, **231**, 561–6.

LANE, D. J., OLSEN, G. J., GIOVANNONI, S. R. & PACE, N. R. (1986). A new device for studying bacterial growth in submarine hydrothermal vent flows *in situ*. *American Society for Microbiology, Abstracts of the Annual Meeting*, p. 171.

LILLEY, M. D., BAROSS, J. A. & GORDON, L. I. (1983). Reduced gases and bacteria in hydrothermal fluids: Galapagos spreading center and 21 °N East Pacific Rise. In *Hydrothermal Processes at Seafloor Spreading Centers*, ed. P. A. Rona, K. Bostrom, L. Laubier & K. L. Smith, Jr, pp. 411–49. New York: Plenum Press.

LISTER, C. R. B. (1980). Heat flow and hydrothermal circulation. *Annual Review of Earth Planetary Science*, **8**, 95–117.

LIU, C. L., HART, N. & PECK, H. D., JR (1982). Inorganic pyrophosphate: energy source for sulfate-reducing bacteria of the genus *Desulfotomaculum*. *Science*, **217**, 363–4.

LONSDALE, P. (1977). Clustering of suspension-feeding macrobenthos near abyssal hydrothermal vents at oceanic spreading centers. *Deep-Sea Research*, **24**, 857–63.

LUPTON, J. E. & CRAIG, H. (1981). A major Helium-3 source at 15°S on the East Pacific Rise. *Science*, **214**, 13–18.

McCONACHY, T. F., BALLARD, R. D., MOTTL, M. J. & VON HERZEN, R. P. (1986). Geologic form and setting of a hydrothermal vent field at lat 10°56′N, East Pacific Rise: a detailed study using *Angus* and *Alvin*. *Geology*, **14**, 295–8.

MACDONALD, K. C., BECKER, K., SPEISS, F. N. & BALLARD, R. D. (1980). Hydrothermal heat flux of the 'black smoker' vents on the East Pacific Rise. *Earth and Planetary Science Letters*, **48**, 1–7.

MACDONALD, K. C. (1982). Mid-Ocean Ridges: Fine scale tectonic, volcanic and hydrothermal processes within the plate boundary zone. *Annual Review of Earth and Planetary Sciences*, **10**, 155–90.

McDUFF, R. E. & EDMOND, J. M. (1982). On the fate of sulfate during hydrothermal circulation at mid-ocean ridges. *Earth and Planetary Science Letters*, **57**, 117–32.

MINAGAWA, M. & WADA, E. (1984). Stepwise enrichment of ^{15}N along food chains: further evidence and the relation between ^{15}N and animal age. *Geochimica et Cosmochimica Acta*, **48**, 1135–40.

MORRIS, I., GLOVER, H. E., KAPLAN, W. A., KELLY, D. P. & WEIGHTMAN, A. L. (1985). Microbial activity in the Cariaco Trench. *Microbios*, **42**, 133–44.

OVERBECK, J. (1979). Dark CO_2 uptake – biochemical background and its relevance to *in situ* bacterial production. *Archiv für Hydrobiologie Beiheft Ergebnisse der Limnologie*, **12**, 38–47.

PARK, R. & EPSTEIN, S. (1960). Carbon isotope fractionation during photosynthesis. *Geochimica et Cosmochimica Acta*, **21**, 110–26.

PAULL, C. K., HECKER, B., COMEAU, R., FREEMAN-LYNDE, R. P., NEUMANN, C., CORSO, W. P., GOLUBIC, S., HOOK, J. E., SIKES, E. & CURRAY, J. (1984). Biological communities at the Florida Escarpment resemble hydrothermal vent taxa. *Science*, **226**, 965–7.

PAULL, C. K., JULL, A. J. T., TOOLIN, L. J. & LINICK, T. (1985). Stable isotope evidence for chemosynthesis in an abyssal seep community. *Nature*, **317**, 709–11.

POWELL, M. A. & SOMERO, G. N. (in press). Hydrogen sulfide oxidation is coupled to oxidative phosphorylation in mitochondria of *Solemya reidi*. *Science*.

RAU, G. H. (1981*a*). Hydrothermal vent clam and tube worm $^{13}C/^{12}C$: further evidence of nonphotosynthetic food sources. *Science*, **213**, 338–40.

RAU, G. H. (1981*b*). Low $^{15}N/^{14}N$ in hydrothermal vent animals: ecological implications. *Nature*, **289**, 484–5.

RAU, G. H. (1985). $^{13}C/^{12}C$ and $^{15}N/^{14}N$ in hydrothermal vent organisms: ecological and biogeochemical implications. *Biological Society of Washington Bulletin*, No. 6, 243–7.

RAU, G. H. & HEDGES, J. I. (1979). Carbon-13 depletion in a hydrothermal vent mussel: suggestion of a chemosynthetic food source. *Science*, **203**, 648–9.

ROMANENKO, V. I. (1964). Heterotrophic assimilation of CO_2 by bacterial flora of water. *Microbiology*, **33**, 679–83.

RUBY, E. G. & JANNASCH, H. W. (1982). Physiological characteristics of *Thiomicrospira* sp. strain L-12 isolated from deep-sea hydrothermal vents. *Journal of Bacteriology*, **149**, 161–5.

RUBY, E. G., WIRSEN, C. O. & JANNASCH, H. W. (1981). Chemolithotrophic sulfur-oxidizing bacteria from the Galapagos Rift hydrothermal vents. *Applied and Environmental Microbiology*, **42**, 317–24.

SACKETT, W. M., BROOKS, J. M., BERNARD, B. B., SCHWAB, C. R., CHUNG, H. & PARKER, R. A. (1979). A carbon inventory for Orca Basin brines and sediments. *Earth and Planetary Science Letters*, **44**, 73–81.

SCHLEGEL, H. G. & JANNASCH, H. W. (1967). Enrichment cultures. *Annual Review of Microbiology*, **21**, 49–70.

SHOKES, R. F., TRABANT, P. K., PRESLEY, B. J. & REID, D. F. (1977). Anoxic, hypersaline basin in the northern Gulf of Mexico. *Science*, **196**, 1443–6.

SIMONEIT, B. R. T., MAZUREK, M. A., BRENNER, S., CRISP, P. T. & KAPLAN, I. R. (1979). Organic geochemistry of recent sediments from Guaymas Basin, Gulf of California. *Deep-Sea Research*, **26**. 879–91.

SPIESS, F. N., MACDONALD, K. C., ATWATER, T., BALLARD, R., CARRANZA, A., CORDOBA, D., COX, C., DIAZ GARCIA, V. M., FRANCHETEAU, J., GUERRERO, J., HAWKINS, J., HAYMON, R., HESSLER, R., JUTEAU, T., KASTNER, M., LARSON, R., LUYENDYK, B., MACDOUGALL, J. D., MILLER, S., NORMARK, W., ORCUTT, J. & RANGIN, C. (1980). East Pacific Rise: hot springs and geophysical experiments. *Science*, **207**, 1421–33.

SUESS, E. (1980). Particulate organic carbon flux in the oceans – surface productivity and oxygen utilization. *Nature*, **288**, 260–3.

SUESS, E., CARSON, B., RITGER, S. D., MOORE, J. C., JONES, M. L., KULM, L. D. & COCHRANE, G. R. (1985). Biological communities at vent sites along the subduction zone off Oregon. In *The Hydrothermal Vents of the Eastern Pacific: An Overview – Bulletin of the Biological Society of Washington*, No. 6, 475–84.

SWINBANKS, D. (1985*a*). Japan finds clams and trouble. *Nature*, **315**, 624.

SWINBANKS, D. (1985*b*). New find near Japan's coast. *Nature*, **316**, 475.

TRENT, J. D., CHASTAIN, R. A. & YAYANOS, A. A. (1984). Possible artefactual basis for apparent bacterial growth at 250 °C. *Nature*, **307**, 737–40.

TUNNICLIFFE, V., BOTROS, M., DE BURGH, M. E., DINET, A., JOHNSON, H. P., JUNIPER, S. K. & McDUFF, R. E. (1986). Hydrothermal vents of Explorer Ridge, northeast Pacific. *Deep-Sea Research*, **33**, 401–12.

TUTTLE, J. H. (1985). The role of sulfur-oxidizing bacteria at deep-sea hydrothermal vents. In *The Hydrothermal Vents of the Eastern Pacific: An Overview – Bulletin of the Biological Society of Washington*, No. 6, 335–43.

TUTTLE, J. H. & JANNASCH, H. W. (1973). Sulfide and thiosulfate-oxidizing bacteria in anoxic marine basins. *Marine Biology*, **20**, 64–70.

TUTTLE, J. H. & JANNASCH, H. W. (1977). Thiosulfate stimulation of microbial dark assimilation of carbon dioxide in shallow marine waters. *Microbial Ecology*, **4**, 9–25.

TUTTLE, J. H. & JANNASCH, H. W. (1979). Microbial dark assimilation of CO_2 in the Cariaco Trench. *Limnology and Oceanography*, **24**, 746–53.

TUTTLE, J. H., WIRSEN, C. O. & JANNASCH, H. W. (1983). Microbial activities in the emitted hydrothermal waters of the Galapagos Rift vents. *Marine Biology*, **73**, 293–9.

VAN ANDEL, T. (1981). *Science at Sea: Tales of an Old Ocean*. San Francisco, Freeman.

VON DAMM, K. L., EDMOND, J. M., GRANT, B., MEASURES, C. I., WALDEN, B & WEISS, R. F. (1985). Chemistry of submarine hydrothermal solutions at 21 °N, East Pacific Rise. *Geochimica et Cosmochimica Acta*, **49**, 2197–220.

WEINER, R. M., DEVINE, R. A., POWELL, D. M., DAGASAN, L. & MOORE, R. L. (1985). *Hyphomonas oceantitis* sp. nov., *Hyphomonas hirschiana* sp. nov., and *Hyphomonas jannaschiana* sp. nov. *International Journal of Systematic Bacteriology*, **35**, 237–43.

WEISS, R. F., LONSDALE, P., LUPTON, J. E., BAINBRIDGE, A. E. & CRAIG, H. (1977). Hydrothermal plumes in the Galapagos Rift. *Nature*, **267**, 600–3.

WELHAN, J. A. & CRAIG, H. (1979). Methane and hydrogen in East Pacific Rise hydrothermal fluids. *Geophysical Research Letters*, **6**, 829–31.

WELHAN, J. A. & CRAIG, H. (1983). Methane, hydrogen and helium in hydrothermal fluids at 21 °N on the East Pacific Rise. In *Hydrothermal Processes at Seafloor Spreading Centers*, ed. P. A. Rona, K. Bostrom, L. Laubier & K. L. Smith, Jr, pp. 391–409. New York: Plenum Press.

WHITE, R. H. (1984). Hydrolytic stability of biomolecules at high temperatures and its implication for life at 250 °C. *Nature*, **310**, 430–2.

WHITICAR, M. J., FABER, E. & SCHOELL, M. (1986). Biogenic methane formation in marine and freshwater environments: CO_2 reduction vs. acetate fermentation – Isotope evidence. *Geochimica et Cosmochimica Acta*, **50**, 693–709.

WHITTENBURY, R. & KELLY, D. P. (1977). Autotrophy: a conceptual phoenix. *Symposium, Society for General Microbiology*, **27**, 121–49.

WILLIAMS, D. L., VON HERZEN, R. P., SCLATER, J. G. & ANDERSON, R. N. (1974). The Galapagos spreading center: lithospheric cooling and hydrothermal circulation. *Geophysical Journal of the Royal Astronomical Society*, **38**, 587–608.

WILLIAMS, P. M., SMITH, K. L., DRUFFEL, E. M. & LINICK, T. W. (1981). Dietary carbon sources of mussels and tubeworms from Galapagos hydrothermal vents determined from tissue ^{14}C activity. *Nature*, **292**, 448–9.

WILLIAMS, S. T., GOODFELLOW, M. & VICKERS, J. C. (1984). New microbes from old habitats? In *The Microbe 1984. Part II, Prokaryotes and Eukaryotes. Symposium of the Society for General Microbiology*, vol. 36, ed. D. P. Kelly & N. G. Carr, pp. 219–56. Cambridge: Cambridge University Press.

WINN, C. D. & KARL, D. M. (1985). Microbiology of geographically distinct hydrothermal systems: Galapagos Rift, East Pacific Rise and Juan de Fuca Ridge. *American Society for Microbiology, Abstracts of the Annual Meeting*, p. 228.

WINN, C. D., KARL, D. M. & MASSOTH, G. J. (1986). Microorganisms in deep-sea hydrothermal plumes. *Nature*, **320**, 744–6.

WIRSEN, C. O., TUTTLE, J. H. & JANNASCH, H. W. (in press). Activities of sulfur-oxidizing bacteria at the 21 °N East Pacific Rise vent site. *Marine Biology*.

BIOFILMS: MICROBIAL INTERACTIONS AND METABOLIC ACTIVITIES

W. ALLAN HAMILTON

Department of Genetics and Microbiology, Marischal College, University of Aberdeen, Aberdeen AB9 1AS, UK

INTRODUCTION

Concepts and methods

As befits an experimental science, conceptual advances in microbiology have been dependent upon the development of novel practical techniques. Equally, however, the extension and further development of these concepts have often been restricted by the limitations inherent in that methodology. One can cite the petri dish and the chemostat in support of this thesis.

The invention of the petri dish and nutrient media – solidified at first with gelatin and later with agar – formed the very foundation of microbiology as an experimental science. From this technique stems our understanding of the chemical and physical characteristics and of the metabolic activities and potential of individual species of microorganisms. Still today it constitutes the critical first step in the identification and isolation of novel organisms genetically engineered by man, if not in his own image then at least with the hope of his further advancement. This pure culture technique has not only been a cornerstone of studies in microbial physiology and genetics. It has also found extensive use in microbial ecology where it has, at least in the case of the bacteria, been essentially the only method available for the identification of the particular organisms, and their unique activities, present in a given ecosystem. For some hundred years, however, the great power of the pure culture technique has served to limit our appreciation that the very essence of microbial ecology and the root of the massive environmental and economic impact of microorganisms in both natural and manmade ecosystems lies in the interactions and interdependencies of communities of organisms in mixed culture, or microbial consortia as they are called.

With the chemostat came, if not a revolution, at least a greatly heightened awareness of the potential of individual species and of the mechanisms for the expression and control of that potential. With its capacity for the maintenance of steady state conditions, low growth rates and limitation by the experimenter's choice of nutrient, the chemostat greatly extended the power of the physiologist to probe the integration of individual biochemical reactions in the phenomenon of cellular growth, and of the ecologists to develop predictive experimental models with respect to specific environmental parameters. Although the chemostat has found its principal use in the analysis of the growth of individual species in pure culture, it has also been of value in elucidating features of the positive interactions among complex communities of microorganisms (see, for example, Slater & Somerville, 1979). The chemostat, like the petri dish, also has a severe limitation, however, which greatly reduces its value as an experimental model for the study of microbial ecosystems. Both in theory and in practice the chemostat demands that the component organisms should be in homogeneous suspension, a condition which in fact applies to only a very restricted range of naturally occurring ecosystems.

Structure and function

The metabolic activity of microorganisms is of central importance to the experimenter, whether he be taxonomist, physiologist or ecologist. It is the extent and diversity of microbial metabolism which both characterises individual species and underlies the enormous capacity of microorganisms to modify their immediate environment. The basis of metabolism is, of course, enzymic activity and it was the solubilisation of the enzymes of sugar fermentation from yeast which gave birth to the sister subject of biochemistry. Although the concept of cellular metabolism being simply the expression of the activities of a 'bag of enzymes' has long since been formally laid to rest, it is still true that implicit in much of both our thinking and our experimental analysis is the idea that the whole is merely the sum of the parts and that the individual parts can be fully understood by isolating the enzyme/cell in aqueous solution/suspension and assaying its activity. This approach is dangerously simplistic, however, and we find in biochemistry at the present time increasing reference to the gel-like state of the cytoplasm and the importance of multi-enzyme complexes, cellular ultrastructure and local concentrations of substrates, cofactors and allosteric effectors. Perhaps the most telling illustration of the limitations of 'solution biochemistry' is to be found in bioenergetics and the mechanism of oxidative phosphorylation. The search for the supposed chemical high-energy intermediate led inexorably to ever more complex experimental artefacts, contradictory hypotheses and general confusion. It was only with the appreciation of the true significance of the localisation of the component reaction sequences within biological membranes and the development of the relevant experimental techniques that the revolutionary chemiosmotic theory could be evolved (see, for example, Harold, 1972).

In the same manner that the integrated functioning of the eukaryotic cell is intimately dependent upon its structural organisation, it seemed reasonable to consider whether the concerted activity of a mixed microbial consortium might also show features equivalent to the group transfer and group translocation coupling reactions which orchestrate the energetics of individual cells. It appears that such a structured approach to the functioning of microbial consortia is valid and offers a deeper understanding of the processes involved (Hamilton, 1984). A similar analysis was presented by Zeikus (1983) in the previous Society for General Microbiology Symposium on 'Microbes in their Natural Environments'. In that Symposium a number of other contributions also stressed the importance to microbial ecology of structure in general, and of surfaces in particular (Wardell, Brown & Flannigan, 1983; White, 1983; Wimpenny, Lovitt & Coombs, 1983). This question of physical heterogeneity within microbial ecosystems will be explored further in a number of papers in this present Symposium, and in this particular contribution I shall focus my attention on the growth and activity of surface-associated consortia, or biofilms, and examine the role of structural heterogeneities on the physiological homogeneity that is so much a feature of such mixed microbial communities. In order to do this it will be necessary to describe the general features of biofilms including their natural occurrence, to discuss the methodology that is being developed for their study and the assay of their biological activities, and to indicate the findings that are extending our understanding of these systems. This last I shall do with particular reference to microbially induced corrosion of metals.

Surfaces

The term biofilm is used to define the discrete aggregation of organisms, generally microorganisms, and their metabolic products at an interface. While the neuston at the air/water interface is an example of such a biofilm, most usually biofilms refer to those depositions of organisms on a solid surface within an aqueous phase. Examples of such biofilms are legion: on the surface of stones and particulate matter in dilute aqueous environments; on soil and sediment particles; in trickling filters; on leaves, roots and germinating seeds of plants; in dental plaque; on intestinal and rumen epithelial tissues in animals; on medical prostheses such as catheters, artificial joints and pace makers; on ships' hulls and the external surfaces of offshore oil production platforms; on internal surfaces of chemical process equipment; in film fermenters (Atkinson & Fowler, 1974; Wimpenny, Lovitt & Coombs, 1983; Costerton, Marrie & Cheng, 1985; Hamilton, 1985). The activities of such biofilms are of fundamental importance in nature, where Costerton & Geesey (1979) have shown that the number of organisms per cm^2 of surface normally exceeds that in $1\,cm^3$ of flowing water by a factor of at least 200. They can also be of direct practical significance, positive and negative, to man: effluent treatment and biotechnological reactors; decreased heat transfer and fluid flow performance; corrosion and general biodeterioration of equipment; dental caries.

There are two striking features of biofilms which fundamentally influence both our experimental approach to their study and our understanding of the significance of our findings: (1) biofilms exist at surfaces; and (2) they are generally composed of a mixed community of organisms and their metabolic products, predominantly extracellular polysaccharides. Although overlapping, these characteristics circumscribe two quite separate conceptual and experimental approaches to the study of biofilms. In the first instance the focus is on the attachment of organisms to the surface, with reference to the nature of the surface, concentration of nutrients, and the relative metabolic activities of attached (sessile) and free (planktonic) organisms. In the second instance the direct role of the surface *per se* is greatly diminished. While the above considerations clearly apply during the initial stages of development of the biofilm, the activities of the mature biofilms owe more to the physicochemical and biological heterogeneity of the biofilm itself, with the surface being reduced to, quite literally, a supporting role. In this respect studies of biofilms have more in common with the theoretical and practical processes relating to research into the behaviour and properties of mixed microbial consortia than to the equivalent techniques required to examine the attachment to surfaces and growth as a cellular monolayer. Although this chapter will be principally concerned with mature, or at least complex, biofilms, it is necessary to consider first at least the major determinants of their initiation and development; a subject which has been extensively covered in a number of recent review articles and books (Costerton, Irving & Cheng, 1981; Fletcher & Marshall, 1982*a*; Characklis & Cooksey, 1983; Wardell, Brown & Flannigan, 1983; Wimpenny, Lovitt & Coombs, 1983; Marshall, 1984).

CELL SURFACE INTERACTIONS

Attachment

The process of cell attachment and biofilm formation can be considered as taking place in a number of discrete phases. (1) A so-called conditioning film is formed by the adsorption of organic molecules, commonly protein, to the surface. (2) This is followed by the reversible adsorption of bacterial cells to the conditioned surface. The time scale of these two processes is measurable in minutes rather than hours. (3) Thereafter the cells become irreversibly bound to the surface, largely through the synthesis of extracellular polymeric material, predominantly polysaccharide.

Both the nature of the surface itself and the identity and history of the bacterial species involved can have a major influence on the extent of these initial adsorption stages. For example, Fletcher & Marshall (1982*b*) have related the adhesion of bacteria to plastic surfaces modified to give a range of wettability, as measured by contact angles. Surface charge is another important parameter (Fletcher & Loeb, 1979). With metal surfaces both the nature of the metal and its electrochemical characteristics are significant. Ford, Walch & Mitchell (in press) have shown that a thermophilic *Thermus* sp. demonstrates a preference for attachment to stainless steel and titanium in contrast to aluminium and copper alloys. The attachment of two marine bacteria to copper surfaces could however be enhanced by polarizing the metal cathodically (Gordon, Gerchakov & Udey, 1981). A number of workers have observed that a high level of attachment to surfaces tends to be a property of bacteria in nature and that this characteristic is greatly diminished after maintenance of a culture under laboratory conditions (Costerton, Irvin & Cheng,

1981). This phenomenon is considered to result from the decreased synthesis of extracellular polysaccharide by laboratory strains. A similar explanation lies behind the observed increase in the rate of adsorption of bacteria after chemostat growth at low dilution rates as compared to batch-grown organisms (Robinson, Trulear & Characklis, 1984; Nelson, Robinson & Characklis, 1985; Dowling, Guezennec & White, in press). The nature of nutrient limitation also appears to be important and Brown, Ellwood & Hunter (1977) have shown that the attachment of cells from a river inoculum onto aluminium foil within a chemostat vessel was slight under nitrogen limitation with glucose in excess, despite evidence of the synthesis of polysaccharide material. Under carbon limitation, on the other hand, they found the development of an extensive mixed population on the surface. Douglas and her colleagues have also shown that the adherence of *Candida* spp. to buccal epithelial cells or acrylic surfaces is severely affected by the sugar used as carbon source for their growth in batch culture (McCourtie & Douglas, 1981; Critchley & Douglas, 1985).

External factors relating to natural conditions or experimental design, for example flow velocity within the liquid phase and surface sheer stress (Characklis & Cooksey, 1983; Duddridge, Kent & Law, 1982), can also markedly affect attachment rates, but such considerations are beyond the province of this paper.

Metabolic activities

The initial rate of cell attachment is a function of the bacterial concentration in the bulk phase (Fletcher, 1977; Characklis, 1981) with the spatial pattern or attachment considered to be random (Characklis, 1984; Nelson, Robinson & Characklis, 1985). Thereafter the further development of the biofilm is essentially nutrient-driven through growth. In considering this point, it is also logical to discuss the relative activities of free and attached organisms and the proposed advantage to the cells of becoming attached in the first place.

The generally held view is that organisms are attracted to surfaces in response to their capacity to concentrate organic materials and so to act as a nutrient sink. This idea stems from the original observation by Zobell (1943) that the growth of bacteria in an oligotrophic environment could be stimulated by the inclusion of a glass surface. In fact the data from an extensive range of experimental studies make it quite clear that it is not possible to draw a simple conclusion regarding the effects of the attachment to a surface on the metabolic and growth activities of bacterial cells. Atkinson & Fowler (1974) reported that the growth of *Escherichia coli* was improved after surface adsorption, but only at nutrient (glucose) concentrations less than 25 ppm. Jeffrey & Paul (1986a) quoted a range of studies, the balance of which also showed increased metabolic activities of surface-associated bacteria at low or zero concentrations of exogenous nutrients. Their own findings, on the other hand, were that free-living cells were more active than those attached in the presence of added nutrients, while the reverse was true with cells metabolising their endogenous reserves, and they suggested that this effect might result from the attached cells having decreased area available for the uptake of exogenous nutrients. Other examples of surfaces stimulating metabolic activity are: a two-fold increase in the rate of growth of a *Pseudomonas* sp. (Ellwood *et al.*, 1982); increases in the growth rate and yield of *E. coli* and of the growth and sporulation rates of *Bacillus subtilis* after adsorption to an anion exchange resin (Hattori & Hattori, 1981); stimulation by clay particles of the rate of ethanol degradation by starved cultures of sulphate-reducing bacteria (Laanbroek & Geerligs, 1983); and a 20–25% increase in the nitrite oxidation rates of a *Nitrobacter* sp. after attachment to glass or an anion exchange resin (Keen, 1984).

On the other hand, adsorption of a marine bacterium to hydroxyapatite gave no stimulation of activity, and respiration was in fact decreased (Gordon, Gerchakov & Millero, 1983). Bright & Fletcher (1983a,b) studied the effects of adsorption of a marine *Pseudomonas* sp. to a range of plastic surfaces and found that the nature of the effect depended upon the substratum, the amino acid and its concentration, and the parameter being measured. Assimilation of amino acid by surface-associated cells was generally greater than, and respiration less than, that by free-living bacteria. In a detailed study of the growth of *Ps. aeruginosa* in steady-state biofilms at various substrate (glucose) loading rates and reactor dilution rates, Bakke *et al* (1984) concluded that the yield and rate coefficients were the same as those in suspended cultures, provided that the film was sufficiently thin ($\leqslant 50\,\mu\mathrm{m}$) to cause negligible diffusional resistance to substrate penetration and uptake. *Nitrosomonas europaea* however, did show a reduction in specific growth rate after becoming attached to glass slides (Powell, 1985).

If the premise that surfaces serve to concentrate nutrients has any validity then one would expect it to confer some advantage

on copiotrophic species that have need of relatively high concentrations of nutrients, rather than on oligotrophs which can grow at
extremely low nutrient availability. Novitsky & Morita (1976, 1977,
1978), Kjelleberg, Humphrey & Marshall (1982, 1983) and Kjelleberg (1984) have studied a particular type of copiotroph which shows
a pattern of cell fragmentation when nutrient supply is decreased,
eventually giving rise to small cocci which may survive for long periods. These cocci can also attach to surfaces and grow on the nutrients
accumulated there. Kepkay *et al.* (1986) have recently developed
a microelectrode technique that allows the increase of metabolic
activity to be measured as these resting coccal forms grow into normal
rod-shaped cells. These authors have further shown that this survival
strategy may also be used by metal-oxidising autrophs.

Thereafter, cellular processes will develop. Consideration of the
complex interactions resulting from the time-dependent absorption
and growth of multiple microbial species, and their containment
within the particular environment created by their own extracellular
polymers, now leads us from discussion of cell-surface attachment
to the study of biofilms *sensu stricto*.

BIOFILMS

In the remainder of this chapter I should like first to discuss the
essential characteristics of biofilms in general terms and, based on
this, to present a conceptual model which can serve to direct the
experimental study of biofilms. I shall then consider certain selected
biofilm systems in greater detail, describing specific methods that
have been developed and discussing the results obtained, with reference to the conceptual model and, where relevant, in terms of their
practical significance.

Glycocalyx

Perhaps the single most striking feature of biofilms is their high
content (50–90%) of extracellular polymer substances (EPS) (Characklis & Cooksey, 1983). This is predominantly polysaccharide and
J. W. Costerton has given it the name glycocalyx, which accurately
implies its essential role as an integral structural and functional component of the biofilm. The glycocalyx is responsible for the irreversible binding of the primary colonising bacteria to the substratum

surface, and also for the binding of primary and later colonisers to each other. The great majority of organisms within a mature biofilm are fixed within the three-dimensional domain of the glycocalyx rather than attached to a two-dimensional surface. In terms of the metabolic activities of these cells therefore, many of the points referring to cell-surface interactions are not strictly relevant.

It has been claimed that the glycocalyx is largely composed of mannans, glucans and uronic acids (Costerton & Geesey, 1979), but Characklis & Cooksey (1983) suggest that the situation may be a great deal more complex with glycoproteins and other heterocopolymers possibly being involved. Uhlinger and White (1983) have studied synthesis of the glycocalyx in the marine bacterium *Pseudomonas atlantica* and have shown that it is stimulated both by the presence of sand (increased surface area) and by the addition of galactose, and that its composition changes during the growth cycle in batch culture. In a similar vein, Christensen, Kjosbakken & Smidsrod (1985) reported that another marine *Pseudomonas* sp. produced more than one polymer, but at different rates and at different stages of the growth cycle.

The polyanionic nature of the glycocalyx dictates that it functions as an ion exchange matrix, serving to concentrate organic nutrients and at the same time to limit the penetration of charged molecules such as cationic biocides (Costerton & Lashen, 1984; Sharma, Battersby & Stewart, in press). As Costerton has pointed out, the glycocalyx also serves to protect cells within a biofilm from other environmental hazards such as surfactants, antibodies and phagocytic amoebae (Costerton, Marrie & Cheng, 1985).

Biofilms may also contain significant amounts of inorganic material adsorbed as silt, clay or general detritus from the environment, or in the form of corrosion products formed from the action of the constitutent bacteria and retained within the glycocalyx (Characklis & Cooksey, 1983; Wimpenny, Lovitt & Coombs, 1983). In a parallel manner, organic acids are retained within dental plaque and so stimulate the onset and progress of caries (Ruseska *et al.*, 1982).

Other less clearly defined roles suggested for the glycocalyx component of biofilms are as electron sink, energy store, and site of increased intracellular communication and genetic transfer (Characklis & Cooksey, 1983; Sharma, Battersby & Stewart, in press).

The thickness of biofilms depends both on environmental factors and on the organisms involved. Bacterial films may grow to a thickness of 100–200 μm. For example, it has been reported that biofilms in an oil-field pipeline were up to 150 μm thick and contained 5×10^7 cells per cm^{-2}. W. G. Characklis (personal communication) has obtained films up to 50 μm in an experimental system with pure cultures of *Pseudomonas aeruginosa*, but in excess of 120 μm with mixed cultures. Similar dimensions of 70–100 μm have also been estimated by Atkinson & Fowler (1974) for 'active' film thickness as measured by oxygen tension and substrate utilisation. Interestingly, it appears that the thickness of a mature biofilm is independent of whether the substratum is steel, wood, rock or plastic (Costerton, Marrie & Cheng, 1985). Larger-dimension biofilms, up to say 100 mm, generally contain a more complex mixture of microorganisms such as, for example, an algal/bacterial mat formed in a slow-moving nutrient-rich stream. Marine biofilms containing higher organisms as well as bacteria may be centimetres or even metres thick, but although still biofilms, these will not concern us directly in the present context.

Temporal heterogeneity

The time course of the development of biofilms is also responsive to environmental variables, principally nutrient availability. In laboratory simulations using natural or artificial fresh or sea water irrigation, biofilm development is observed to plateau after 10–15 days, and experiments requiring mature biofilms are generally carried out after 3 weeks. With a richer supply of nutrients, however, it is possible to obtain maximum biofilm development within 1–3 days. As has been summarised by Characklis & Cooksey (1983), the initial phase of microbial attachment and growth involves copiotrophic species such as *Pseudomonas* spp. (copious polysaccharide producers), followed by various oligotrophs. It is important to appreciate however, that biofilm development does not only involve phases of attachment, growth and polysaccharide production, but also a phase of detachment. This last may be a normal feature of a dynamic equilibrium resulting, for example, from nutrient or oxygen depletion arising within the film (Howell & Atkinson, 1976), or it may arise from a specific experimental manipulation. Characklis & Cooksey (1983) reported a particularly intriguing situation where increasing the nutrient supply to a pure culture biofilm of *Pseudomonas aeruginosa* resulted in the immediate loss of a significant proportion of the biofilm. As no decrease in cell numbers was detected, it was proposed that the material lost was extracellular polymer. Clearly, therefore, biofilms demonstrate a degree of temporal heterogeneity, with regard to both the numbers and species of organisms present, and the amount and chemical composition of the attendant glycocalyx polymers.

Structural heterogeneity

The spatial organisation of biofilms is also subject to variation. Characklis & Cooksey (1983) and Characklis (1984) have stressed the likely occurrence of 'patchiness' in biofilms due to uneven distribution of cells and/or polymer across the substratum surface. With its effect of creating concentration cells this can have a significant influence on the microbially induced corrosion of metals (Hamilton, 1985; Nivens et al., 1986) or on hydrogen uptake with the consequential embrittlement or blistering of steels (Walch & Mitchell, in press).

More clearly defined, however, and of more direct significance to the biological activities of biofilms are spatial variations in the vertical dimension. In response to nutrient supply a biofilm will grow with an increase in both cell numbers and extracellular polymer. Substrate utilisation by the developing biofilm increases proportionally with biofilm thickness up to a critical point where nutrient diffusion through the biofilm becomes limiting (Characklis, 1984). This situation has been subjected to detailed theoretical analysis by Pirt (1973). Of particular importance amongst such nutrient gradients through the thickness of the biofilm is the oxygen gradient, which can lead to anoxic and reducing conditions being established within the biofilm and close to the substratum surface. Various estimates are available as to the thickness of biofilm necessary to demonstrate this effect. Cox, Bazin & Gull (1980) suggested that oxygen became limiting for the activity of nitrifying bacteria on the surfaces of glass beads at a film thickness of $8.8\,\mu m$. In studies of dental plaque, Coulter & Russell (1976) concluded that $12\,\mu m$ was sufficient to give reduced levels of Eh at the colonized surface. It is of the greatest significance that these conditions can lead to the growth and activity of anaerobic species within the depths of a biofilm, even though it may itself be sited in a highly aerobic bulk environment (Gibbons & van Haute, 1975; Hamilton, 1985; Paerl, 1985).

That is to say, the structure of biofilms shows a marked degree of heterogeneity which relates both to the physical organisation of the individual cellular and chemical components and to the array of biochemical activities consequent upon this organisation. An appreciation of these temporal and structural heterogeneities of biofilms is fundamental to our understanding of their biological functioning with its often major ecological and economic implications. To return to my statement at the beginning of this chapter, the extension and full development of this concept will depend totally on our ability to generate the appropriate experimental approaches and methodology.

Functional homogeneity

As a counter to the heterogeneities of biofilms it is germane to point out that biofilms demonstrate significant homogeneity of function. The individual organisms do not operate in isolation but rather show significant interaction and even interdependency. This is particularly marked in the case of anaerobic organisms. The generation of reducing conditions within the biofilm is dependent upon aerobic and facultative species utilising oxygen at a faster rate than it can diffuse to the underlying regions of the biofilm. There is also an element of nutrient succession in such interrelationships as is evident, for example in dental caries or corrosion. Wimpenny and colleagues (1983) have summarised the succession of organisms and climax community associated with dental plaque formation, with first aerobic species capable of growth on salivary glycoproteins, followed by facultative and anaerobic organisms utilising the metabolic products of the primary colonisers and producing the deleterious organic acids as their own fermentation products. With biofilms causing corrosion in oil-carrying pipelines, the hydrocarbon-degrading bacteria are required both to create the anaerobic conditions and to produce the necessary nutrients for the sulphate-reducing bacteria which are the principal causative organisms of anaerobic corrosion (Hamilton, 1985).

These examples illustrate a point of the greatest importance in the study of biofilms. The biofilm community is capable of activities other than those only of the individual organisms, and these activities are dependent on the structural integrity of the biofilm *in toto*.

It is these characteristics which have led Costerton to suggest that a biofilm might be considered as a 'quasi-tissue' with measurable rates of respiration and nutrient uptake (Costerton, Marrie & Cheng, 1985). These authors discuss the rumen in these terms and point out that the microbial population is predominantly in the form of a biofilm associated with the epithelial tissue of the rumen wall.

The biofilm contains some twenty-three bacterial species and possesses broad proteolytic and urolytic activities.

In this regard biofilms have a closer similarity with mixed microbial consortia than with surface-attached organisms which were the subject matter of the earlier parts of this chapter. The cynobacterial mats, for example, (see Ward, this Symposium), could be legitimately considered under either 'Biofilms' or 'Microbial Consortia'.

Jeffrey & Paul (1986b) for example, have noted that complex populations (including diatoms) which attached to polystyrene petri dishes from an estuarine environment showed higher activities than the free organisms. They put this down to cross-feeding and the development of suitable microenvironments within the adherent biofilm. Similarly, Murray, Cooksey and Priscu (in press) working with model films containing both a diatom and a bacterium have shown higher rates of metabolic activity with the combined community than with either organisms singly. Most notably, after 70 hours of constant illumination [^{3}H]-thymidine incorporation (generally considered a measure of bacterial activity) was 16-fold greater than with bacteria alone. Haack and McFeters (1982) extend our understanding of the reactions involved in this type of situation from their studies of natural algal/bacterial biofilms in nutrient-deficient stream water. There appears to be a direct flux of soluble algal products to the bacteria, with little or no heterotrophic utilisation of dissolved organic material. Phototrophic productivity and bacterial utilisation of algal products both peak at approximately the same time of year. Activity of the diatom-dominated algal population declined as silica concentrations in the stream water dropped, leading to a situation in which the sessile bacteria were substrate-limited.

Biofilm models and non-destructive methodology

The factual information presented in the preceding sections can be summarised with the help of a diagram (Fig. 1). The biofilm represented in this instance would be one occurring naturally within a stream and involving algal and bacterial components. The diagram suggests both the physical and chemical organisation of the biofilm and its physiological interactions. In particular it stresses the integration of these separate factors in such a manner that the biofilm as a whole has a clearly identifiable biological activity that is more than simply the sum of the activities of the constituent organisms.

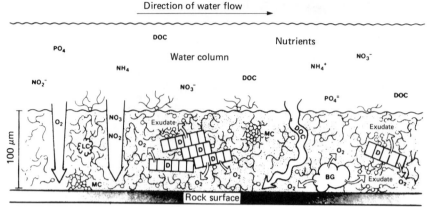

Fig. 1. Diagrammatic representation of a natural adherent biofilm in which bacteria (open circles) live within a continuous matrix of exopolysaccharide made by themselves and by their algal symbionts. The diagram speculates concerning processes within this microbial biofilm, where diatoms and blue-green bacteria (cyanobacteria) are physiologically integrated with the adherent bacteria. BG, blue-green bacteria; D, diatoms; DOC, dissolved organic carbon; LC, lysed cyanobacteria; MC, microcolony. (From Costerton, Marrie & Cheng (1985), with permission.)

Yet it is totally dependent on these activities and their expression as dictated by the physicochemical characteristics of individual microenvironments within the biofilm.

Stemming directly from this last point, we see clearly defined the need for non-destructive assay procedures which are capable of measuring activities representative of the undisturbed biofilm with its physical, chemical and biological interactions intact. Further to this, the inherent limitations of conventional techniques, such as enrichment culture when applied to complex ecosystems such as consortia or biofilms (see Parkes and Ward, this Symposium), must also be taken into consideration.

A key figure in both the projection of these ideas and the development of the appropriate methodology has been David White (1983). In an extended series of publications, White and his colleagues have stressed the value of the use of chemical signatures in obtaining valid measurements of the biomass and metabolic activities of microbial ecosystems. Principal amongst these has been the identification of species-specific fatty acids from the membrane phospholipid, a technique that has been used to good effect by John Parkes (Taylor & Parkes, 1985; Parkes, this Symposium). More recently the repertoire has been extended with the development of Fourier transform–infrared spectroscopy (Nichols *et al.*, 1985). Examples of the power of these techniques in the present context are the identification of the quantitative importance of genera of sulphate-reducing bacteria other than the classical lactate-utilising *Desulfovibrio* in marine sediments and corroding steel (Taylor & Parkes, 1985; Dowling, Guezennec & White, in press), and the demonstration of facilitation of the corrosion of stainless steel by deposition of extracellular polymers from a marine bacterium (Nivens *et al.*, 1986).

Other techniques that have been widely applied in dissecting the chemical and biological activities of microbial consortia and biofilms involve the use of microprobes and radiotracers, often coupled with the application of specific inhibitors. Microelectrodes have been used to determine values of pH, and oxygen and sulphide concentrations within sediment and biofilm communities (Jørgensen, 1982; Revsbech & Ward, 1983). Paerl (1985) has discussed the use of tetrazolium dyes to gain a measure of Eh profiles. Jørgensen and his colleagues (1978, 1982) have used [35S]-sulphate turnover to [35S]-sulphide to assay *in situ* activity of sulphate-reducing bacteria in sediments, while addition of the inhibitor molybdate has allowed them to identify hydrogen, acetate, propionate and butyrate as the major energy substrates of these bacteria in nature (Sørensen, Christensen & Jørgensen, 1981).

In our own studies, which have focussed on the role of the sulphate-reducing bacteria in anaerobic corrosion as experienced, for example, in the offshore oil industry (Hamilton, 1983), we have presented a less generalised model defined within narrower environmental limits (Hamilton, 1985) and have modified the [35S] radiotracer method so that it can more readily be applied in field situations (Rosser & Hamilton, 1983; Maxwell & Hamilton, in press *a*).

The biofilm model illustrated in Fig. 2 concentrates on the central importance of the sulphate-reducing bacteria. On the one hand, they are dependent upon other organisms to generate both the nutrient and the physicochemical conditions for their growth on hydrogen, acetate and related compounds in anaerobic microenvironments within the depths of the biofilm. On the other hand, they are the major causative organism responsible for the metal corrosion which can occur under these circumstances, with hydrogen oxidation and sulphide production likely to be key factors in the mechanism.

First with sediment samples (Rosser & Hamilton, 1983) and later with the inclusion of a metal coupon (Maxwell & Hamilton, in press, *a*), we have adapted the radiorespirometric assay so that the entire assay, including periods of both incubation and of equilibrium, can be carried out within a single test tube. This ensures that through

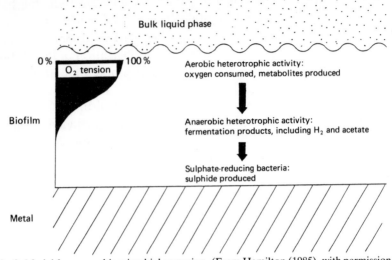

Fig. 2. Model for anaerobic microbial corrosion. (From Hamilton (1985), with permission.)

the choice of suitable incubation media, corrosion coupons can be retrieved from whatever environment they may have been in, and the activity of the sulphate-reducing bacterial population within the undisturbed biofilm can be measured under conditions relating directly to that environment.

Experimental systems for the study of biofilms

(i) Characklis (1984) approaches biofilm development as a series of physical, chemical and biological processes which must be analysed in terms of the stoichiometry and kinetics of each individual process, making every possible use of quantitative data and mathematical modelling. His experimental systems are centred on the chemostat growth of a pure culture, normally *Pseudomonas aeruginosa*.

Robinson, Trulear & Characklis (1984) noted that the synthesis of extracellular polmer was inversely related to growth rate, and that the majority of the glucose consumed under non-growing conditions was utilised for expolymer synthesis. Similarly, Bryers (1984) concluded from a detailed analysis of the effects of biofilm formation on chemostat dynamics in pure and mixed culture that the so-called maintenance requirement observed at low dilution rate may be largely the result of biofilm formation. These data are consistent with the finding of Nelson, Robinson & Characklis (1985) that bacterial adsorption to glass surfaces decreased more or less linearly with increasing growth rate.

Direct comparisons have been made between steady state biofilm populations and suspended cells under steady state growth in the chemostat (Bakke et al., 1984; Turahhia, 1986). It appears that the activities of the two populations are equivalent with regard to specific growth rate (as a function of substrate concentration), cell yield, and stoichiometry of glucose conversion to biomass.

More recently, Characklis (in press) has been examining in detail the processes of transport, adsorption, growth, detachment, cell separation, and desorption that represent the initial events on biofilm formation. The use of a rectangular capillary as a reaction vessel, in association with high-quality microscopy and image analysis, allows these events to be followed at the level of individual cells.

(ii) Wimpenny, Lovitt & Coombs (1983) have described a wide range of film fermenters which have been designed for, and used in, different environmental conditions, notably waste water treatment plants and experimental studies of dental plaque. They have identified some of the questions that need to be answered in order to advance our understanding of biofilms, and suggested the features necessary in an experimental system designed to examine these questions. Perhaps not too surprisingly many of these features are incorporated in the laboratory model thin film fermenter that Wimpenny's group have developed and built themselves (Coombe, Tatevoision & Wimpenny, 1982). The geometry of the fermenter is such that a large number of identical films can be obtained at constant depth (and therefore in a quasi-steady state) at any point predetermined within a range of depths up to 300 μm. Growth of the film can therefore be measured readily by time-dependent removal of individual films and determination of their weight, nitrogen content, etc. The inoculum (pure culture or mixed enrichment from a natural environment), nutrient composition and concentration, gas phase, temperature, presence of illumination, etc., can all be readily altered to suit experimental design. After removal of individual films from the fermenter, oxygen concentration can be measured at different depths through the film with the help of a micromanipulator controlled by a microcomputer (J. W. T. Wimpenny, personal communication). Present plans include a project in which the PTFE base to the film pan will be replaced by steel, and sulphate-reducing bacteria will be introduced with the inoculum. It will then be possible to monitor corrosion of the steel and relate that to the activity of the biofilms, with particular reference to the development of anaerobic conditions as measured both by microprobe analysis and by the growth of sulphate-reducing bacteria within the film.

(iii) The microbial corrosion problems experienced in chemical process industries, and particularly associated with water flood treatments in the oil industry, have stimulated much work in recent years on the effective use of biocides to reduce microbial numbers and so minimise the consequential engineering problems. The pioneering work of J. W. Costerton and his colleagues has greatly heightened the general awareness that such analyses can only be meaningful if they are carried out using sessile organisms within a biofilm rather than the planktonic population freely suspended in the bulk aqueous phase. In keeping with the general thesis underlying this chapter, the application of such a philosophy demanded the development of suitable methodology. This was achieved with the manufacture of the Robbins' device (McCoy et al., 1981). In essence the device consists of a spool tube section, originally made from admiralty brass but now more commonly from mild steel, from which sample pieces may be withdrawn, either in the form of replaceable studs or as sections of the tube itself. The spool section can be incorporated into a laboratory circulating loop fed with a bacterial inoculum and nutrient supply, and the biofilm which develops can then be examined directly by removing a representative section of it on the surface of one of the replaceable studs. The most up-to-date modifications to the system are described by Tanner et al. (1985). Ruseska et al. (1982) and Costerton & Lashen (1984) have made extensive use of the Robbins' device to assay the effectiveness of biocide treatments directed principally against sulphate-reducing bacteria in oilfield situations. It is important to note in this regard that the Robbins' device can be incorporated into an actual water flood system so that biocide and other treatments can be directly monitored in situ.

My own group are also involved in a series of studies involving the use of Robbins' devices. These concern tests on the efficacy of a range of biocides both in situ and in laboratory simulations, and more fundamental studies on the biofilm itself. These experiments, and those of Tanner et al. (1985), demonstrate that the time period for development can be reduced to only a few days in response to increased biofouling pressure in the form of bacterial inoculum and, more particularly, nutrient concentration. What remains to be established are: the time course of development of active metabolism of sulphate-reducing bacteria within the film and the effect thereon of the presence of oxygen in the circulating medium; and the physiological activities of the biofilm in terms of its metabolic interactions and nutrient succession. The rationale of these studies will be to extend the scope of preliminary findings obtained with natural biofilms formed in marine and process environments directly associated with the offshore oil industry.

(iv) The siting of corrosion coupons in environments relevant to the offshore oil industry for periods of up to 2 years raises many problems, of both of logistics and experimental design; the latter must include a significant element of allowance for material that cannot be retrieved in a usable form. The merits of such an experimental approach, however, are two-fold. Data are obtained that are of direct relevance to problems, actual or potential, experienced in environments that are often unique and always interesting. These problems afford the opportunity to work at an interface between biology and engineering, and to contribute some practical advice as to their solution. Secondly, the findings gained from such studies can form the basis for more detailed analyses using model systems capable of a greater degree of experimental control.

Metal test coupons (1.5×3.0 cm) are generally made of 50D structural steel. Where required, cathodic protection against corrosion is achieved by coupling to a sacrificial zinc anode. The mechanism of this measure is that the steel is held at the non-corroding potential of -950 mV (with respect to a Ag/AgCl reference electrode) and the zinc corrodes by dissolution. Exposure of such coupons in a mildly polluted harbour location (Maxwell, 1984) led to high numbers of attached organisms within 1 day; 10^7–10^8 by epifluorescent direct count and 10^4–10^5 by culturing aerobic bacteria. These numbers reached plateaux by about 20 days, of 10^9 and 10^7–10^8, respectively. Sulphate-reducing bacteria, on the other hand, could not be cultured from the biofilms until after 3 days and took longer to reach their plateau value of 10^2–10^3, all figures being expressed per coupon. Sulphate reduction activity, as measured by [^{35}S]-sulphate turnover, developed more slowly, however, and was not measurable until until after 20 days. Apart from their indication of the time course of the development of anaerobic conditions within the biofilm, these data reinforce the point made above (Costerton & Geesey, 1979) that biofilms characteristically have higher numbers of organisms and greater metabolic activity than the bulk aqueous phase in which they are situated. Further demonstration of this important point was gained from coupons sited in the entrained sea water inside the drilling leg of a concrete offshore platform, or in the water injection system of another production platform (Maxwell & Hamilton, in press, Sanders & Hamilton, in press). After a period of 72 days exposure in the drilling leg, assay of coupons gave 10^8 sulphate-reducing bacteria per m^{-2} and a sulphate reduction rate of 0.5 mmoles $m^{-2} day^{-1}$. The corresponding figures for the bulk aqueous phase were 10^3 cells l^{-1} and below detection, respectively.

The microbiological data relating to the effects of cathodic protection remain equivocal at present time. While short-term (up to 123 days) studies suggest that cathodic protection severely reduces sulphate reduction (Maxwell, 1984; Maxwell & Hamilton, 1985), longer-term (up to 2 years) exposures in marine environments give a range of data that in general do not support this contention. What is clear, however, is that cathodic protection, as measured by anode life, is reduced in effectiveness in the presence of high sulphate reduction rates. Coupons sited in sea-bed deposits of discarded drill cuttings arising from the use of diesel-based drilling muds show high rates of sulphate reduction, and although sacrificial zinc anodes can protect against corrosion, under these conditions anodes designed for periods of up to 8 years are becoming exhausted in less than 2 years (Sanders & Tibbetts, in press).

The high rates of sulphate reduction in diesel-based drill cuttings further illustrate the major point made above that biofilms are complex communities of microorganisms and that the creation of the necessary physicochemical and nutrient conditions for the sulphate-reducing bacteria requires the active metabolism of other bacterial species, in this case aerobic hydrocarbon-degrading heterotrophs. It is interesting, however, to seek to identify the carbon and energy sources generated by the various heterotrophic species that may serve to drive the growth and physiological activity of the sulphate-reducing bacteria. It has been found, for example, that films formed on coupons after 64 days' exposure in a harbour location showed substrate-limitation when the [^{35}S]-sulphate assay was incubated for 24 h. This substrate-limitation could be overcome, however, by the addition of either acetate of lactate to the incubation (Hamilton & Maxwell, in press).

Particularly with the help of laboratory systems such as chemostat-linked biofilm growth, film fermenters, and the Robbins' device it will be fascinating to extent these findings. By relating them to the organisms present, the physicochemical conditions prevailing, and the corrosion resulting, a more complete understanding of the totality of activities associated with biofilms will be gained.

REFERENCES

ATKINSON, B. & FOWLER, H. W. (1974). The significance of microbial film in fermenters. *Advances in Biochemical Engineering*, **3**, 221–7.

BAKKE, R., TRULEAR, M. G., ROBINSON, J. A. & CHARACKLIS, W. G. (1984). Activity of *Pseudomonas auruginosa* in biofilms: steady state. *Biotechnology and Bioengineering*, **26**, 1418–24.

BRIGHT, J. J. & FLETCHER, M. (1983a). Amino acid accumulation and electron transport system activity in attached and free-living marine bacteria. *Applied and Environmental Microbial*, **45**, 818–25.

BRIGHT, J. J. & FLETCHER, M. (1983b) Amino acid assimilation and respiration by attached and free-living populations of a marine *Pseudomonas* sp. *Microbial Ecology*, **9**, 215–26.

BROWN, C. M., ELLWOOD, D. C. & HUNTER, J. R. (1977). Growth of bacteria at surfaces: influence of nutreint limitation *FEMS Microbiology Letters*, **1**, 163–6.

BRYERS, J. D. (1984). Biofilm formation and chemostat dynamics: pure and mixed culture considerations. *Biotechnology and Bioengineering*, **26**, 948–58.

CHARACKLIS, W. G. (1981). Fouling biofilm development: a process analysis *Biotechnology and Bioengineering*, **23**, 1923–60.

CHARACKLIS, W. G. (1984). Biofilm development: a process analysis. In *Microbial Adhesion and Aggregation*, ed. K. C. Marshall, pp. 137–97. Berlin, Springer-Verlag.

CHARACKLIS, W. G. (in press). Microbial fouling: a fundamental approach. In *Marine Biodeterioration*, ed. R. Nagabhushanam & R. D. Turner. New Delhi, Oxford and IBH Publishing Co.

CHARACKLIS, W. G. & COOKSEY, K. E. (1983). Biofilms and microbial fouling. *Advances in Applied Microbiology*, **29**, 93–138.

CHRISTENSEN, B. E., KJOSBAKKEN, J. & SMIDSROD, O. (1985). Partial chemical and physical characterization of two extracellular polysaccharides produced by marine, periphytic *Pseudomonas* sp. strain NCMB 2021. *Applied and Environmental Microbiology*, **50**, 837–45.

COOMBE, R. A., TATEVOISION, A. & WIMPENNY, J. W. T. (1982). Bacterial thin films as *in vitro* models for dental plague. In *Surface and Colloid Phenomena in the Oral Cavity: Methodological Aspects*, ed. R. M. Frank & S. A. Teach, pp. 239–49. London, IRL Press.

COSTERTON, J. W. & GEESEY, G. C. (1979). Microbial contamination of surfaces. In *Surface Contamination*, ed. K. L. Mittal, pp. 211–21. New York, Plenum Publishing Corporation.

COSTERTON, J. W., IRVIN, R. T. & CHENG, K.-J. (1981). The bacterial glycocalyx in nature and disease. *Annual Review of Microbiology*, **35**, 299–324.

COSTERTON, J. W. & LASHEN, E. S. (1984). The inherent biocide resistance of corrosion-causing biofilm bacteria. *Materials Performance*, **23**(2), 13–16.

COSTERTON, J. W., MARRIE, T. J. & CHENG, K.-J. (1985). Phenomena of bacterial adhesion. In *Bacterial Adhesion: Mechanisms and Physiological Significance*, ed. D. C. Savage & M. Fletcher, pp. 3–43. New York, Plenum Press.

COULTER, W. A. & RUSSELL, C. (1976). pH and Eh in single and mixed culture bacterial plaque in an artificial mouth. *Journal of Applied Bacteriology*, **40**, 73–81.

Cox, D. J., Bazin, M. J. & Gull, K. (1980). Distribution of bacteria in a continuous-flow nitrification column. *Soil Biology and Biochemistry*, **12**, 241–6.

Critchley, I. A. & Douglas, L. J. (1985). Differential adhesion of pathogenic *Candida* species to epithelial and inert surfaces. *FEMS Microbiology Letters*, **28**, 199–203.

Dowling, N. J. E., Guezennec, J. & White, D. C. (in press). Facilitation of corrosion of stainless steel exposed to aerobic seawater by microbial biofilms containing both facultative and absolute anaerobes. In *Microbial Problems in the Offshore Oil Industry*, ed. E. C. Hill. London, Wiley.

Duddridge, J. E., Kent, C. A. & Laws J. F. (1982). Effect of surface sheer stress on the attachment of *Pseudomonas fluorescans* to stainless steel under defined flow conditions. *Biotechnology and Bioengineering*, **24**, 153–64.

Ellwood, D. C., Keevil, C. W., Marsh, P. D., Brown, C. M & Wardell, J. N. (1982). Surface-associated growth. *Philosophical Transactions of the Royal Society of London, Series B*, **297**, 517–32.

Fletcher, M. (1977). The effects of culture concentration and age, time and temperature on bacterial attachment to polystyrene. *Canadian Journal of Microbiology*, **23**, 1–6.

Fletcher, M. & Loeb, G. I. (1979). Influence of substratum characteristics on the attachment of a marine pseudomonad to solid surfaces. *Applied and Environmental Microbiology*, **37**, 67–72.

Fletcher, M. & Marshall, K. C. (1982*a*). Are solid surfaces of ecological significance to aquatic bacteria? *Advances in Microbial Ecology*, **6**, 199–236.

Fletcher, M. & Marshall, K. C. (1982*b*). Bubble contact angle method for evaluating substratum interfacial characteristics and its relevance to bacterial attachment. *Applied and Environmental Microbiology*, **44**, 184–92.

Ford, T. E., Walch, M. & Mitchell, R. (in press). Corrosion of metals by thermophilic microorganisms. *Materials Performance*.

Gibbons, R. J. & van Haute, J. (1975). Bacterial adherence in oral microbial ecology. *Annual Review of Microbiology*, **29**, 19–44.

Gordon, A. S., Gerchakov, S. M. & Millero, F. J. (1983). Effects of inorganic particles of metabolism by a periphytic marine bacterium. *Applied and Environmental Microbiology*, **45**, 411–17.

Gordon, A. S., Gerchakov, S. M. & Udey, L. R. (1981). The effect of polarization on the attachment of marine bacteria to copper and platinum surfaces. *Canadian Journal of Microbiology*, **27**, 698–703.

Haack, T. K. & McFeters, G. A. (1982). Nutritional relationships among microorganisms in an epilithic biofilm community. *Microbial Ecology*, **8**, 115–26.

Hamilton, W. A. (1983). Sulphate-reducing bacteria in the offshore oil industry. *Trends in Biotechnology*, **1**, 36–40.

Hamilton, W. A. (1984). Energy sources for microbial growth: an overview. In *Aspects of Microbial Metabolism and Ecology*, ed. G. A. Codd, pp. 35–57. London, Academic Press.

Hamilton, W. A. (1985). Sulphate-reducing bacteria and anaerobic corrosion. *Annual Review of Microbiology*, 39, 195–217.

Hamilton, W. A. & Maxwell, S. (in press). Biological and corrosion activities of sulphate-reducing bacteria within natural biofilms. In *Biologically Induced Corrosion*, ed. S. C. Dexter. Houston, National Association of Corrosion Engineers.

Harold, F. M. (1972). Conservation and transformation of energy by bacterial membranes. *Bacteriological Reviews*, **36**, 172–230.

Hattori, R. & Hattori, T. (1981). Growth rate and molar growth yield of *Escherichia coli* adsorbed on an anion-exchange resin. *Journal of General and Applied*

Microbiology, **27**, 287–98.

HOWELL, J. A. & ATKINSON, B. (1976). Sloughing of microbial film on trickling filters. *Water Research*, **18**, 307–15.

JEFFREY, W. H. & PAUL, J. H. (1986*a*). Activity of an attached and free-living *Vibrio* sp. as measured by thymidine incorporation, *p*-iodonitrotetrazolium reduction and ATP/DNA ratios. *Applied and Environmental Microbiology*, **51**, 150–6.

JEFFREY, W. H. & PAUL, J. H. (1986*b*). Activity measurements of planktonic microbial and microfouling communities in a eutrophic estuary. *Applied and Environmental Microbiology*, **51**, 157–62.

JØRGENSEN, B. B. (1978). A comparison of methods for the quantification of bacterial sulphate reduction in coastal marine sediments. 1. Measurement of radiotracer techniques. *Geomicrobiology Journal*, **1**, 11–27.

JØRGENSEN, B. B. (1982). Ecology of the bacteria of the sulphur cycle with special reference to anoxic–oxic interface environments. *Philosophical Transactions of the Royal Society, London, Series B*, **298**, 543–61.

KEEN, G. A. (1984). Nitrification in continuous culture: the effect of pH and surfaces. Ph.D. thesis, University of Aberdeen.

KEPKAY, P. E., SCHWINGHAMER, P., WILLAR, T. & BOWEN, A. J. (1986). Metabolism and metal binding by surface-colonizing bacteria: results of microgradient measurements. *Applied and Environmental Microbiology*, **51**, 163–70.

KJELLEBERG, S. (1984). Effects of interfaces on survival mechanisms of copiotrophic bacteria at low-nutrient habitats. In *Current Perspectives in Microbial Ecology*, ed. M. J. Klug & C. A. Reddy, pp. 151–9. Washington, American Society for Microbiology.

KJELLEBERG, S., HUMPHREY, B. A. & MARSHALL, K. C. (1982). The effects of interfaces on small, starved bacteria. *Applied and Environmental Microbiology*, **43**, 1166–72.

KJELLEBERG, S., HUMPHREY, B. A. & MARSHALL, K. C. (1983). Initial phases of starvation and activity of bacteria at surfaces. *Applied and Environmental Microbiology*, **46**, 978–84.

LAANBROEK, H. J. & GEERLIGS, H. J. (1983). Influence of clay particles (illite) on substrate utilization by sulphate-reducing bacteria. *Archives of Microbiology*, **134**, 161–3.

McCOURTIE, J. & DOUGLAS, L. J. (1981). Relationship between cell surface composition of *Candida albicans* and adherence to acrylic after growth on different carbon sources. *Infection and Immunity*, **32**, 1234–41.

McCOY, W. F., BRYERS, J. D., ROBBINS, J. & COSTERTON, J. W. (1981). Observations of fouling biofilm formation. *Canadian Journal of Microbiology*, **27**, 910–17.

MARSHALL, K. C. (1984) (ed.). *Microbial Adhesion and Aggregation*. Berlin, Springer-Verlag.

MAXWELL, S. (1984). Ecological studies of sulphate-reducing bacteria on metal surfaces. Ph.D. thesis, University of Aberdeen.

MAXWELL, S. & HAMILTON, W. A. (1985). Effect of cathodic protection on the activity of microbial biofilms. *U. K. Corrosion '85*, 281–91.

MAXWELL, S. & HAMILTON, W. A. (1986*a*). Modified radiorespirometric assay for determining the sulfate reduction activity of biofilms on metal surfaces. *Journal of Microbiological Methods*, **5**, 83–91.

MAXWELL, S. & HAMILTON, W. A. (in press). Activity of sulphate-reducing bacteria on metal surfaces in an oilfield situation. In *Biologically Induced Corrosion*, ed. S. C. Dexter. Houston, National Association of Corrosion Engineers.

MURRAY, R. E., COOKSEY, K. E. & PRISCU, J. C. (in press). Algal-bacterial syntrophy in laboratory produced biofilms. *Journal of Phycology*.

NELSON, C. H., ROBINSON, J. A. & CHARACKLIS, W. G. (1985). Bacterial adsorption to smooth surfaces: rate, extent, and spatial pattern. *Biotechnology and Bioengineering*, **27**, 1662–7.

NICHOLS, P. D., HENSEN, J. M., GUCHERT, J. B., NIVENS, D. E. & WHITE, D. C. (1985). Fourier transform–infrared spectroscopic methods for microbial ecology: analysis of bacteria, bacteria–polymer mixtures and biofilms. *Journal of Microbiological Methods*, **4**, 79–94.

NIVENS, D. E., NICHOLS, P. D., HENSON, J. M., GEESEY, G. G. & WHITE, D. C. (1986). Reversible acceleration of the corrosion of A1S1 304 stainless steel exposed to seawater induced by growth and secretions of the marine bacterium *Vibrio natriegens*. *Corrosion*, **42**, 204–10.

NOVITSKY, J. A. & MORITA, R. Y. (1976). Morphological characterization of small cells resulting from nutrient starvation of a psychrophilic marine vibrio. *Applied and Environmental Microbiology*, **32**, 617–22.

NOVITSKY, J. A. & MORITA, R. Y. (1977). Survival of a psychrophilic marine vibrio under long-term nutrient starvation. *Applied and Environment Microbiology*, **33**, 635–41.

NOVITSKY, J. A. & MORITA, R. Y. (1978). Possible strategy for the survival of marine bacteria under starvation conditions. *Marine Biology (Berlin)*, **48**, 289–95.

PAERL, M. V. (1985). Influence of attachment on microbial metabolism and growth in aquatic acosystems. In *Bacterial Adhesion: Mechanisms and Physiological Significance*, ed. D. C. Savage & M. Fletcher, pp. 363–400. New York, Plenum Press.

PIRT, S. J. (1973). A quantitative theory of the action of microbes attached to a packed column: relevant to trickling filter effluent purification and to microbial action in soil. *Journal of Applied Chemistry and Biotechnology*, **23**, 389–400.

POWELL, S. J. (1985) Inhibition of *Nitrosomonas europaea* by nitrapyrin: the role of surfaces. Ph.D. thesis, University of Aberdeen.

REVSBECH, N. P. & WARD, D. M. (1983). Oxygen microelectrode that is insensitive to medium chemical composition: use in an acid microbial mat dominated by *Cyanidium caldarium*. *Applied and Environmental Microbiology*, **45**, 755–9.

ROBINSON. J. A., TRULEAR, M. G. & CHARACKLIS, W. G. (1984). Cellular reproduction and extracellular polymer formation by *Pseudomonas aeruginosa* in continuous culture. *Biotechnology and Bioengineering*, **26**, 1409–17.

ROSSER, H. R. & HAMILTON, W. A. (1983). Simplified assay for accurate determination of [^{35}S] sulfate reduction activity. *Applied and Environmental Microbiology*, **45**, 1956–9.

RUSESKA, I., ROBBINS, J., COSTERTON, J. W. & LASHEN, E. S. (1982). Biocide testing against corrosion-causing oil-field bacteria helps control plugging. *Oil and Gas Journal*, **10**, 253–8.

SANDERS, P. F. & HAMILTON, W. A. (in press). Biological and corrosion activities of sulphate-reducing bacteria in industrial process plant. In *Biologically Induced Corrosion*, ed. S. C. Dexter. Houston, National Association of Corrosion Engineers.

SANDERS, P. F. & TIBBETS, P. J. C. (in press). Effects of discarded drill muds on microbial populations. *Proceedings of the Royal Society, London, Series B*.

SHARMA, A. P., BATTERSBY, N. S. & STEWART, D. J. (in press). Techniques for the evaluation of biocide activity against sulphate-reducing bacteria. In *Preservatives in the Food, Pharmaceutical and Environmental Industries*. Society of Applied Bacteriology Technical Series, No. 22, ed. J. G. Barks, R. G. Board & M. C. Allwood. Oxford, Blackwell Scientific Publications.

SLATER, J. H. & SOMERVILLE, H. J. (1979). Microbial aspects of waste treatment with particular attention to the degradation of organic compounds. In *Microbial*

Technology: Current State, Future Prospects, ed. A. T. Bull, D. C. Ellwood & C. Ratledge, pp. 221–61. Cambridge, Cambridge University Press.

SØRENSEN, J., CHRISTENSEN, D. & JØRGENSEN, B. B. (1981). Volatile fatty acids and hydrogen as substrates for sulfate-reducing bacteria in anaerobic marine sediment. *Applied and Environmental Microbiology*, **42**, 5–11.

TANNER, R. S., HAACK, T. K., SEMET, R. F. & GREENLEY, D. E. (1985). A mild steel tubular flow system for biofilm monitoring. *U.K. Corrosion '85*, 259–69.

TAYLOR, J. & PARKES, R. J. (1985). Identifying different populations of sulphate-reducing bacteria within marine sediment systems, using fatty acid biomarkers. *Journal of General Microbiology*, **131**, 631–42.

TURAHHIA, M. H. (1986). The influence of calcium on biofilm processes. Ph.D. thesis, Montana State University.

UHLINGER, D. J. & WHITE, D. C. (1983). Relationship between physiological status and formation of extracellular polysaccharide glycocalyx in *Pseudomonas atlantica. Applied and Environmental Microbiology*, **45**, 64–70.

WALCH, M. & MITCHELL, R. (in press). Microbial influence in hydrogen uptake. In *Biological Induced Corrosion*, ed. S. C. Dexter, Houston, National Association of Corrosion Engineers.

WARDELL, J. N., BROWN, C. M. & FLANNIGAN, B. (1983). Microbes and surfaces. In *Microbes in their Natural Environments*, ed. J. H. Slater, R. Whittenbury & J. W. T. Wimpenny, pp. 351–78. Cambridge, Cambridge University Press.

WHITE, D. C. (1983). Analysis of microorganisms in terms of quantity and activity in natural environments. In *Microbes in their Natural Environments*, ed. J. H. Slater, R. Whittenbury & J. W. T. Wimpenny, pp. 37–66. Cambridge, Cambridge University Press.

WIMPENNY, J. W. T., LOVITT, R. W. & COOMBS, J. P. (1983). Laboratory model systems for the investigation of spatially and temporally organised microbial eco-systems. In *Microbes in their Natural Environments*, ed. J. H. Slater, R. Whittenbury & J. W. T. Wimpenny, pp. 67–117. Cambridge, Cambridge University Press.

ZEIKUS, J. G. (1983). Metabolic communication between biodegradative populations in nature. In *Microbes in their Natural Environments*, ed. J. H. Slater, R. Whittenbury & J. W. T. Wimpenny, pp. 423–62. Cambridge, Cambridge University Press.

ZOBELL, C. E. (1943). The effect of solid surfaces upon bacterial activity. *Journal of Bacteriology*, **46**, 39–56.

MODELLING THE GROWTH AND DISTRIBUTION OF MARINE MICROPLANKTON

PAUL TETT

School of Ocean Sciences, University College of North Wales, Marine Science Laboratories, Menai Bridge, Gwynedd LL59 5EH, UK

INTRODUCTION

The marine *microplankton* may arbitrarily be defined as free-living waterborne protists less than 200 μm in size. They include unicellular or colonial autotrophs from several divisions of the eukaryotic microalgae, autotrophic cyanobacterial and heterotrophic eubacterial prokaryotes, and several types of heterotrophic protozoans. Whereas classical descriptions of planktonic food webs (e.g. Hardy, 1924) stressed the central role of invertebrate zooplankton feeding on diatoms and phytoflagellates, recent studies have suggested that a large proportion of photoassimilated carbon is rapidly recycled either through dissolved organic matter (DOM), bacteria and protozoans, or by protozoans feeding directly on the smaller algae. Some of these studies are reviewed by Williams (1981) and Fasham (1985), the latter in the context of a model of carbon and nitrogen flow through the marine pelagic food web.

Although nutrition is important, microplankton are also strongly affected by water movements. Such movements, which redistribute microorganisms in relation to food, predators, mineral nutrients or light, may be directional ('currents') or chaotic ('turbulence'). The former have, except in upwelling regions, generally little effect on microplankton communities, whose rates of turnover are rapid compared with the speed of most mean currents in the sea. Turbulent diffusion is more important. Although a stochastic process, its scale is greater in the horizontal then in the vertical, where turbulent eddies must work against gravity. Nevertheless, the existence of density layering in the sea, the rapid attenuation of photosynthetically effective light with depth, and the consequent strong vertical gradients of microplankton biomass and plant nutrients, suggest that vertical rather than horizontal diffusion is the dominant physical transport of microplankton (Tett & Edwards, 1984).

The main aim of this chapter is therefore to discuss the way in which vertical turbulence interacts with density and illumination gradients to determine the distribution of dissolved nutrients and microplankton biomass in the sea. This approach has so far been most successful for phytoplankton, perhaps because of the ease with which its biomass can be estimated from measurements of photosynthetic pigments. The larger part of the review thus deals with published 'vertical-process' numerical models for the growth and distribution of generalized phytoplankton and, especially, with the physical and biological background necessary to understand these models. The subsequent part is more speculative, and discusses the way in which models might describe interactions amongst specified algae or protozoans, or between generalized autotrophic and heterotrophic processes, in a water column with depth-varying mixing. I concentrate on temperate coastal seas with strong tidal mixing.

SOME VERTICAL DISTRIBUTIONS OF MICROPLANKTON AND NUTRIENTS

Simpson *et al.* (1982) investigated the physical and biological effects of enhanced tidal stirring in the water surrounding the Scilly Isles, SW England, during July 1979. Figure 1 shows profiles from the outer part of the region influenced by island mixing. Points to note include the temperature layering, the depletion of dissolved inorganic nitrogen in near-surface waters, and the occurrence of high concentrations of chlorophyll at depths of 10–20 m below the surface. These high concentrations were dominated by the dinoflagellate *Gyrodinium aureolum*, giving algal biomasses and productivities up to ten times those in the undisturbed water further away from the islands. Holligan *et al.* (1984*a*) describe similar profiles, and dominance by *G. aureolum*, in frontal regions elsewhere in the western English Channel.

Enclosed water columns allow some of the assumptions of vertical-process models to be tested accurately. Fig. 2 shows profiles in an abandoned slate quarry, now flooded by the sea, on a small island

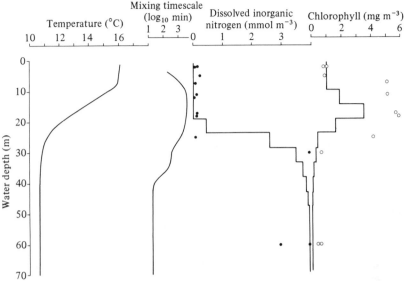

Fig. 1. Profiles for a frontal station near the Scilly Isles, SW England, July 1979, based on the mean of stations FG2 and FG3 of Simpson *et al.* (1982). Observed temperatures are shown by a (smoothed) continuous line. Mixing timescales are shown for layers of thickness 4.74 m, calculated from turbulent diffusivities, and hence from the temperature profile, assuming a vertical heat flux of 700 kcal m^{-2} day^{-1}. Observed dissolved inorganic nitrogen (DIN: nitrate plus nitrite measured) and chlorophyll-a concentrations are shown by open and closed circles. (One chlorophyll concentration of 31 mg m^{-3} at 12 m is omitted (see Fig. 5 and p. 405).) Continuous lines show the steady-state predictions of the model, and standard parameter values, of Table 2; the steps indicate layers in the simulation.

on the west coast of Scotland. The quarry maintains intermittent contact with the sea through a narrow, tidally flushed channel, and each day exchanges about 10% of its contents. Profiles presented are for late May 1984 and are based on Grantham (1984), Weeks (1984) and unpublished observations. The microplankton at this time contained a mixture of diatoms (dominated by *Rhizosolenia*), gonyalacoid dinoflagellates and protozoans (including oligotrich ciliates). Particular points to note are the thermal layering, the midwater maxima of chlorophyll, ATP concentration and numbers of dinoflagellates, and the minimum of dissolved inorganic nitrogen corresponding to these maxima. Some of the subsequent discussion will concern turnover of materials at a depth of 20 m, at the bottom of the photic zone and in the upper part of the thermocline, during June, July and August 1984. By this time the microplankton consisted of diatoms, silicoflagellates, phytoflagellates, and zooflagellates and heterotrophic dinoflagellates.

VERTICAL STRUCTURE AND VERTICAL PROCESSES IN COASTAL WATER COLUMNS

Tidal currents are strong in many north-west European and other coastal seas, and their friction with the sea-bed generates turbulence, which is intensified around islands and headlands. Wind drag results in near-surface turbulence. These stirring actions are opposed by solar heating, and by freshwater from rain or rivers, both of which reduce the density of seawater and can be thought of as inputting buoyancy to near-surface waters. In many shelf seas in summer the inputs of stratification-stabilizing buoyancy and stratification-disrupting turbulence are roughly in balance, resulting in a thermally stratified water column such as that shown in Fig. 1. Warmer, less dense, wind-mixed water near the sea surface is here separated from deeper, colder, denser, tide-mixed water by a region of temperature (and density) gradient, the *thermocline*, extending from 8 m to 28 m depth. Turbulence in the wind-mixed and tide-mixed layers is generally such that a passive microplankter may expect to find itself in any part of the relevant layer during a 24 h period. These layers may thus be considered well mixed in relation to the growth of eukaryote microplankton. This is often not true of the thermocline, where density gradients oppose vertical mixing. Bowden (1983) provides an introduction to the physics of these processes.

Photosynthetically effective irradiance (p.e.i.) is rapidly attenuated in coastal seas, falling to 1% of sea-surface values at depths of 40 m or less (Jerlov, 1968). This relative illumination, and these depths, roughly correspond to the bottom of the *photic* zone, to which photoautotrophic growth is confined. In the case of the water column shown in Fig. 1, the photic zone is 25 m thick and includes the wind-mixed layer and most of the thermocline. The mean illumination available to algae in the tide-mixed layer between 28 m and 90 m is only about 0.001% of surface irradiance. Deep water is thus, in this and many other cases, a zone dominated by heterotrophic processes, and is therefore likely to remain nutrient-rich.

Should the turbulent energy injected into the water column be sufficiently great (as for example where tidal currents are strong around islands), or the buoyancy input be weak (as happens in most temperate seas during winter), stratification will break down completely. Except in very shallow waters the mean illumination received by phytoplankters mixed between sea surface and sea-bed (instead of being confined to better illuminated waters above the thermocline) is then likely to be too little for growth (Sverdrup, 1953). Such complete mixing enhances sea-surface nutrient concentrations locally in summer, and generally restores photic zone nutrients in winter.

Stratified conditions are, however, more typical of coastal seas in summer, and the wind-mixed layer is often nutrient-depleted at this time (e.g. Fig. 1 and Pingree, Maddock & Butler, 1977), following the growth and subsequent grazing or sinking down of the spring phytoplankton bloom. Summer production thus depends ultimately on the rate at which the wind-mixed layer exchanges with nutrient-rich deep water. Although it is not easy to give a simple and precise description of the physical processes involved in exchange, it is possible to conceive of their resulting in a flow of dissolved nutrient from a deep-water source to the sink provided by photoassimilation near the sea surface. Such a flow must pass through the thermocline, and Pingree et al. (1975), seeking to explain midwater chlorophyll maxima, pointed to the advantages for phytoplankton growth of thermoclines overlapping the photic zone. Thermocline mixing timescales are of the order of weeks, allowing an algal cell to divide several times before any of its descendants are dispersed into the low-nutrient wind-mixed layer above or the low-light tide-mixed layer below. Such a *light–nutrient–mixing* hypothesis is implicit in many ecological explanations of phytoplankton distribution in the sea. Vertical migration may increase the ability of microalgae to exploit the thermoclinic niche, and the enhanced plant biomass may support a passive or active concentration of heterotrophs here (e.g. Fig. 2 and Holligan et al., 1984b).

Because they experience little tidal stirring, the deep oceans and some shelf seas are mixed mainly from the surface. Thus, although they show a well-delineated wind-mixed layer, the lower edge of the thermocline is often poorly defined (Pickard & Emery, 1982). This was also the case in the flooded quarry whose temperature profile is shown in Fig. 2. During the summer, water entering from the sea was less dense than the enclosed deep water, and so the turbulence resulting from its flow through a narrow channel was largely confined to the top 18 m of the water column. No tidal energy was available to mix from the bottom upwards, and the deepest water became stagnant and ultimately anoxic. Nevertheless, some exchange did take place between water within the thermocline and that above, since observations at this time (Fig. 8) showed a slow

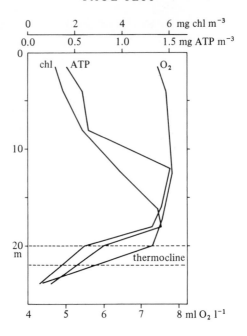

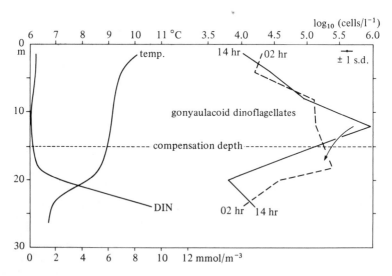

Fig. 2. Profiles in a flooded quarry on Easdale Island, Scotland, May 21–22, 1984. Except for the profiles of gonyaulacoid dinoflagellates where the downward shift (arrow) of peak numbers suggests vertical migration, the profiles are means of three to five taken during a 24 h period. The (algal) compensation depth is taken from Fig. 7. DIN includes nitrate, nitrite and ammonium concentrations. The depths relate to an arbitrary datum, roughly mean high water spring tide. Low water in the quarry was 1.4 m below this.

warming of water between 18 m and 25 m. Mixing and regenerative processes in this part of the quarry may bear a close resemblance to those in the deep water of some fjords (e.g. Edwards & Grantham, 1986).

NUTRIENT LIMITATION AND NUTRIENT CYCLING

The growth of axenic cultures of planktonic microalgae is limited by a range of elements and organic compounds, and the nutrient kinetics of nitrogen, phosphorus, silicon and vitamin B12 are well, although perhaps not perfectly, understood (Rhee, 1980; Droop, 1983). The uptake of these nutrients can often be represented by Michaelis–Menten or similar saturation kinetics and, following the work of Monod (1942) with bacterial cultures, algal ecologists (such as Dugdale, 1967) initially assumed that phytoplankton growth could similarly be described as the product of the uptake kinetics and a constant yield. The suggestion (see Redfield, 1958) that the elemental composition of plankton, and the stoichiometry of mean planktonic processes, tended to a constant ratio of (by atoms) 276 O: 106 C: 16 N: 1 P: encouraged this view.

There was evidence, however, (reviewed by Eppley & Strickland, 1968) that microalgae could assimilate nutrient in excess of immediate need, thus altering their gross chemical composition. Although some authors saw such variation in composition as a *result* of variations in growth, others suggested that it *controlled* growth rate. Formal *cell quota* models for the control of steady-state microalgal growth by the intracellular concentration of limiting nutrient were proposed for vitamin B12 by Droop (1968) and for nitrogen by Caperon (1968) and Caperon & Meyer (1972). Subsequent physiological studies (reviewed by Droop, 1983) have shown the applicability of the hypothesis of internal-nutrient control to a range of nutrients and types of microalgae, and to transient as well as steady-state growth. The fully developed cell-quota models reviewed by Rhee (1980) and Droop (1983) embody several important principles that it seems desirable to include in ecological models of microplankton growth.

(1) Growth rate depends on the internal concentration, or cell quota, Q, of limiting nutrient, and is some saturating function of the ratio, Q^*, of Q to the minimum cell content or subsistence

quota, k_Q, of that nutrient. Droop (1968, 1979) defined the quota as the cellular total of all forms of the nutrient in question, divided by cell biomass. Schemes for partitioning cellular nutrient pools (e.g. Maske, 1982; Davis, Breitner & Harrison, 1978, discussed by Droop, 1978) seem insufficiently well established to be of use in ecological models.

(2) The limiting nutrient is the factor in least supply relative to need, and is thus that with the lowest value of Q^*. The switching of control of growth predicted by this *threshold hypothesis* is well established for vitamin B12 and phosphorus (Droop, 1974, 1975) and for nitrogen and phosphorus (Rhee, 1974). Droop *et al.* (1982) have shown that the interaction between light and B12 limitation may be described analagously.

(3) Nutrient uptake can be partially uncoupled from growth, and depends on internal as well as external concentrations of the nutrient in question, and on the cell quota of the limiting nutrient (or on the concentration analogue for light) if this is different. Various formulae have been proposed (see Rhee, 1980; Droop, 1983; Tett & Droop, in press). Since the uptake kinetics of microalgae are also likely to be affected by diffusion gradients near cells (Pasciak & Gavis, 1974) and by the short-term kinetics of surface adsorption and, in some cases, by the effects of extra-cellular nutrient binding (Droop, 1968), the use in ecological models of simple Michaelis–Menten uptake equations can represent reality only approximately (see Goldman & Gilbert, 1983).

It is generally believed, following Dugdale (1967) and Ryther & Dunstan (1971), that nitrogen is the main potentially limiting chemical resource in the sea. Dugdale & Goering (1967) distinguished between 'new' and 'regenerated' forms of combined inorganic nitrogen, and the photoautotrophic production associated with each of these. Although cyanobacterial nitrogen fixation may be locally important, new nitrogen is mainly nitrate resulting from oxidative remineralization by deep-water or benthic organisms. Its supply by mixing or upwelling (McCarthy & Carpenter, 1983) is the main long-term constraint on the export of production from the superficial layers of the sea (Eppley & Peterson, 1979). Photic zone primary production may however be enhanced by locally excreted ammonium, and the importance of recycling, indicated by the reciprocal of the *f-ratio* of new to total primary production, has been much discussed, most recently by Platt and Harrison (1985).

There are three routes by which nitrogen (and indeed phosphorus) may be recycled in the photic zone. Best understood is that resulting from the grazing of larger phytoplankton by mesozooplankton such as calanoid copepods, which excrete as ammonium or urea about a third to a half of the organic nitrogen assimilated during their lifetimes (Corner, Cowey & Marshall, 1967; Corner & Davies, 1971; Bidigare, 1983). Given their lower overall growth efficiency (Scott, 1985), planktonic protozoans probably excrete a higher proportion of assimilated nitrogen. Bacteria may play an important role in remineralizing faecal organic nitrogen or dissolved organic nitrogen (DON) produced by cell leakage during zooplankton grazing (Hollibaugh, 1980), or in oxidizing ammonium to nitrate (Kaplan, 1983). The relative importance of each group of organisms is much debated and may vary between environments. Microorganisms supplied with particulate or dissolved organic material of low nitrogen content may excrete little ammonium, and bacteria may in these circumstances take up nitrate or ammonium in competition with algae.

MODELS AND MODELLING

A *mathematical model* is one or more equations describing the functional dependence of one or more quantities on time, space, and each other. The quantities are generally called *state variables*, and the equations may also contain *rate variables* and locally constant *parameters*. Models that contain one or more random terms are *stochastic*; those that do not are *deterministic*.

General *analytical solutions* to the equations of a model may sometimes be reached by exact or approximate symbol manipulation (e.g. Burghes & Borrie, 1981). Particular *numerical* solutions are obtained by substituting numbers for some of the terms, perhaps after analytical solution of the model. There is, however, a narrow sense in which the term numerical solution is reserved for models that cannot be solved analytically. *Numerical simulation* generally implies the use of a (digital) computer program to substitute numerical values repeatedly in the model equations, so representing the changes taking place during small steps in time or space. Parameter values are generally supplied to such simulations, and the aim is to follow changes in the state variables. In other cases, values of the state variables are known from observation and the task of *parameterization* involves the evaluation of these (local) constants. *Sensitivity analysis* concerns the response of a model to changes in one or more parameters. In an *empirical* model the functional relations are arbitrary, perhaps taking the form of correlations, regressions or time-series, and parameters are evaluated by purely statistical techniques. *Theoretical* models assume some insights into the nature of the functional relationships and their parameter values. Tett & Droop (in press) distinguished between *ecological frameworks*, which are sets of differential equations in the state and rate variables, and *physiological equations* expressing rate variables as functions of parameters and state variables.

In oceanographic terminology (Pond & Pickard, 1978), the *Lagrangian* perspective travels with a moving particle, whereas the *Eulerian* view involves a fixed spatial grid. The latter may have *one*, *two* or *three dimensions*, or may be collapsed to a *point model*. Numerical solutions of Eulerian models must be cast in terms of a finite number of grid-points, equivalent in the case of one-dimensional, vertical-process models to slicing the water columns of Fig. 1 and 2 into a number of layers, assumed to be horizontally infinite and well mixed internally. The model represents processes within, and exchange between, each layer (Figure. 3).

BOUNDARY CONDITIONS AND POINT MODELS

Analytical solutions often involve integration between defined limits in space and time. The temporal limits correspond to the initial and final states of a numerical simulation, and it is sometimes useful to solve the equations analytically for, or arrive numerically at, a steady state, in which there are no further changes with time and which may be independent of initial conditions. The spatial limits correspond to the *boundary conditions* of a numerical solution, and to real-world limits such as the air–sea and sediment–water interfaces. Model solutions are generally strongly influenced by assumptions about boundaries, as can be demonstrated by the framework equations for a well-mixed, continuously diluted, nutrient-limited microbial culture (Table 1).

This is a point model. The only exchanges are those across a single boundary and they describe dilution by fresh medium which contains nutrient but no microbes. The state variables are microbial biomass concentration, X, microbial nutrient quota, Q, and medium nutrient concentration, S. The three framework equations specify the rates of change of these variables as functions of the boundary exchange

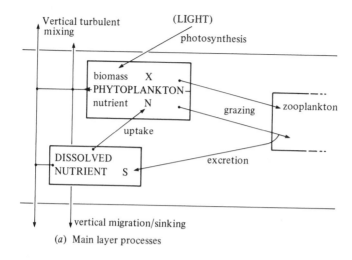

(a) Main layer processes

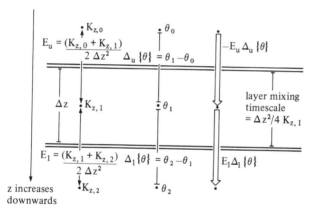

(b) finite difference exchange terms for temperature θ

Fig. 3. Processes described by one-dimensional models for each (non-boundary) layer of a water column. The derivation of a finite-difference equation for exchange between layers is illustrated for temperature, θ. The relation between specific exchange rates (E, day^{-1}) and turbulent diffusivities (K_z, m^2 day^{-1}) is shown.

and the intrinsic processes of growth and nutrient uptake. The latter terms are expanded in the physiological equations 1.4 and 1.5, which assume cell-quota growth and Michaelis–Menten uptake kinetics.

In a numerical solution the framework equations would be replaced by finite-difference approximations, one of the simplest forms being given in eqn 1.6. If the framework equations are simplified by removal of the boundary/dilution term, their numerical integration will decribe the time course of a batch culture. In eqn 1.4,

Table 1. *A diluted algal culture*

	Rate of change	Boundary exchange	Growth	Uptake	
Biomass	$\dfrac{dX}{dt}$	$= -DX$	$+\mu X$		(1.1)
Cell quota	$\dfrac{dQ}{dt}$	$=$	$-\mu Q$	$+u$	(1.2)
Dissolved nutrient	$\dfrac{dS}{dt}$	$= D(S_0 - S)$		$-uX$	(1.3)

Rate equations
Growth: $\mu = \mu'_m(1 - k_Q/Q)$ (1.4)
Uptake: $u = u_m \cdot S/(k_u + S)$ (1.5)

Finite-difference equation
Biomass: $X_{t+\Delta t} = X_t[1 + \Delta t(\mu - D)]$ (valid when $\Delta t(\mu - D) \ll 1$)   (1.6)

Variables
X: microbial biomass concentration
Q: microbial nutrient quota
S: medium nutrient concentration
t: time

Parameters

D: dilution rate $\left(\dfrac{dV}{dt}\dfrac{1}{V}\text{where } V \text{ is culture volume}\right)$
k_u: nutrient concentration for $\frac{1}{2}$ maximum uptake
k_Q: subsistence quota
S_0: nutrient concentration in new medium
u_m: maximum biomass-related nutrient uptake rate
μ'_m: maximum specific growth rate (when $k_Q/Q = 0$)

replacing μ by uy, where y is a yield constant, gives Monod growth. Setting dX/dt, dQ/dt and dS/dt to zero defines the conditions for a chemostat. Analytical solutions can be derived for this steady state for Monod and cell-quota growth (Herbert, Ellsworth & Telling, 1956; Tempest, 1970; Droop, 1976) and depend on (1) the values of the parameters in the rate equations 1.4 and 1.5, and (2) the boundary conditions D and S_0.

Jannasch (1974) has discussed the application of chemostat theory to natural microplanktonic populations. Despite his caution, equations analogous to those in Table 1 may be useful; Tett (1986) was able to deduce net growth rates for bulk phytoplankton in small fjordic estuaries treated as single, well-mixed, continuously diluted basins. 'Point-model' frameworks have been used with some success to examine competition amongst microbial species in continuous culture (e.g. Curds, 1974; Taylor & Williams, 1975; Tilman, 1977) and were employed by Riley (1946) in what was probably the first theoretically based numerical model of phytoplankton growth in the sea. But since the argument of this review is that water column physical structure is an important determinant of microplankton growth and distribution, I shall concentrate, in what follows, on models that allow for imperfect mixing and the exploration of the interaction between depth-varying mixing and depth-varying microbial processes.

THE DESCRIPTION OF MIXING

Smith (1975) provides an introduction to the complex and difficult topic of mixing in fluids. One approach is to view the process as the sum of the 'random walks' of each particle of the fluid. On all scales larger than a few millimetres the appropriate unit is a packet of water and associated dissolved and particulate tracers. Most models assume that rates of turbulent mixing deduced from observations of temperature change or dye dispersion are directly applicable to dissolved nutrients or non-motile organisms.

Woods & Onken (1982) modelled individual 'random walks' of phytoplankton through vertical light gradients, and summed these walks to obtain a statistical description of primary production in the wind-mixed layer of the ocean. Although such a stochastic, Lagrangian approach seems likely to prove fruitful in ecological modelling (see Tett & Edwards, 1984), it is presently expensive in computer time. Most models instead employ simple deterministic statistics in an Eulerian framework to provide a rough but numerically efficient representation of the effects of turbulent mixing on the distribution of tracers.

One such approach describes the flow of a tracer along a gradient as the product of the local gradient in the tracer and a *turbulent* or *eddy diffusivity*, which is a property of the water. Rates of change of concentration of a tracer at a point are then given by the rate at which the product of diffusivity and gradient changes with distance along the gradient. This approach is exemplified in Tables 2 to 4. The diffusivities can be parameterized from observations on any conservative tracer whose flux is known. King & Devol (1979) estimated vertical eddy diffusivities in an oceanic thermocline from data on nitrate gradients and fluxes, assuming that the only source of nitrate was deep-water mineralization and the only sink was wind-mixed-layer uptake by algae. Such assumptions may be questionable in seas with shallow thermoclines, and it is more usual to employ temperature as a tracer unaffected by biological activity. Pingree & Pennycuick (1975) used temperature time-series and gradients to estimate heat flows, and therefore diffusivities and nutrient mixing rates, in a shelf-sea thermocline. General values of diffusivity may be obtained from the literature or calculated from observations of wind speed, tidal current speed and density stratification (Bowden, 1983).

Vertical diffusivity, K_z, is in principle a continuous function of depth. A somewhat different approach can be adopted in numerical models where the water column is divided into discrete layers of thickness Δz. The relevant statistics are then those of (1) mixing timescale within a layer, calculable from $\Delta z^2/4K_z$, and (2) relative rates of exchange between layers, given by $K_z/\Delta z^2$. The first diffusivity refers to the mean value within a layer, which the numerical model will treat as well mixed. The limiting turbulence is thus that across the layer boundaries; the second diffusivity refers to the mean value between the centre of the layer in question and that above or below. The net change due to mixing is the sum of the products, at the upper and lower boundaries of the layer, of exchange rate and interlayer concentration differences (Fig. 3).

Exchange rate may be computed directly from a time series of concentrations of conservative tracers in adjacent layers, as in the case of the quarry data analysed below. Simpson & Bowers (1984) describe a two-layer model which predicts heat exchange between near-surface and deep layers as a function of tidal speed and meterological data. Such a model, which avoids the difficulties associated with the description of diffusivity, might provide a basis for a simple biological model of a two-layered shelf-sea water column. It does not, however, provide a detailed description of thermocline structure and cannot thus be used to predict microplankton distributions in seas with mid-water chlorophyll maxima.

The physical model of Gill & Turner (1976) describes the seasonal evolution of the wind-mixed layer of the ocean and predicts thermocline depth; although it was used by Kiefer & Kremer (1981) as the basis for a phytoplankton model it does not describe mixing within the thermocline. The model of James (1977) for the seasonal development of a shelf-sea thermocline, predicts diffusivity as a function of depth and thus could be (but has not yet been) used as the physical basis for several of the models discussed below.

A BULK PHYTOPLANKTON MODEL

One-dimensional (1-D), vertical-mixing-driven, numerical models with Monod-type growth have been described by Radach & Maier-Reimer (1975); Steele & Henderson (1976); Jamart *et al.* (1977); Jamart, Winter & Banse (1979); Fasham, Holligan & Pugh (1983). The 1-D model of Kiefer & Kremer (1981) is anomalous in that it does not allow exchange between layers. Like that of Walsh (1975), the upwelling model of Howe (1979, 1982) has a more complex physical and biological framework than the 1-D models but, unlike the other models, it uses cell-quota equations. A vertical-process model with cell-quota growth has been described by Tett (1981) and by Tett, Edwards & Jones (1986). Of all these modellers, only Fasham, Howe and Tett attempt detailed comparison of model predictions with particular observations. Tett, Edwards & Jones (1986) discuss the success of these comparisons, and Tett & Droop (in press) review in detail the nutrient-uptake and nutrient-controlled-growth equations in the models of Howe & Tett. I thus propose to concentrate here on the assumptions, structure and results of my own model, which is summarized in Table 2. This version differs slightly from that of Tett, Edwards & Jones (1986) in respect of irradiance attenuation, nutrient uptake and lower boundary conditions. The values given in the Table are mostly taken from that paper and are those appropriate to the mean Scilly Isles station in Fig. 1.

The ecological framework contains equations for four state variables. With the exception of eqn 2.1 for irradiance, each equation contains terms for changes due to turbulent diffusion, algal processes and zooplankton grazing.

Dissolved nutrient, S, (eqn 2.4) does not discriminate between oxidized and reduced forms of nitrogen. It is, nevertheless, possible to estimate f-ratios (eqn 2.9) from predictions of zooplankton excretion (corresponding to recycled nitrogen) and total phytoplankton nutrient uptake. Howe (1979, 1982) distinguishes ammonium, nitrate and silicate as potential limiting nutrients.

The model assumes that the phytoplankton is dominated by, or behaves as, a single species (see Jones *et al.*, 1978). Biomass and particulate nitrogen are thus measured by the bulk variables X and

Table 2. *Bulk phytoplankton model*

	Rate of change	Irradiance	Diffusion	Algal processes	Grazing	
Irradiance $(\mathrm{W\,m^{-2}})$	$\dfrac{dI}{dz}$	$= -I(\lambda_0$		$+ \varepsilon X)$		(2.1)
Biomass $(\mathrm{mg\,chl\,m^{-3}})$	$\dfrac{dX}{dt}$	$=$	$\dfrac{d}{dz}\!\left(K_z\dfrac{dX}{dz}\right)$	$+\mu X$	$-gX$	(2.2)
Algal nitrogen, $(\mathrm{mmol\,m^{-3}})$	$\dfrac{dN}{dt}$	$=$	$\dfrac{d}{dz}\!\left(K_z\dfrac{dN}{dz}\right)$	$+uX$	$-gN$	(2.3)
Dissolved inorganic nitrogen $(\mathrm{mmol\,m^{-3}})$	$\dfrac{dS}{dt}$	$=$	$\dfrac{d}{dz}\!\left(K_z\dfrac{dS}{dz}\right)$	$-uX$	$+egN$	(2.4)

Rate equations
Specific growth rate $(\mathrm{day^{-1}})$ $\mu = \mathrm{L}\{\mu'_m(1 - k_Q/Q), (\alpha \bar{I} - r^B) \cdot {}^{\mathrm{chl}}q\}$ (2.5)
Uptake $(\mathrm{mmol\,DIN\,mg\,chl})^{-1}\,\mathrm{day^{-1})}$ $u = [u_m(1 - Q/Q_m) \cdot S/(k_u + S)]$
$ + (\mu Q : \mu < 0; 0 : \mu \geqslant 0)$ (2.6)
$\bar{I}$: layer average irradiance

Boundary conditions
Irradiance $(\mathrm{W\,m^{-2}})$ $I = I_0 : z = 0$ (2.7)
Fluxes   ⎧ biomass ${}^X\phi = 0 : z = 0$ and $z = z_m$
$(\mathrm{m^{-2}\,day^{-1}})$ ⎨ algal N ${}^N\phi = 0 : z = 0$ and $z = z_m$
        ⎩ DIN ${}^S\phi = 0 : z = 0; S = S_m : z = z_m$
Related equations
Net production $(\mathrm{mgC\,m^{-3}\,day^{-1}})$ $p = \mu X / {}^{\mathrm{chl}}q$ (2.8)
Ratio 'new' : total production $f = 1 - (eg/\mu)$ (2.9)

Standard parameter values (Figs 1 and 4)
e: 0.5 (recycled proportion of grazed nutrient)
g: $0.12\,\mathrm{day^{-1}}$ (grazing impact)
I_0: $63\,\mathrm{W\,m^{-2}}$ (24 h mean irradiance)
k_Q: $0.2\,\mathrm{mmol\,N\,(mg\,chl)^{-1}}$
k_u: $0.3\,\mathrm{mmol\,DIN\,m^{-3}}$
Q_m: $1.0\,\mathrm{mmol\,N\,(mg\,chl)^{-1}}$
${}^{\mathrm{chl}}q$: $0.02\,\mathrm{mg\,chl\,(mg\,C)^{-1}}$
r^B: $3.5\,\mathrm{mgC\,(mg\,chl)^{-1}\,day^{-1}}$ (respiration)
S_m: $4.0\,\mathrm{mmol\,DIN\,m^{-3}}$
u_m: $2.0\,\mathrm{mmol\,DIN\,(mg\,chl)^{-1}\,day^{-1}}$
z_m: 90 m (depth of sea-bed)
α: $4.1\,\mathrm{mg\,C\,(mg\,chl)^{-1}\,day^{-1}\,(W\,m^{-2})^{-1}}$
ε: $0.012\,\mathrm{m^2\,(mg\,chl)^{-1}}$
λ_0: $0.14\,\mathrm{m^{-1}}$ (seawater attenuation)
μ'_m: $1.2\,\mathrm{day^{-1}}$
$\mathrm{L}\{\,\}$ refers to least of values in braces

N (eqns 2.2 and 2.3), with the cell quota defined as the ratio N/X. The question of the preferred units for biomass is discussed by Tett & Droop (in press). Cell number does not provide a good basis for standardizing many physiological parameters (Banse, 1976; Droop, 1979). Algal carbon is difficult to measure chemically because of the problem of detrital corruption (Banse, 1977) and is tedious to determine from microscopical measurements. Chlorophyll is thus the practical choice, and has advantages in relation to photosynthetic and self-shading terms in the model. It does, however, necessitate information on the cell chlorophyll quota, or chlorophyll:carbon ratio, which I treat as a fixed although perhaps a species-dependent parameter ^{chl}q. The term enters into eqn 2.5 and into the calculation of some cell-quota parameters. Fasham, Holligan & Pugh (1983), Jamart, Winter & Banse (1979) and Tett, Edwards & Jones (1986) report the sensitivity of model predictions to variation in assumed chlorophyll:carbon ratio. Howe's (1979, 1982) model allows for variation in cell chlorophyll content and size, as well as in nitrogen and silicon quotas. This would seem a satisfactory alternative if we were better informed on the control of cell chlorophyll content by light and nutrient; see for example Laws & Bannister (1980).

Phytoplankton growth is predicted (eqn 2.5) by a threshold cell-quota hypothesis, the rate being the lower of the alternatives deduced from the cell nitrogen quota or the net rate of photosynthesis. Because light limitation occurs when the mean irradiance experienced by algae is low, it is possible to use a simple linear function for photosynthesis. The value chosen for the coefficient, α, of this function, corresponded to the maximum possible quantum yield (Droop et al., 1982; Tett & Droop, in press; cf. Platt, 1986). The nutrient uptake expression (eqn 2.6) has been modified from Tett, Edwards & Jones (1986) to correspond better to the results of Droop (1974, 1975) and Rhee (1973). The modified equation causes uptake to saturate as the cell quota approaches a maximum value, Q_m, as well as when seawater nutrient concentration exceeds the half-saturation constant for uptake, k_u. In addition, negative growth rate, the result of an excess of respiration over gross photosynthesis, now causes excretion of nitrogen.

Biomass is removed in the model, and some of its nutrient recycled, by a term g, referring to the proportion of phytoplankton crop grazed in unit time. This proportion can be computed from data on zooplankton biomass and their volume clearance rate (see below), but in the absence of such data Tett, Edwards & Jones

(1986) estimated the grazing impact as 0.12 day^{-1} by fitting the model to data from another of the Scilly Isles stations described by Simpson *et al.* (1982). The fraction, *e*, of grazed nutrient which is recycled was taken as a half. The model is sensitive to the value chosen for grazing; similar conclusions were reached by Howe (1982) and Jamart, Winter & Banse (1979).

Boundary conditions are such that there are no flows of nutrients or biomass across the sea surface, nor of particulates at the sea-bed. Deep-water nutrient concentration is, in the present version, held constant at the highest observed value at the relevant stations. The result is a simulated flow of nutrients across the lower boundary of the model which may be ascribed to benthic regeneration.

The model is driven by the depth distribution of turbulent diffusivity, K_z, which was estimated from observed temperature gradients at the relevant stations on the assumption of a constant vertical heat flux of 700 kcal m^{-2} day^{-1} (Pingree *et al.*, 1975). Diffusivities thus estimated were about 20 m^2 day^{-1} in the wind-mixed layer, less than 5 m^2 day^{-1} in the main part of the thermocline, and at least 500 m^2 day^{-1} in the tide-mixed layer. These latter values are probably too low, since calculations (e.g. Pingree & Griffiths, 1977) from tidal flows suggests values of 10^3 to 10^4 m^2 day^{-1}. The discrepancy may have arisen from difficulties in observing accurately the very small temperature gradients found in the deep water.

The model does not allow for diel variation in illumination, stratification or grazing, and so the boundary condition, I_0, for photosynthetically effective irradiance, I, is the 24 h mean sea-surface value, corrected for surface losses. The present version allows for self-shading, and so irradiance changes with depth (eqn 2.1) at an exponential rate partly determined by the local concentration of chlorophyll. The value for the pigment absorption cross-section, ε (also called the *in vivo* specific extinction), was taken from Tett & Droop (in press) and λ_0, the residual attenuation of downwelling light by water and nonalgal particulates, was deduced from measurements or submarine irradiance at the relevant stations.

Numerical simulations were run from arbitrary but realistic initial distributions of the state variables. After 80 simulated days a steady state had been reached. Results are shown in Fig. 1 and 4 for the standard values of Table 2. With the exception of the diffusivities, deep nutrient concentration and residual optical attenuation, all parameters were derived from the literature or from fitting the model to data at a different station.

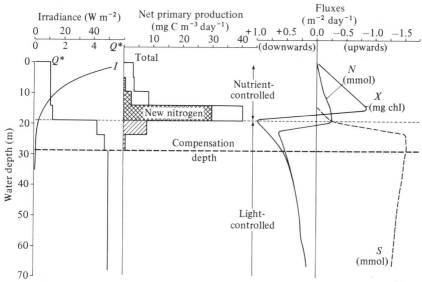

Fig. 4. Steady-state predictions of the model of Table 2, with standard parameter set, for the mean Scilly Isles station of Fig. 1. Continuous profiles are shown for photosynthetically available irradiance (I, Wm^{-2}) and for vertical (turbulent) fluxes of biomass (X, $mg\,chl\,m^{-2}\,day^{-1}$) and nutrient (particulate, N, and dissolved, S, $mmol\,N\,m^{-2}\,day^{-1}$). Note that with depth, z, increasing downwards, a negative flux indicates an *upwards* movement. The stepped profiles show values for each layer of standardized cell quota ($Q^* = Q/k_Q$, where $Q = N/X$) and net primary production ($mg\,C\,m^{-3}\,day^{-1}$). The euphotic zone above the compensation depth is divided into regions in which production is nutrient- or light-controlled. In the former region the part of production not supportable by recycled nitrogen is shown hatched.

A model is most conveniently assessed by comparing its predictions with observations of the state variables. Tett, Edwards & Jones (in press) found that 35% of observed variations in chlorophyll concentration was predicted by the model after *a priori* parameterization for a group of Scilly Isles. This could be increased to 71% by varying the grazing term so as to fit the model to data from a particular station. In the present case the standard simulation predicts 53% of the variation in dissolved inorganic nitrogen and 37% of the variation in chlorophyll (omitting the highest value, which can probably only be explained by vertical migration: see below).

The rate predictions are of some interest. Fig. 4 shows carbon production and diffusive fluxes of the state variables as functions of depth. These graphs show the following:

(1) Predicted compensation depth occurs at 28 m. This provides a better-defined estimate of the thickness of the *euphotic* zone. Note that, because of the discrete depth-step in the simulations,

compensation depth is a discontinuous variable. It is the bottom of the deepest layer in which light-limited growth exceeds zero.

(2) The euphotic zone consists of a lower region in which growth is limited by light and an upper region in which it is limited by nutrient. In this upper region the proportion of growth supportable by uptake of recycled nutrients can be computed (eqn 2.9). Limitation in a particular layer can be determined by substituting values of layer average irradiance, $\bar{I}$, and Q^* into eqn 2.5, to see which predicts the lower rate of growth.

(3) The biomass peak (Fig. 1) occurs in the light-limited region of the thermocline and absorbs most of the upwards flux of dissolved nutrient. It is additionally supported by a flux of particulate nutrient, to be interpreted as a mixing upwards of phytoplankton made nutrient-rich by luxury uptake in deeper water. Biomass is dispersed from the maximum by the same turbulence but this removes cells with a lower nutrient content.

Finally, depth averaging of the results of the simulation allows the construction of budgets. Predicted euphotic zone carbon production is $300\,\mathrm{mg\,m^{-2}\,day^{-1}}$, of which 272 is immediately grazed and 28 exported downwards by turbulence. Algal respiration in the aphotic zone accounts for $9\,\mathrm{mg\,m^{-2}\,day^{-1}}$, and the remaining 19 is taken by zooplankton. As discussed by Tett, Edwards & Jones (1986) the production rates predicted by the model are lower than those measured by carbon-14 uptake at similar stations in the English Channel, perhaps because the model underestimates chlorophyll concentrations.

Predicted euphotic zone nitrogen uptake is $2.50\,\mathrm{mmol\,N\,m^{-2}}$ $\mathrm{day^{-1}}$, of which 1.00 can be met from local excretion by zooplankton and 1.50 by turbulent diffusion from deep water. The overall f-ratio is thus 0.67. Zooplankton and algal excretion in the aphotic zone provide $0.34\,\mathrm{mmol\,N\,m^{-2}\,day^{-1}}$ of the 'new' nitrogen, and the remainder, 1.20, must be supplied by implied benthic regeneration. This is within the range of $1\text{--}4\,\mathrm{mmol\,N\,m^{-2}\,day^{-1}}$ given by Lancelot & Billen (1985) for coastal sediments. Other fluxes are, however, lower than those estimated by Holligan, et al. (1984b) for their equivalent station F in the English Channel and by Harrison et al. (1983) for the Middle Atlantic Bight. The explanation may lie partially in underestimation of zooplankton excretion, the water column total of which Holligan et al. (1984b) estimated as $8.4\,\mathrm{mmol\,N\,m^{-2}\,day^{-1}}$, and partly in underestimation of turbulence in the tide-mixed layer.

MICROPLANKTON DIVERSITY: VERTICAL MIGRATION

Questions relating to the diversity of the plankton have been long discussed. In 'The paradox of the plankton', Hutchinson (1961) pointed out that there seemed to be insufficient ecological niches for the observed number of species of plankton. Several freshwater ecologists have investigated dominance relationships amongst a few, usually closely related, species, modelling this in terms of nutrient-growth kinetics (e.g. Tilman, 1977) or photosynthetic properties (e.g. Talling, 1957). In contrast, marine phytoplanktonologists have concentrated on identifying ecologically relevant differences, such as those in photosynthetic properties resulting from pigment differences (e.g. Ryther, 1956; Atlas & Bannister, 1980), amongst the main taxonomic groups of algae. Clear-cut distinctions in physiological parameters are, however, not always evident. Tett & Droop (in press) could find no systematic differences amongst published values for nitrogen and phosphorus subsistence quotas for several classes of eukaryotic algae. An alternative approach is to model differences between functional groups based on (1) interaction with vertical turbulence (Margalef, 1978; Bowman, Esaias & Schnitzer, 1981), and (2) size (Banse, 1976; Shuter, 1978).

Algae can interact with turbulence, or the water column structure resulting from turbulence, through vertical movement. The model in Table 3 is derived from that in Table 2 with the addition of such a movement term. The function $f(Q)$ (eqn 3.5) is such as to result in simulated upwards swimming or floating when cells are nutrient-replete, and downwards swimming or sinking when the cell quota is close to the subsistence quota. Although not representing the diel component of vertical migration observed by Cullen & Horrigan (1981) and Heaney & Eppley (1981), some such function would seem a reasonable interpretation of the net result of nutrient limitation on the dinoflagellate migrations studied by these authors. They found typical swimming rates of up to $1\,m\,h^{-1}$, and so a maximum rate of $5\,m\,day^{-1}$ would seem plausible for migration net of cyclic diel movements.

Although such vertical migration can sharpen a midwater chlorophyll maximum, it cannot alone bring about the large biomasses observed in blooms. Reduced grazing is also required (Tett, 1984) to predict a large midwater chlorophyll maximum, as in Fig. 5. The model predicts some interesting differences from the state of affairs shown in Fig. 4:

Table 3. *Phytoplankton model with vertical movement*

	Rate of change		Irradiance	Diffusion	Algal processes	Algal movement	Grazing	
Irradiance (W m^{-2})	$\dfrac{\mathrm{d}I}{\mathrm{d}z}$	$=$	$-I(\lambda_0)$		$+\varepsilon X)$			(3.1)
Biomass (mg chl m^{-3})	$\dfrac{\mathrm{d}X}{\mathrm{d}t}$	$=$		$\dfrac{\mathrm{d}}{\mathrm{d}z}\!\left(K_z\dfrac{\mathrm{d}X}{\mathrm{d}z}\right)$	$+\mu X$	$-w\dfrac{\mathrm{d}X}{\mathrm{d}z}$	$-gX$	(3.2)
Algal nitrogen (mmol m^{-3})	$\dfrac{\mathrm{d}N}{\mathrm{d}t}$	$=$		$\dfrac{\mathrm{d}}{\mathrm{d}z}\!\left(K_z\dfrac{\mathrm{d}N}{\mathrm{d}z}\right)$	$+uX$	$-w\dfrac{\mathrm{d}N}{\mathrm{d}z}$	$-gN$	(3.3)
Dissolved inorganic nitrogen, (mmol m^{-3})	$\dfrac{\mathrm{d}S}{\mathrm{d}t}$	$=$		$\dfrac{\mathrm{d}}{\mathrm{d}z}\!\left(K_z\dfrac{\mathrm{d}S}{\mathrm{d}z}\right)$	$-uX$		$+egN$	(3.4)

Rate equation

$$w = w_m f(Q): \quad f(Q) = 2(Q_w - Q)/(Q_m - k_Q): \quad \begin{array}{l}\text{dinoflagellates}\\ 1 \quad : \quad \text{diatoms}\\ 0 \quad : \quad \text{others}\end{array} \qquad (3.5)$$

Parameter values (Fig. 5)

g : 0.01 day^{-1}

r_B : 3.5 mg C (mg chl)$^{-1}$ day^{-1} (diatoms); 10 mg C (mg chl)$^{-1}$ day^{-1} (dinoflagellates and others)

Q_w : 0.6 mmol N (mg chl)$^{-1}$

w_m : 5 m day^{-1} (dinoflagellates); 2 m day^{-1} (diatoms); 0 m day^{-1} (others)

(See Table 2 for other equations and parameters)

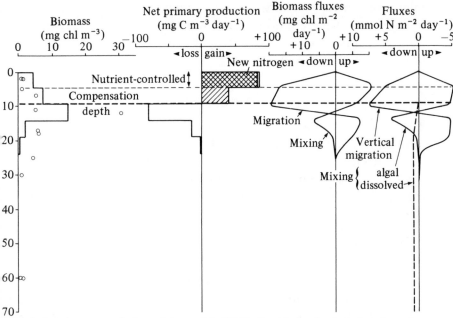

Fig. 5. Steady-state predictions of the model of Table 3, with simulated nutrient-dependent vertical migration, for the mean Scilly Isles station of Fig. 1. Circles show all observed chlorophyll concentrations. Migration and turbulent mixing fluxes are shown. See also legend to Fig. 4.

(1) Compensation depth now occurs, because of self-shading, at 9.5 m. The biomass peak lies in the aphotic zone, maintained by vertical migration from the euphotic zone.

(2) Although predicted euphotic zone production is, at 594 mg C m^{-2} day^{-1}, higher than that for the standard simulation in Fig. 4, and is enhanced by turbulent diffusion from the biomass peak, most of the new organic carbon is taken into the aphotic zone by vertical migration and respired at the biomass peak. Predicted water column production is thus only 137 mg C m^{-2} day^{-1}.

(3) Because grazing is low, little nitrogen is recycled and the euphotic zone f-ratio is 0.97. Vertical turbulent fluxes of dissolved inorganic nitrogen are less than in the standard simulation, and most of the nitrogen required to support predicted production comes to the euphotic zone in nutrient-rich algae mixed from the biomass peak.

Points 1 and 3 above describe the respective roles of active vertical migration and passive turbulent transport at this station. Vertical migration conveys algal carbon (and hence photosynthetically fixed energy) out of the euphotic zone, whereas turbulence carries back into the euphotic zone the particulate nitrogen which that migrated biomass has assimilated.

MICROPLANKTON DIVERSITY: SIZE AND GRAZING

The model in Table 4 is a further development of the single-species model in Table 3. It allows for differences between the main functional groups of algae resulting from different patterns of vertical movement and from differences in size. Simulations have been run for a microplankton consisting of three types of alga and one protozoan.

The simulated algae may be equated with dinoflagellates, diatoms and small flagellates. Dinoflagellates are given the swimming behaviour described above. Diatoms are allowed to sink at a fixed rate of $2 \, \text{m day}^{-1}$. It is assumed that small flagellates have no powers of vertical movement. Cell-quota parameters keep the standard values of Table 2 for all algae. It seems that dinoflagellates and small flagellates may have a higher biomass-related respiration rate than diatoms (see Tett & Droop, in press), and the rate given to dinoflagellates in this model and that of Table 3 derives from observations made by Holligan et al. (1984b) and D. Purdie (personal communication) on blooms of Gyrodinium aureolum.

The importance of surface processes argues that the maximum biomass-related rate of nutrient uptake should decrease with increasing cell size, since the ratio of surface area to cell volume is inversely proportional to the cell radius. Pasciak & Gavis (1974) have shown that, at low external concentrations of nutrient, molecular diffusion can substantially limit nutrient transport near cells larger than (my calculations) about 5 μm diameter. Small flagellates were thus given more efficient nutrient uptake kinetics than diatoms or dinoflagellates. Other differences between algal groups are detailed in Table 4 and derive from points discussed by Tett & Droop (in press).

Grazing is seen mainly as a consequence of size. Diatoms and protozoans are thus subject to constant removal by mesozooplankton. The impact of mesozooplankton on dinoflagellates is set low in view of the apparent unpalatability of species such as Gyrodinium aureolum. In order to simplify this early stage in the development of a multiple-group model it is assumed that mesozooplankton do not feed directly on small flagellates, and that protozoans eat nothing else. No account is taken of the production of particles or dissolved organics during feeding or by defaecation, nor of the heterotrophic bacteria that may be supported by this material (see Fasham, 1985).

The protozoan component (eqn 4.5) of the model includes terms for growth (eqn 4.7) at a rate determined by the assimilation of carbon or nitrogen, whichever is in least supply, and for excretion of nitrogen (eqn 4.6) at such a rate as to maintain a constant cellular nitrogen:carbon ratio. The assumption of constant composition thus renders simulated protozoans fundamentally different from algae as described by the cell-quota model, and requires further laboratory investigation. The rate at which protozoans obtain food is described by the product of specific clearance rate and phytoplankton biomass. Calculations made from Heinbokel (1978) suggests that the tintinnid ciliate *Tintinnopsis* sweeps clear about $60 \, l \, day^{-1} \, mg^{-1}$ tintinnid carbon. Fenchel (1982) gives a mean clearance rate for small zooflagellates equivalent to $24 \, l \, day^{-1} \, mg^{-1} \, C$. Laboratory investigations of one species of heterotrophic dinoflagellate and two of ciliates feeding on algae in pure culture give clearance rates of $5–33 \, l \, day^{-1} \, mg^{-1} \, C$ (A. Fuller, personal communication). These rates contrast with values for crustacean zooplankton, which are typically about $0.5 \, l \, day^{-1} \, mg^{-1} \, C$, according to data in Parsons & Lebrasseur (1970).

Numerical work with this model has only just begun, and adequate physiological data for parameterization are not fully available. Nevertheless, results in Fig. 6 show that the phased oscillations that might be expected to result from Lotka–Volterra interactions between protozoans and their small flagellate prey, and the effects of these interactions via nutrient supply on the other algae, are damped to give a roughly stable community which still contains the original three kinds of alga even after 80 days. Dinoflagellates have come to dominate algal biomass by this time, as observed in nature, but some diatoms and small flagellates remain. This persistence of diversity needs further exploration, but seems likely to be at least partly the result of interactions between the behaviour of each type of alga and the vertical turbulent structure of the environment.

A BULK MODEL FOR AUTOTROPHIC AND HETEROTROPHIC PROCESSES

The model of Table 4 might be further developed by the addition of more kinds of microplankton, perhaps including algal and bacterial picoplankton (see Fogg, 1986), and by the separation of

Table 4. *Microplankton model*

	Rate of change	=	Irradiance	Diffusion	Algal processes	Vertical movement	Protozoan processes	Mesozooplankton grazing	
Irradiance (W m⁻²)	$\dfrac{dI}{dt}$	$=$	$-I(\lambda_0$		$+^i\varepsilon^i X$				(4.1)
Biomass of alga i (mg chl m⁻³)	$\dfrac{d^iX}{dt}$	$=$		$M_z\{^iX\}$	$+^i\mu^i X$	$-^iw\dfrac{d^iX}{dz}$	$-kP^ia_1^iX$	$-g^ia_2^iX$	(4.2)
Nitrogen of alga i (mmol N m⁻³)	$\dfrac{d^iN}{dt}$	$=$		$M_z\{^iN\}$	$+^iu^iX$	$-^iw\dfrac{d^iN}{dz}$	$-kP^ia_1^iN$	$-g^ia_2^iN$	(4.3)
Dissolved inorganic nitrogen (mmol N m⁻³)	$\dfrac{dS}{dt}$	$=$		$M_z\{S\}$	$-\hat{\Sigma}^iu^iX$		$+\eta P$	$+\hat{\Sigma}eg^ia_2^iN$	(4.4)
Protozoan biomass, (mg C m⁻³)	$\dfrac{dP}{dt}$	$=$		$M_z\{P\}$	$+eg^Pa_2^PNq^P$	$-^Pw\dfrac{dP}{dz}$	$+^P\mu P$	$-g^Pa_2P$	(4.5)

Rate equations

Protozoan nitrogen excretion (mmol N $(mg\,C)^{-1}\,day^{-1}$) $\eta = \hat{\Sigma} k^i a_1^i N - {}^P \mu^N q^P$ (assimilation – growth requirement)   (4.6)

Protozoan growth rate (day^{-1}) $^P \mu = L\{\hat{\Sigma} k^i a_1^i X / {}^{chl} q^i - r^P\}$, $(\hat{\Sigma} k^i a_1^i N / {}^N q^P)$, $^P \mu_m\}$ (4.7)

Boundary condition

Protozoan biomass flux $(mg\,C\,m^{-2}\,day^{-1})$ $^P \phi = 0 : z = 0$ and $z = z_n$ (4.8)

Parameter values (for Fig. 6)

a_1 : 1 small flagellates; 0 (others)

a_2 : 0 (small flagellates); 0.2 (dinoflagellates); 1 (others)

g : 0.07 day^{-1} (mesozooplankton grazing)

k : 0.030 m^3 $(mg\,protozoan\,C)^{-1}\,day^{-1}$

k_u : 0.1 (small flagellates), 0.3 (others) mmol DIN m^{-3}

i : 1 (diatoms), 2 (dinoflagellates), 3 (small flagellates)

n : 3 (algal types)

^{chl}q : 0.033 (diatoms), 0.02 (dinoflagellates), 0.040 (small flagellates) mg chl $(mgC)^{-1}$

$^N q^P$: 0.0108 mmol N $(mg\,protozoan\,C)^{-1}$

r^P : 0.4 day^{-1} (protozoan specific respiration rate)

u_m : 1 (diatoms, dinoflagellates), 5 (small flagellates) mmol DIN $(mg\,chl)^{-1}\,day^{-1}$

$^P \mu_m$: 1.0 day^{-1} (protozoan maximum specific growth rate)

$M_z\{\ \}$: is the vertical diffusion operator; for example

$$M_z\{S\} = \frac{d}{dz}\left(K_z \frac{dS}{dz}\right)$$

(see Tables 2 and 3 for other values and equations)

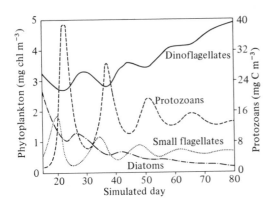
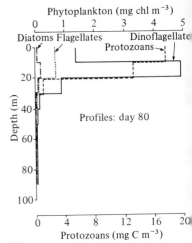

Fig. 6. Prediction of the model of Table 4 for the Scilly Isles mean station of Fig. 1. Depth increment is 10 m. Time-series are for the layer from 10 m to 20 m depth, starting from arbitrary initial conditions at day 0. Vertical profiles shown for simulated day 80.

nutrient species (as in the model of Howe, 1979, 1982). Such developments are, however, limited by the extent that complex models can be parameterized, programmed, and tested against observation. Until new optical techniques (Yentsch & Yentsch, 1984) become routinely available, observations of the distribution of individual kinds of microplankton must be based on time-consuming microscopy. The advantage of the bulk model of Table 2 is that its predictions of algal biomass distribution may be tested by simple, routine, chemical or fluorescence measurements of chlorophyll concentration. There may be equal advantage in extending the bulk model approach to include other components of the microplankton.

Table 5 describes such a model. In essence, photoautotrophic processes are made simple functions of chlorophyll concentration, representing algal biomass, X; heterotrophic processes are made simple functions of ATP concentration, representing total microplankton biomass, B, and *including the algal component*. The framework equations include radiant energy, turbulent diffusion, algal photoassimilation, microplankton mineralization, and microplankton vertical movement terms. Dissolved nutrient, S, refers to all forms of dissolved inorganic nitrogen. Temperature, θ, is included as an explicit state variable, allowing either for the prediction of thermal layering given data on diffusivities, or for the parameterization of diffusivity or layer exchange given a temperature time-series. Oxygen concentration, O, is included on analogous grounds: either oxygen changes

	Rate of change	Irradiance	Diffusion	Autotrophic processes	Heterotrophic processes	Vertical movement	
Heterotrophic biomass (mg ATP m^{-3})	$\dfrac{dB}{dt}$		$= \quad M_z\{B\}$	$+pX^Bq$	$-rB^Bq$	$-w\dfrac{dB}{dz}$	(5.1)
Irradiance (W m^{-2})	$\dfrac{dI}{dt}$	$= \quad -\lambda I$					(5.2)
Dissolved oxygen ($\text{ml O}_2\,\text{m}^{-3}$)	$\dfrac{dO}{dt}$		$= \quad M_z\{O\}$	$+pX^Oq$	$-rB^Oq$		(5.3)
Dissolved inorganic nitrogen (mmol N m^{-3})	$\dfrac{dS}{dt}$		$= \quad M_z\{S\}$	$-pX^Nq$	$+rB^Nq$		(5.4)
Autotrophic biomass (mg chl m^{-3})	$\dfrac{dX}{dt}$		$= \quad M_z\{X\}$	$+pX^Xq$	$-rB^Xq$	$-w\dfrac{dX}{dz}$	(5.5)
Temperature (°C)	$\dfrac{d\theta}{dt}$	$= \quad \lambda I/c$	$+M_z\{\theta\}$				(5.6)

Rate equation

Gross photosynthetic rate ($\text{mg C (mg chl)}^{-1}\,\text{hr}^{-1}$) $p = p_m I/(I_k^2 + I^2)^{\frac{1}{2}}$ where $I_k = p_m/\alpha$ (5.7)

Finite-difference equations (see Fig. 3)

Temperature (°C) $\dfrac{\Delta\theta}{\Delta t} = I_u(1 - e^{-\Lambda\Delta z})/c - E_u\Delta_u\{\theta\} + E_e\Delta_e\{\theta\}$ (5.8)

Dissolved oxygen, ($\text{ml O}_2\,\text{m}^{-3}$) $\dfrac{\Delta O}{\Delta t} = -E_u\Delta_u\{O\} + E_e\Delta_e\{O\} + pX^Oq - rB^Oq$ (5.9)

Parameter values (for Fig. 7 and 9)

C : $4290\,\text{kJ m}^{-3}\text{°C}^{-1}$ (specific heat of seawater)

P_m	: $2.2–3.6\,\text{mg C (mg chl)}^{-1}\,\text{hour}^{-1}$	X_q : $0.028\,\text{mg chl (mg C)}^{-1}$
B_q	: $0.0060\,\text{mg ATP (mg C)}^{-1}$	r : $60\,\text{mg C (mg ATP)}^{-1}\,\text{day}^{-1}$
N_q	: $0.0126\,\text{mmol N (mg C)}^{-1}$	α : $0.11–0.18\,\text{mg C (mg chl)}^{-1}\,\text{hour}^{-1}$
O_q	: $2.4\,\text{ml O}_2$	λ : $0.20–0.30\,\text{m}^{-1}$
		Δz : $4\,\text{m}$

(Boundary conditions unspecified, except for I. See Table 4 for $M_z\{\ \}$ and Fig. 3 for $\Delta\{\ \}$.)

can be predicted, or data on oxygen time-series can be used to estimate the respiration-mineralization parameter, r. The rate, p, of carbon photoassimilation is calculated from irradiance, I, using the equation of Smith & Talling (Talling, 1957). In the interests of simplicity the model employs fixed ratios of organic carbon to ATP biomass, free oxygen, organic nitrogen, and chlorophyll.

This model is in a preliminary form, and may be oversimplified. Its main purpose is, however, to examine the interaction between vertical turbulent exchange, light penetration and microbial processes, and it thus seems important to describe the latter as simply as possible. Its use to estimate parameters and construct budgets for parts of the water column will be briefly illustrated with data from the flooded quarry described above (Fig. 2).

Microscopical estimates of phytoplankton and microheterotroph biomass allowed estimates of $167 \, g \, C \, g^{-1}$ ATP and $36 \, g \, C \, g^{-1}$ chl a for these conversion ratios. The atomic ratio of carbon to nitrogen was taken as $6.6:1$, on the basis of the Redfield (1958) stoichiometry, and the molar ratio of oxygen to carbon as $1:1$ (ammonium-dominated) or $1.3:1$ (nitrate-dominated). Photosynthetic parameters were estimated from light-gradient incubator measurements of carbon-14 uptake rates. α ranged from 0.11 to $0.18 \, mg \, C \, mg^{-1}$ chl a h^{-1} $(W \, m^{-2})$, and p_m from 2.2 to $3.6 \, mg \, C \, mg^{-1}$ chl $a \, h^{-1}$ (K. Jones, personal communication). Optical attenuation was calculated from profiles of downwelling irradiance in the quarry, and used with hourly surface irradiances to predict submarine photosynthetically effective irradiance. Profiles of photosynthesis, computed from eqn 5.7, are shown in Fig. 7 for 2 days in May 1984. The first of these days was exceptionally sunny, and the second quite overcast. Microplankton respiration rates of $0.14 \, g \, O_2 \, mg^{-1}$ ATP day^{-1} were estimated from bottle experiments (K. Jones, personal communication) and of $0.14-0.18 \, g \, O_2 \, mg^{-1}$ ATP day^{-1} from deep-water changes in oxygen concentration (Fig. 8 and see below). An algal respiration of $0.027 \, g \, O_2 \, mg^{-1}$ chl $a \, day^{-1}$ was assumed for the purpose of other calculations in Fig. 7. The profiles of photosynthesis and respiration show that microplankton heterotrophic processes dominated photoautotrophy at depths below about 11 m.

As already mentioned, the near-surface waters of the quarry were stirred twice daily by tidal inflow. Attention was thus turned to the water near the top of the thermocline, which was less vigorously mixed. Rates of vertical exchange in this region were calculated from the temperature time-series (A. Edwards personal communica-

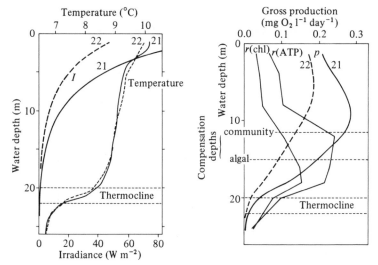

Fig. 7. Further profiles from Easdale Quarry, May 1984. Photosynthetically effective irradiances (I, W m^{-2}) are 24 h means for a sunny (21) and a cloudy (22) day. Profiles of gross, biomass-related photosynthesis were computed from hourly irradiances (using eqn 5.7) and observed depth-varying photosynthetic parameters, and were multiplied by the appropriate chlorophyll profile to give the gross production (p, mg$O_2 l^{-1}$day^{-1}). Phytoplankton respiration (r(chl)) was calculated from the mean chlorophyll profile (Fig. 2) using a respiration rate of 27 mg O_2 mg chl^{-1} day^{-1}. Microplankton community respiration (r(ATP)) was computed from the mean ATP profile and a respiration rate of 16 mg O_2 mg ATP^{-1} day^{-1}.

tion) in Fig. 8 using a finite-difference form (eqn 5.8) of eqn 5.6. Water at 20 m exchanged with that at 16 m at rates falling from 0.093 day^{-1} in early May to 0.044 day^{-1} in August. Exchange with 24 m was 0.021 day^{-1} in early May, falling to 0.003 day^{-1} in August. These are equivalent to diffusivities of 0.05–1.5 m^2 day^{-1}, and imply mixing timescales for the region between 18 m and 22 m of 5–20 days.

Figure 9 summarizes budgets of oxygen, dissolved inorganic nitrogen (DIN), ATP-carbon and algal biomass calculated for 20 m depth during June, July and August using finite-difference approximations to eqns 5.1, 5.3, 5.4 and 5.5. Photosynthesis rates were calculated, as before, from eqn 5.7 and predicted irradiance. Rates of oxygen change were corrected for exchange with surrounding waters, using the time-series data (A. Edwards & K. Petrie, personal communication) in Fig. 8, and for photosynthesis, and then related to mean ATP concentrations in order to estimate microplankton respiration rates (the remaining unknown in eqn 5.3 and its finite-difference equivalent in eqn 5.9). Although nitrate and ammonium concentrations were generally summed, their rates of turbulent diffusion

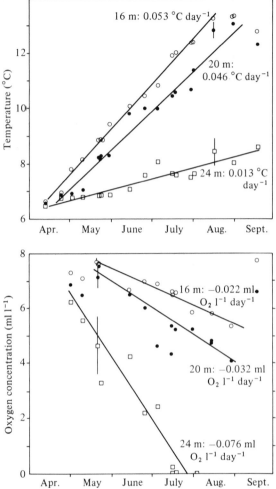

Fig. 8. Temperature and oxygen time-series for three depths in Easdale Quarry, 1984. The lines have been fitted by least-squares regression, omitting early or late points from periods when hydrographic or biological conditions were unstable. Vertical lines drawn about some points show variability of measurements during 24 h, resulting mainly from an imprecision of about 20 cm in sampling in relation to the isotherm.

were treated separately in solving eqn 5.4 because of their crossed gradients: nitrate dominated dissolved nitrogen at 20 m, but ammonium was more important at 24 m. Finally, the equations for biomass were balanced by assuming a sedimentation input; such an input, when divided by observed vertical gradients, gave a sedimentation rate for algae and microplankton.

Figure 9 shows that the budgets for microplankton and for phytoplankton were dominated at this depth by sinking and by respiration

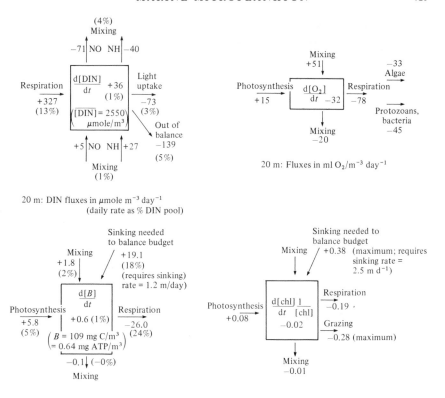

Fig. 9. Flux budgets at 20 m depth in Easdale Quarry, June–August, 1984. See text for explanation.

and grazing. The latter losses consumed each day a large fraction of the 20 m pool of biomass. In the budgets for dissolved oxygen and nitrogen, however, mixing processes were also important. Exchange should affect all tracers equally; the differences between dissolved and particulate components must thus have been due to differences in their vertical gradients. The nitrogen budget reveals an imbalance (Grantham, 1984) which might be explained if the assumed stoichiometry were wrong or if denitrification (Hattori, 1983) had been taking place. In the case of the sedimentation required to balance the biomass budgets the computed sinking rates of 1.2 m day^{-1} for total microplankton and 2.5 m day^{-1} for phytoplankton, seem plausible.

These preliminary results suggest that the model of Table 5 may provide realistic simulations of the distribution of autotrophic and heterotrophic processes in a partially mixed water column. The importance of the mixing terms in the budget for dissolved inorganic nitrogen at 20 m, where the calculated diffusivities were at least as low as those in the upper part of the oceanic thermocline (King & Devol, 1979), suggests that chemical/microbiological models, such as that of Olson (1981) for depth-dependent nitrification, should include the effects of vertical diffusion.

CONCLUSIONS

The theme of this review has been the importance of the interaction between vertical turbulence and depth-dependent microbial processes. 'Vertical-process' models for bulk phytoplankton may now be considered to be well established, although it seems likely that they can be improved by better descriptions of marine turbulence. Work on mixing-driven models involving several groups of microplankton, and including heterotrophs as well as autotrophs, is much less advanced. Nevertheless, the preliminary results reported here suggests that further study of the role of turbulence, in relation to the light- and nutrient-dependent behaviour of the main groups of microplankton, should prove enlightening.

ACKNOWLEDGEMENTS

The studies in the flooded quarry were carried out while the author was on the staff of the Scottish Marine Biological Association, which is grant-aided by the Natural Environment Research Council. The manuscript was prepared whilst in receipt of an SMBA funded research fellowship at UCNW. I am grateful to my SMBA colleagues, and to others acknowledged in the text, for data used in the models.

REFERENCES

ATLAS, D. & BANNISTER, T. T. (1980). Dependence of mean spectral extinction coefficient of phytoplankton on depth, water color, and species. *Limnology and Oceanography*, **25**, 157–9.

BANSE, K. (1976). Rates of growth, respiration and photosynthesis of unicellular algae as related to cell size – a review. *Journal of Phycology,* **12**, 135–40.

BANSE, K. (1977). Determining the carbon : chlorophyll ratio of natural phytoplankton. *Marine Biology*, **41**, 199–213.

BIDIGARE, R. R. (1983). Nitrogen excretion by marine zooplankton. In *Nitrogen in the Marine Environment*, ed. E. J. Carpenter & D. G. Capone, pp. 385–409. New York, Academic Press.

BOWDEN, K. F. (1983). Physical oceanography of coastal waters. Chichester, Ellis Horwood Ltd.

BOWMAN, M. J., ESAIAS, W. E. & SCHNITZER, M. B. (1981). Tidal stirring and the distribution of phytoplankton in Long Island and Block Island Sounds. *Journal of Marine Research*, **39**, 587–603.

BURGHES, D. N. & BORRIE, M. S. (1981). *Modelling with Differential Equations*. Chichester, Ellis Horwood Ltd.

CAPERON, J. (1968). Population growth response of *Isochrysis galbana* to nitrate variation at limiting concentration. *Ecology*, **49**, 866–72.

CAPERON, J. & MEYER, J. (1972). Nitrogen limited growth of marine phytoplankton. I. Changes in population characteristics with steady-stage growth rate. *Deep-Sea Research*, **19**, 601–18.

CORNER, E. D. S., COWEY, C. B. & MARSHALL, S. M. (1967). On the nutrition and metabolism of zooplankton, V. Feeding efficiency of *Calanus finmarchicus*. *Journal of the Marine Biological Association of the United Kingdom*, **47**, 259–70.

CORNER, E. D. S. & DAVIES, A. G. (1971). Plankton as a factor in the nitrogen and phosphorus cycles in the sea. *Advances in Marine Biology*, **9**, 101–204.

CURDS, C. R. (1974). Computer simulations of some complex microbial food chains. *Water Research*, **8**, 769–80.

CULLEN, J. J. & HORRIGAN, S. G. (1981). Effects of nitrate on the diurnal vertical migraton, carbon to nitrogen ratio, and the photosynthetic capacity of the dinoflagellate *Gymnodinium splendens*. *Marine Biology*, **62**, 81–9.

DAVIS, C. O., BREITNER, N. F. & HARRISON, P. J. (1978). Continuous culture of marine diatoms under silicon limitation. 3. A model of Si-limited diatom growth. *Limnology and Oceanography*, **23**, 41–52.

DROOP, M. R. (1968). Vitamin B12 and marine ecology. IV. The kinetics of uptake, growth and inhibition in *Monochrysis lutheri*. *Journal of the Marine Biological Association of the United Kingdom*, **48**, 689–73.

DROOP, M. R. (1974). The nutrient status of algal cells in continuous culture. *Journal of the Marine Biological Association of the United Kingdom*, **54**, 825–55.

DROOP, M. R. (1975). The nutrient status of algal cells in batch culture. *Journal of the Marine Biological Association of the United Kingdom*, **55**, 541–55.

DROOP, M. R. (1976). The chemostat in mariculture. In *Proceedings of the 10th European Marine Biology Symposium*, Vol. 1, ed. G. Personne & E. Jaspers, pp. 71–93. Wetteren, Universa Press.

DROOP, M. R. (1978). Comments on the Davis/Breitner/Harrison model for silicon uptake and utilization by diatoms. *Limnology and Oceanography*, **23**, 383–5.

DROOP, M. R. (1979). On the definition of X and of Q in the cell quota model. *Journal of Experimental Marine Biology and Ecology*, **39**, 203.

DROOP, M. R. (1983). 25 years of algal growth kinetics, a personal view. *Botanica Marina*, **26**, 99–112.

DROOP, M. R., MICKELSON, M. J., SCOTT, J. M. & TURNER, M. F. (1982). Light and nutrient status of algal cells. *Journal of the Marine Biological Association of the United Kingdom*, **62**, 403–34.

DUGDALE, R. C. (1967). Nutrient limitation in the sea: dynamics, identification and significance. *Limnology and Oceanography*, **12**, 196–206.

DUGDALE, R. C. & GOERING, J. J. (1967). Uptake of new and regenerated forms of nitrogen in primary productivity. *Limnology and Oceanography*, **12**, 196–206.

EDWARDS, A. & GRANTHAM, B. E. (1986). Inorganic nutrient regeneration in Loch Etive Bottom water. In *The Role of Freshwater Outflow in Coastal Marine Ecosystems*, ed. S. Skreslet, pp. 195–204. Berlin, Springer-Verlag.

EPPLEY, R. W. & PETERSON, B. J. (1979). Particulate organic matter flux and planktonic new production in the deep ocean. *Nature*, **282**, 677–80.

EPPLEY, R. W. & STRICKLAND, J. D. H. (1968). Kinetics of marine phytoplankton growth. *Advances in Microbiology of the Sea*, **1**, 23–62.

FASHAM, M. J. R. (1985). Flow analysis of materials in the marine euphotic zone. *Canadian Bulletin of Fisheries and Aquatic Science*, **213**, 139–61.

FASHAM, M. J. R., HOLLIGAN, P. M. & PUGH, P. R. (1983). The spatial and temporal development of the spring phytoplankton bloom in the Celtic Sea, April 1979. *Progress in Oceanography*, **12**, 87–145.

FENCHEL, T. (1982). Ecology of heterotrophic microflagellates. IV. Quantitative occurrence and importance as bacterial consumers. *Marine Ecology – Progress Series*, **9**, 35–42.

FOGG, G. E. (1986). Picoplankton. *Proceedings of the Royal Society of London*, **B, 228**, 1–30.

GILL, A. E. & TURNER, J. S. (1976). A comparison of seasonal thermocline models with observations. *Deep-Sea Research*, **23**, 391–401.

GOLDMAN, J. C. & GILBERT, P. M. (1983). Kinetics of inorganic nitrogen uptake by phytoplankton. In *Nitrogen in the Marine Environment*, ed. E. J. Carpenter & D. G. Capone, pp. 233–74. New York, Academic Press.

GRANTHAM, B. E. (1984). 'C/N Flux Project: Easdale Quarry nutrient results.' Scottish Marine Biological Association, internal report no. 125, 52 pp.

HARDY, A. C. (1924). The herring in relation to its animate environment. Part I. The food and feeding habits of the herring. *Fisheries Investigations, London*, Series II, **7**, no. 3.

HARRISON, W. G., DOUGLAS, D., FALKOWSKI, P., ROWE, G. & VIDAL, J. (1983). Summer nutrient dynamics of the Middle Atlantic Bight: nitrogen uptake and regeneration. *Journal of Plankton Research*, **5**, 539–56.

HATTORI, A. (1983). Denitrification and dissimilatory nitrate reduction. In *Nitrogen in the Marine Environment*, ed. E. J. Carpenter & D. G. Capone, pp. 191–232. New York, Academic Press.

HEANEY, S. I. & EPPLEY, R. W. (1981). Light, temperature and nitrogen as interacting factors affecting diel vertical migrations of dinoflagellates in culture. *Journal of Plankton Research*, **3**, 331–44.

HEINBOKEL, J. F. (1978). Studies on the functional role of tintinnids in the Southern Californian Bight. I. Grazing and growth rates in laboratory cultures. II. Grazing rates of field populations. *Marine Biology*, **47**, 177–89, 191–7.

HERBERT, D., ELSWORTH, R. & TELLING, R. C. (1956). The continuous culture of bacteria; theoretical and experimental study. *Journal of General Microbiology*, **14**, 601–22.

HOLLIBAUGH, J. T. (1980). Amino acid fluxes in marine plankton communities contained in CEPEX bags. In *Fjord Oceanography*, ed. H. J. Freeland, D. M. Farmer & C. D. Levings, pp. 439–45. New York, Plenum Press.

HOLLIGAN, P. M., HARRIS, R. P., NEWELL, R. C., HARBOUR, D. S., HEAD, R. N., LINLEY, E. A. S., LUCAS, M. I., TRANTER, P. R. G. & WEEKLY, C. M. (1984a). Vertical distribution and partitioning of organic carbon in mixed, frontal and stratified waters of the English Channel. *Marine Ecology – Progress Series*, **14**, 111–27.

HOLLIGAN, P. M., WILLIAMS, P. J. LE B., PURDIE, D. & HARRIS, R. P. (1984b). Photosynthesis, respiration and nitrogen supply of plankton populations in stratified, frontal and tidally mixed shelf waters. *Marine Ecology – Progress Series*, **17**, 201–13.

HOWE, S. O. (1979). 'Biological consequences of environmental change related to coastal upwelling: a simulation study.' Brookhaven National Laboratory BNL 51069, 123 pp.

HOWE, S. O. (1982). A simulation study of biological responses to environmental changes associated with coastal upwelling off Northwest Africa. *Rapport et Procès-Verbaux, Conseil International pour l'Exploration de la Mer*, **180**, 135–47.

HUTCHINSON, G. E. (1961). The paradox of the plankton. *American Naturalist*, **95**, 137–45.

JAMART, B. M., WINTER, D. F. & BANSE, K. (1979). Sensitivity analysis of a mathematical model of phytoplankton growth and nutrient distribution in the Pacific Ocean off the Northwestern United States coast. *Journal of Plankton Research*, **1**, 267–90.

JAMART, B. M., WINTER, D. F., BANSE, K., ANDERSON, G. C. & LAM, R. K. (1977). A theoretical study of phytoplankton growth and nutrient distribution in the Pacific Ocean off the North Western U.S. Coast. *Deep Sea Research*, **24**, 753–73.

JAMES, I. D. (1977). A model of the annual cycle of temperature in a frontal region of the Celtic Sea. *Estuarine and Coastal Marine Science*, **5**, 339–53.

JANNASCH, H. W. (1974). Steady state and the chemostat in ecology. *Limnology and Oceanography*, **19**, 716–20.

JERLOV, N. G. (1968). *Optical Oceanography*. New York, Elsevier.

JONES, K. J., TETT, P., WALLIS, A. C. & WOOD, B. J. B. (1978). Investigation of a nutrient-growth model using a continuous culture of natural phytoplankton. *Journal of the Marine Biological Association of the United Kingdom*, **58**, 923–41.

KAPLAN, W. A. (1983). Nitrification. In *Nitrogen in the Marine Environment*, ed. E. J. Carpenter & D. G. Capone, pp. 139–90. New York, Academic Press.

KIEFER, D. A. & KREMER, J. N. (1981). Origins of vertical patterns of phytoplankton and nutrients in the temperate, open ocean: a stratigraphic hypothesis. *Deep-Sea Research*, **28A**, 1087–105.

KING, F. D. & DEVOL, A. H. (1979). Estimates of vertical eddy diffusion through the thermocline from phytoplankton nitrate uptake rates in the mixed layer of the eastern tropical Pacific. *Limnology and Oceanography*, **24**, 645–51.

LANCELOT, C. & BILLEN, G. (1985). Carbon–nitrogen relationships in nutrient metabolism of coastal marine ecosystems. *Advances in Aquatic Microbiology*, **3**, 263–321.

LAWS, E. A. & BANNISTER, T. T. (1980). Nutrient- and light-limited growth of *Thalassiosira fluviatilis* in continuous culture, with implications for phytoplankton growth in the ocean. *Limnology and Oceanography*, **25**, 457–73.

McCARTHY, J. J. & CARPENTER, E. J. (1983). Nitrogen cycling in near-surface waters of the open ocean. In *Nitrogen in the Marine Environment*, ed. E. J. Carpenter & D. G. Capone, pp. 487–512. New York, Academic Press.

MARGALEF, R. (1978). Life-forms of phytoplankton as survival alternatives in an unstable environment. *Oceanologica Acta*, **1**, 493–509.

MASKE, H. (1982). Ammonium-limited continuous culture of *Skeletonema costatum* in steady and transitional state: experimental results and model simulations. *Journal of the Marine Biological Association of the United Kingdom*, **62**, 919–43.

MONOD, J. (1942). *Recherche sur la Croissance des Cultures Bacteriennes*. Paris, Herman et Cie.

OLSON, R. J. (1981). Differential photoinhibition of marine nitrifying bacteria: a possible mechanism for the formation of the primary nitrate maximum. *Journal of Marine Research*, **39**, 227–38.

PARSONS, T. R. & LEBRASSEUR, R. J. (1970). The availability of food to different trophic levels in the marine food chain. In *Marine Food Chains*, ed. J. H. Steele, pp. 323–43. Edinburgh, Oliver & Boyd.

PASCIAK, W. J. & GAVIS, J. (1974). Transport limitation of nutrient uptake in phytoplankton. *Limnology and Oceanography*, **19**, 881–8.

PICKARD, G. L. & EMERY, W. J. (1982). *Descriptive Physical Oceanography*, 4th edn. Oxford, Pergamon.

PINGREE, R. D. & GRIFFITHS, D. K. (1977). The bottom mixed layer on the continental shelf. *Estuarine and Coastal Marine Science*, 5, 399–413.

PINGREE, R. D., MADDOCK, L. & BUTLER, E. I. (1977). The influence of biological activity and physical stability in determining the chemical distributions of inorganic phosphate, silicate and nitrate. *Journal of the Marine Biological Association of the United Kingdom*, 57, 1065–73.

PINGREE, R. D. & PENNYCUICK, L. (1975). Transfer of heat, fresh water and nutrients through the seasonal thermocline. *Journal of the Marine Biological Association of the United Kingdom*, 55, 261–74.

PINGREE, R. D., PUGH, P. R., HOLLIGAN, P. M. & FORSTER, G. R. (1975). Summer plankton blooms and red tides along tidal fronts in the approaches to the English Channel. *Nature, (London)*, 258, 672–7.

PLATT, T. (1986). Primary production of the ocean water column as a function of surface light intensity: algorithms for remote sensing. *Deep-Sea Research*, 33, 149–63.

PLATT, T. & HARRISON, W. G. (1985). Biogenic fluxes of carbon and oxygen in the ocean. *Nature*, 318, 55–8.

POND, S. & PICKARD, G. L. (1978). *Introductory Dynamic Oceanography*. Oxford, Pergamon.

RADACH, G. & MAIER-REIMER, E. (1975). The vertical structure of phytoplankton growth dynamics: a mathematical model. *Mémoires de la Société Royale des Sciences de Liège, 6e série*, 7, 113–46.

REDFIELD, A. C. (1958). The biological control of chemical factors in the environment. *American Scientist*, 46, 205–21.

RHEE, G.-Y. (1973). A continuous culture study of phosphate uptake, growth rate and polyphosphate in *Scenedesmus* sp. *Journal of Phycology*, 9, 495–506.

RHEE, G.-Y. (1974). Phosphate uptake under nitrogen limitation by *Scenedesmus* and its ecological implications. *Journal of Phycology*, 10, 470–5.

RHEE, G.-Y. (1980). Continuous culture in phytoplankton ecology. *Advances in Aquatic Microbiology*, 2, 151–203.

RILEY, G. A. (1946). Factors controlling phytoplankton populations on Georges Bank. *Journal of Marine Research*, 6, 54–73.

RYTHER, J. H. (1956). Photosynthesis in the ocean as a function of light intensity. *Limnology and Oceanography*, 1, 61–70.

RYTHER, J. H. & DUNSTAN, W. M. (1971). Nitrogen, phosphorus and eutrophication in the inshore marine environment. *Science*, 171, 1008–13.

SCOTT, J. M. (1985). The feeding rates and efficiencies of a marine ciliate, *Strombidium* sp., grown under chemostat steady-state conditions. *Journal of Experimental Marine Biology and Ecology*, 90, 81–95.

SHUTER, B. J. (1978). Size dependence of phosphorus and nitrogen subsistence quotas in unicellular microorganisms. *Limnology and Oceanography*, 23, 1248–55.

SIMPSON, J. H. & BOWERS, D. G. (1984). The role of tidal stirring in controlling the seasonal heat cycle in shelf seas. *Annales Geophysicae*, 2, 411–16.

SIMPSON, J. H., TETT, P. B., ARGOTE-ESPINOZA, M. L., EDWARDS, A., JONES, K. J. & SAVIDGE, G. (1982). Mixing and phytoplankton growth around an island in a stratified sea. *Continental Shelf Research*, 1, 15–31.

SMITH, I. R. (1975). Turbulence in lakes and rivers. *Freshwater Biological Association, Scientific Publication* no. 29, 79 pp.

STEELE, J. H. & HENDERSON, E.-W. (1976). Simulation of vertical structure in a planktonic ecosystem. *Scottish Department of Agriculture and Fisheries, Fisheries Research Reports*, 5, 1–27.

SVERDRUP, H. U. (1953). On conditions for the vernal blooming of phytoplankton.

Journal du Conseil Permanent International pour l'Exploration de la Mer, **18**, 287–95.

TALLING, J. F. (1957). Photosynthetic characteristics of some freshwater plankton diatoms in relation to underwater radiation. *New Phytologist*, **56**, 29–50.

TAYLOR, P. A. & WILLIAMS, P. J. LE B. (1975). Theoretical studies on the coexistence of competing species under continuous-flow conditions. *Canadian Journal of Microbiology*, **21**, 90–8.

TEMPEST, D. W. (1970). The continuous cultivation of microorganisms: 1. Theory of the chemostat. *Methods in Microbiology*, **2**, 259–76.

TETT, P. (1981). Modelling phytoplankton production at shelf-sea fronts. *Philosophical Transactions of the Royal Society of London, Series A*, **302**, 605–15.

TETT, P. (1984). The ecology of exceptional blooms. International Commission for the Exploration of the Sea, Special Meeting on the Causes, Dynamics and Effects of Exceptional Marine Blooms and related events. Copenhagen, 4–5 October 1984: B1, 17 pp.

TETT, P. (1986), Physical exchange and the dynamics of phytoplankton in Scottish sea-lochs. In *The Role of Freshwater Outflow in Coastal Marine Ecosystems*, ed. S. Skreslet, pp. 205–18. Berlin, Springer-Verlag.

TETT, P. & DROOP, M. R. (in press). Cell quota models and planktonic primary production. In *Handbook of Laboratory Model Systems for Microbial Ecosystem Research*, ed. J. W. T. Wimpenny. Florida, CRC Press.

TETT, P. & EDWARDS, A. (1984). Mixing and plankton: an interdisciplinary theme in oceanography. *Oceanography and Marine Biology, Annual Review*, **22**, 99–123.

TETT, P., EDWARDS, A. & JONES, K. (1986). A model for the growth of shelf-sea phytoplankton in summer. *Estuarine, Coastal and Shelf Science*, **23** (in press).

TILMAN, D. (1977). Resource competition between planktonic algae: an experimental and theoretical approach. *Ecology*, **58**, 338–48.

WALSH, J. J. (1975). A spatial simulation model of the Peru upwelling system. *Deep-Sea Research*, **22**, 201–36.

WEEKS, A. (1984). 'The vertical distribution of phytoplankton and related variables in Easdale Quarry, 21–22 May, 1984.' Scottish Marine Biological Association, internal report no. 118, 32 pp.

WILLIAMS, P. J. LE B. (1981). Incorporation of microheterotrophic processes into the classical paradigm of the planktonic food web. *Kieler Meeresforschungen*, **5**, 1–28.

WOODS, J. D. & ONKEN, R. (1982). Diurnal variation and primary production in the ocean – preliminary results of a Lagrangian ensemble model. *Journal of Plankton Research*, **4**, 735–56.

YENTSCH, C. M. & YENTSCH, C. S. (1984). Emergence of optical instrumentation for measuring biological properties. *Oceanography and Marine Biology, Annual Review*, **22**, 55–98.

INDEX